Real-World Electrical Engineering Case Studies

# Real-World Electrical Engineering Case Studies

*Manikandan Palanichamy*
*Institute of Engineering, Østfold University of Applied Sciences, Fredrikstad Campus, Norway*

IEEE*Press*
WILEY

Published by John Wiley & Sons, Inc., Hoboken, New Jersey.



***Library of Congress Cataloging-in-Publication Data:***

Names: Palanichamy, Manikandan, 1980- author
Title: Real-world electrical engineering case studies / Manikandan Palanichamy, Institute of Engineering, Østfold University of Applied Sciences, Fredrikstad Campus, Norway.
Description: Hoboken, New Jersey: John Wiley & Sons, Inc, [2026] | Includes bibliographical references and index.
Identifiers: LCCN 2026006157 | ISBN 9781394359431 Hardback | ISBN 9781394359455 ePDF | ISBN 9781394359448 ePub | ISBN 9781394359462 oBook
Subjects: LCSH: Electrical engineering–Study and teaching | Electrical engineering–Environmental aspects | Electrical engineering–Materials | Renewable energy sources | Power (Mechanics) | LCGFT: Case studies
Classification: LCC TK165.P35 2026
LC record available at https://lccn.loc.gov/2026006157

Cover Design: Wiley
Cover Image: © ABCDstock/stock.adobe.com

*To all the future engineers of this world.*

## Contents

## About the Author

**Dr. Palanichamy** is an Associate Professor at Østfold University of Applied Sciences with over 12 years of post-PhD teaching and research experience. Specializing in green transitions, smart energy systems, and emerging ICT technologies, he supervises theses across Bachelor's, Master's, and PhD levels while leading multiple externally funded research projects. His forward-thinking research focuses on Industry 5.0, applied artificial intelligence, circular economy, and information technology security, grounded in the belief that today's research lays the foundation for tomorrow's knowledge.

He has contributed over 90 academic publications and reports and collaborated with diverse research groups and laboratories across the globe. His work has achieved an H-index of 21 and i10-index of 49. He actively collaborates with multinational industries and serves as an industrial consultant for custom software development, and Industrial IoT integration projects, bridging academic expertise with practical industry applications.

ORCID: 0009-0006-6582-3480
Google Scholar: https://scholar.google.com/citations?hl=en&authuser=4&user=awzJenIAAAAJ
Affiliation: https://www.hiof.no/iio/ing/english/people/aca/manikanp/index.html

*Manikandan Palanichamy*
*Associate Professor*
*Østfold University of Applied Sciences, HiØF, Norway*

## Acknowledgements

I am deeply grateful to everyone who supported me throughout the journey of writing this book.

First and foremost, I would like to express my heartfelt gratitude to my parents, my wife, my sisters, and my daughter for their endless love, patience, and unwavering support. Their encouragement has been my constant source of strength.

I am profoundly thankful to the Almighty for blessing me with this wonderful opportunity and guiding me through this endeavor.

My sincere appreciation goes to the entire Wiley team for their professionalism and dedication. I am particularly grateful to Brett for believing in the idea of this book, and to his colleague Nandhini for her exceptional support from the beginning through the entire publication process.

To all my colleagues, friends, reviewers and contributors who provided valuable insights and feedback – thank you all for enriching this work with your expertise and encouragement.

# Introduction

## Bridging Theory and Practice: The Power of Engineering Case Studies in Real-World Problem Solving

Every engineer shares a fundamental common ground, whether a novice student drawing their first circuit diagram or an experienced professional designing a multi-billion-dollar infrastructure project, because engineering is about problem solving. It is about transforming theoretical concepts and making them into real-time implementations. Engineering overcomes limitations and develops solutions that shape our world. But a critical gap exists between the academic theory and the complex, often unpredictable demands of real-world engineering challenges. How do we bridge this gap? How do we prepare aspiring engineers not just to understand theories, but to apply them effectively when faced with unpredictable challenges, tight deadlines, and unforeseen variables?

The solution, as history and experience have time and again indicated, is found in stories. More specifically, in the power of narratives found within engineering case studies. Just as a pilot trains on flight simulators and a surgeon gains practice through clinical cases, engineers cultivate their analytical thinking and problem-solving skills by analyzing real-life cases. It is part of our human nature to gravitate toward stories, and it is through the accomplished projects that we are best able to see the lessons that shape theoretical learning into pragmatic understanding.

## The Engineering Gap

This book, *Real-World Electrical Engineering Case Studies* is designed as an essential guide to transition from theory to practical engineering feats. Unlike traditional textbooks that seek to teach fundamental theory and complex ideas (which are still essential), this book takes a different yet complementary approach. The primary purpose is not to repeat the "what" about engineering, but rather the "how" and "why." How are fundamental theories, ideas, and equations used in real-world situations? And why particular implementations succeed? How and why new challenges occur?

The book is intended to make difficult engineering concepts easier to understand by demonstrating where they are used. With theories and textbook learning in real context, *Real-World Electrical Engineering Case Studies* provides students with a realistic methodology and better equips them for the complex issues they will likely be faced within the engineering profession. To educators, this book provides a comprehensive and practical guide to teaching with cases, enabling active learning in which students can simulate or solve real engineering problems. The application of this book is broader than classroom use. To the non-engineers, whether for policymakers, investors, or simply for the curious minded, this book provides a window into the engineering technology that drives modern life.

For anyone who is interested in understanding the mechanics of the global transition to green energy or to decode the tiny world of semiconductors and digital connectivity, these case studies make the knowledge more digestible without the need for a background in advanced calculus or engineering. To those interested in cultivating a forward-leaning problem mentality on any level will also find immense value in these pages.

## Motivation and Learning Outcomes

Each case study in this series is carefully designed to provide real context balanced with the technical depth required. With this immersive style, readers can acquire five key learning outcomes: 1. **Applied Knowledge**, which closes the gap between abstract theory and the physical world. 2. **Critical Thinking Skills,** for assessing complex, and frequently contradictory information. 3. **Problem-Solving Intuition**, to help deal with the uncertainty of real-world situations and challenges. **4. Contextual Thinking**, to understand the socioeconomic and regulatory context that surrounds engineering projects; and **5. Best Practices**, which helps engineers to develop a mental library of various experiences derived from the successes and failures of others.

1. **Applied Knowledge**

   *Learn how engineering principles, theories, and methods are applied to real projects and real problems.*

   This book bridges the gap between abstract theory and physical reality. Traditional lecturing often leads to passive absorption. In contrast, research confirms that active learning methods, particularly through case studies, can significantly improve information retention and student engagement [1]. This pedagogical shift is not unique to engineering. This has long been the gold standard in medical and business education. For example, Herreid et al. highlights how case-based learning in the sciences compels students to confront ambiguity, thereby promoting deeper conceptual understanding than textbook lessons [2].

Similarly, in the discipline of business management, the case method study is essential for bridging the "knowing-doing gap" as it permits students to engage in decision-making in a risk-free environment [3]. This pedagogical approach directly addresses the phenomenon described by Pfeffer and Sutton (1999) [4]. They reasoned that the "knowing-doing gap" exists partly because organizations and educational institutions have historically rewarded complex talk over simple action. This results in a form of paralysis where individuals fear the risks associated with implementation. Therefore, their theory legitimizes the case method as a crucial mechanism to overcome the disconnect between intellectual understanding and practical execution. It serves to break this cycle of hesitation, empowering students to transform concepts from their textbook concepts and numbers from equations into plans of deeds and action-based solutions.

In engineering this leads to what is called as the "real performance." Such proficiency is acquired not just by knowing equations, but with a fundamental understanding of the social and physical context within which those variables exist [5]. This concept is grounded in the theory of "situated learning," originally conceptualized by Lave and Wenger (1991). They explained that learning is not merely the transmission of abstract concepts from teacher to student, but a social process of "legitimate peripheral participation," where novices gradually move toward full participation in the actual work of that organization or community.

Johri and Olds (2011) reinforce this within the technical discipline by stating that engineering knowledge is inherently social and material. It is not only in the thinking of individuals, but is distributed through tools, representations and collaborative conversations [6]. Instead of directing students on mere memorization of disjointed facts, this approach ties abstract rules to the real world. When a student can observe thermodynamic laws governing the cooling systems of an enormous infrastructure project like the grid connection system of BorWin2 or the quantum tunneling constraints dictating the microscopic design of a 3nm chip, the knowledge becomes "indexed" to the real world. In this way, the technical ideas are not only understood in the classroom vacuum, but they can be retrieved and used when confronted with the tangled challenges in real professional practice.

**2. Critical Thinking Skills**

*Develop the ability to analyze complex situations, identify root causes, evaluate alternate solutions, and understand the consequences of engineering decisions.*

Beyond simple calculation, engineers must evaluate conflicting data and navigate the gray areas of design trade-offs. This requirement mirrors the basic principle of "thinking reflectively" as put forth by John Dewey (1910).

The concept is defined as the active, persistent, and careful consideration of any belief or form of knowledge in light of the grounds that support it [7]. Although this approach was originally applied to the humanities, it has become the basis of all modern technical competence.

In the specific context of engineering education, Yadav et al. (2007) found that the use of case studies into the curriculum is specifically beneficial to improving higher-order thinking skills. This progression is best understood through the framework of Bloom's Taxonomy, which organizes cognitive objectives into levels of complexity [8]. This means that the students move up the hierarchy of Bloom's Taxonomy from simple application to complex analysis and evaluation. Knowledge of facts and basic comprehension are utilized in lower-order thinking, while learners need to break down information and make judgments based on evidence in higher-order thinking. As Anderson and Krathwohl (2001) discussed in their revision to the taxonomy, this cognitive level cannot be realized through passive learning, as it requires "meaningful learning," in which students are expected to actively relate abstract ideas to real world events to tackle new situations [9].

However, the driving force for this pedagogical shift extends beyond the classroom; it is a direct response to the evolving demands of the engineering profession. This is an industry driven change. A recent review by Ahern et al. (2019) emphasizes that while technical competence is assumed, the modern workforce requires professionals who can critically assess the validity of the problems they are asked to solve, rather than just solving them blindly [9]. To that end, as Paul and Elder have observed, this form of critical thinking is essential for moving from being a passive consumer of information to active questioners of the validity of engineering, so that our decisions are based on rigorous reasoning rather than convenient assumptions [10].

Bridging the gap between these educational theories and professional realities requires a deliberate change in how technical content is presented. Consequently, this book challenges the reader to think through complex situations, identify root causes, and assess options even when data is incomplete or even during conditions of uncertainty.

3. **Problem-Solving Intuition**

   *Engage with open-ended scenarios that mimic the unpredictable nature of real-world engineering, enhancing your capacity for innovative and practical solutions.*

   Solving a textbook problem is like driving with GPS on a straight road with clear weather. The destination is set, the route is calculated, and the variables are controlled. Such configurations are considered to be "well-structured" with clearly defined boundaries. However, the real world resembles driving off-road terrain where the map is outdated and the destination itself might shift. This is aligned with what Jonassen (2000) describes as "ill-structured problems" [10]. This means problems involve conflicting objectives, multiple answers and many uncertain factors. In such situations, the "right" engineering answer is rarely a simple "right/wrong" decision, but a calculated balancing act involving cost, efficiency, and safety.

To deal with these uncertainties, engineers can no longer rely on "routine expertise," or the ability to carry out well-practiced tasks quickly and accurately. Rather, they need to develop "adaptive expertise," a term coined by Hatano and Inagaki (1986) [11]. Unlike routine experts who adhere to established routines, adaptive experts have a more flexible understanding of their field that enables them to adapt their knowledge and develop new procedures when existing ones do not work.

This is more meaningful in engineering today, when problems are becoming systemic, not isolated. Professional success in engineering is related to the ability to integrate multiple competing constraints, e.g., thermodynamic efficiency with materials cost, rather than utilizing a cookbook method [12]. This actually reflects a broader economic pattern. According to the World Economic Forum's *Future of Jobs Report 2023*, creative thinking and analytical thinking top the list of skills and outranking more specific technical skills [13].

By engaging through open-ended situations that feature this unpredictability of scenarios, like grid stability challenges in the Tesla Hornsdale case, readers can enhance the kind of novel, adaptive logic required to tackle issues that have never been seen before.

4. **Contextual Thinking**

*Appreciate the broader factors – industry trends, regulatory environments, stakeholder interests that influence engineering projects.*

Engineering does not happen in a vacuum; it is essentially a social-technical activity. This reality is illustrated in every chapter of this book. Starting from Chapter 1, the BorWin2 HVDC project shows that transmitting offshore power is not just about technical voltage conversion challenge but also involves regulatory harmonization and infrastructure planning. Likewise, Chapter 3 examines Ørsted's "Viking Longship" strategy to show how the "Triple Helix of Innovation" compels engineers to coordinate technical innovations with government policy and economic market drivers.

This view has been derived from the Socio-Technical Systems (STS) theory proposed by Eric Trist (1981). The theory advocates that the technical performing system cannot be designed or improved in isolation from the social system within which it is embedded. Ignoring one of the two systems ultimately results in the defeat of the other [14]. Specifically, within the boundaries of engineering design, Bucciarelli (1994) prominently suggested that engineering is not the isolated exercise of a calculative mind but rather a "social process of negotiation," to which engineers must also somehow negotiate competing "object worlds" of stakeholders [15].

More recent research validates and enhances this perspective. For instance, Sovacool et al. (2020) point out that today's industrial transitions, including the green hydrogen and wind transitions explored in this book are more than mere technological substitutions but complex "socio-technical transitions." It requires engineers to contend with profound changes in user practices, cultural norms, and institutional logic in addition to the design of hardware [15].

**5. Best Practices**

*Extract valuable insights from both successful implementations and challenging situations, understanding what works and why.*

Best practice is the embodiment of industry knowledge. These are the standardized techniques that have been tested in the real world. In high-risk industries such as renewable energy or semiconductor manufacturing, these tried-and-true templates help engineers adapt to the unknown without reinventing the wheel. This method is a derivative of "Case-Based Reasoning" (CBR). As Kolodner (1992) notes, human thinking is largely driven by previous experiences applied to new problems, we think about the future by thinking through the past [16]. By extracting valuable insights from both successful implementations, such as TSMC's 3nm breakthrough (Chapter 6) and difficult pivots such as Vestas' journey from farm implement to a "Smart Wind" leader (Chapter 5), readers can create a mental library of proxy experiences.

The approach is consistent with "experiential learning theory" developed by David Kolb (1984), which explains that transformative learning occurs through a cycle of reflective observation (analyzing case studies) as a source of "Abstract Conceptualization or developing novel engineering solutions" [17]. Studying and analyzing challenging cases are important. As Henry Petroski (1985) famously pointed out, the history of engineering is essentially the history of learning from failure, since good engineering conceals inherent weaknesses while bad engineering reveals the boundaries of the possible [18].

Recent research reinforces there is a need for structured reflection. Modern engineering demands "double-loop learning" [19]. Instead of simply just correcting surface-level mistakes or fixing obvious errors, this method forces teams to question the underlying logic of a design. Love et al. (2019) showed that without this careful approach, complex projects can inevitably face repeated failures that affect the entire system [20]. As Flyvbjerg (2006) also reasoned, it is precisely such close-up, context-dependent knowledge that makes the beginner, who follows rules, different from the expert, who has an intuitive understanding of past realities [21].

Taken together, these five key learning outcomes create an all-encompassing framework that provides more than just a review of history. By internalizing these proxy experiences through case studies, engineers can foresee potential problems, turning the hindsight of others into their own strategic advantage. This book, therefore, should not be viewed as simply a series of cases, but as a conceptual toolkit. Ultimately, this establishes a foundation where professional judgment is not a product of guessing, but the result of rigorous, evidence-based synthesis.

## Methodology and Chapter Structure

Standard engineering textbooks often present problems as isolated mathematical variables, stripping away the complexity of the real world. But professional engineering in reality requires the determination to deal with uncertain data, competing demands from stakeholders, and sometimes complex ethical trade-offs.

To bridge this gap, this book employs Case-Based Reasoning (CBR) as its core methodological engine. Drawing upon the "double-loop learning" framework established in the previous section, this book does not treat case studies as static historical accounts. Rather, it treats the historical context as a simulation.

Each chapter is designed to guide the reader out of passive consumption (learning the theory) and into active production (applying judgments in context). Each chapter is structured to lead the reader along a particular cognitive trajectory: the incentive to define the context, the risk to confront the conflict, the obligation to analyze the variables, and the duty to synthesize the solution.

While the format of the chapters is standardized, the analysis within them is unique. Rather than concentrating on technical failure or systems theory analysis like other traditional approaches, this book evaluates each case through three integrated perspectives to reflect the actual decision-making matrix of a professional engineer.

First, the **Technical Lens (feasibility)** grounds the case on hard science. Under this lens, the relevant engineering considerations are restated in terms of known physics, mathematical models, and the logic of materials science [22]. For example, in the BorWin2 case (Chapter 1), this lens examines a particular Modular Multilevel Converter (MMC) technology to stabilize the voltage over a hundred kilometers of subsea cable. It poses the basic question: "Can we make it and will it work?"

However, technical feasibility is almost never the sole limitation. The analysis then extends to the **Techno-Economic Lens (Strategic Engineering)**. This perspective looks at the competitive advantage and productivity gains that drive the choice of technology. In Chapter 1, this lens explains why Siemens switched to "HVDC Plus" technology as a means to challenge ABB's market control. It reveals that engineering choices are frequently in line with strategic business decisions as well. In this example, trading off maximizing transmission capacity against ensuring long-term viability.

Finally, the **Operational and Environmental Lens (Context)** expands the perspective further. It investigates how outside influences, such as government regulation or physical geography, define the limits of design. Through the BorWin2 example, this lens places engineering in the context of Germany's Energiewende policy as well as the physical realities of the North Sea, where installations and equipment must endure 14-meter waves. By triangulating these three perspectives, the narratives prove that a "right" calculation can be wrong if strategic or environmental factors are overlooked.

To ensure consistency across all case study, whether the topic is microelectronics or civil infrastructure, every case study follows a rigorous, six-part narrative structure. This format is designed to facilitate Double-Loop Learning by encouraging readers to question underlying assumptions rather than just correcting immediate errors.

1. **Synopsis and Context** – The chapter opens with an introductory summary that sets the context. This section identifies the specific engineering subdiscipline, the key stakeholders, and the industry context. This allows readers to quickly grasp the scope of the challenge before diving into the details.
2. **Background and Case Scenario** – Here, the text establishes the narrative context. This section contains a chronological account of the project history or the origin of the technology and the specific "Decision Point" or moment of crisis. This is the crucial part of the case study where it introduces the problem or challenges. The engineers often had to work with incomplete and sometimes conflicting information.
3. **Analysis and Problem Statement** – This is the core of the case study where the "Three Lenses" framework is used. Instead of just narrating the outcome or results, this section breaks down the key point of friction. It uses engineering principles to identify the problem, considers alternative solutions that were proposed, and discusses the effects of the decisions made.
4. **Conclusion and Key Takeaways** – This section, synthesis from analysis, and integrates the key findings. Instead of categorizing projects as simply success or failures, it looks at the engineering trade-offs and strategic pivots that shaped the result. It shows how competing complex physical, financial, and regulatory constraints were managed to result in a breakthrough. There is certainly still a focus on that transferability. It demonstrates how a specific solution to a problem in one field, for example, the design of a subsea cable, can yield universal principles that can be applied to later engineering problems.
5. **Discussion Points** – To support educators, this section transitions from reading to active debate. It provides open-ended questions and classroom topics designed to spark discussion on ethics, risk management, and alternative history ("What would you have done differently in such specific case or situation?").

6. **Limitations of the Case Study** – Finally, to maintain academic rigor, the chapter concludes by acknowledging the specific constraints of the case. This section addresses hindsight bias and the unique variables of the situation, ensuring that readers understand the boundaries of the case and do not over-generalize the findings.

## Scope and Limitations

This book focuses on case studies within the disciplines of Electrical, Electronics and Computer Engineering. Cases included in this book are not necessarily the most commercially well-known or historically significant. Rather, they were selected because of the specific engineering narratives they provide. Every case is a distinct vehicle to illustrate or explain a fundamental concept. While there are many other viable examples to explore in the field, this selection is curated as the "first set" of case studies limited by the physical constraints of this volume. In that sense, this book is more complementary than a classical primary textbook. It provides the link between academic research and industrial practice, focusing on engineering and strategic decisions while limiting coverage of commercial or financial metrics.

To ensure the integrity of these narratives, the analysis relies on a robust triangulation of open-source data. Information is drawn from official project documentation, corporate reports, peer-reviewed academic publications, and high-quality media analysis. Although the intention is to achieve maximum accuracy through this cross-verification, the depth of the case studies is inevitably limited by the availability of public records. Therefore, any proprietary information, or internal business data that is not in the public domain are excluded from the analysis.

Finally, just as the analysis is bound by data availability, it is also defined by time. Engineering projects are inherently dynamic and subject to constant updates. The narratives here capture the status of these projects and the state of knowledge at the time of writing. Unannounced modifications, or software updates may occur after the date of this publication. Therefore, no responsibility is assumed for any discrepancies arising from subsequent developments or information that are not publicly available. The reader should treat the text as an investigation into discrete engineering moments, rather than a live operational manual. Finally, to address unique constraints relevant to specific technologies, detailed limitations are also provided within each individual chapter.

## Organization of the Book

The choice of case studies in this volume is not random. It traces a purposeful course from the megawatt power grids of large-scale projects, down to the nanometer world of integrated circuits. This allows readers to appreciate the distinct yet related issues in managing electrons for power or electrons for information. As a result, the book is organized into two separate sections, reflecting the physical duality between the bulk infrastructure of the energy transition and the precision hardware of the digital economy.

**Part 1: Renewable Energies and Power Infrastructure** deals with the macrophysics that encompasses energy generation, energy transmission, and energy storage. The narrative starts with the problem of transmission, taking **BorWin2** as an example for the HVDC technology needed to bring power from offshore to the onshore. From there it's on to power storage, and strategic reorientation with **GreenHyScale**'s hydrogen electrolysis and **Ørsted**'s move away from fossil fuels to renewables. Next up is the physics of generation itself, **First Solar**'s thin-film diffusion and **Vestas**' smart turbine engineering. These generator technologies are enabled by **Nexans**' subsea cables material science. This section concludes with **Tesla**'s Hornsdale project with the critical solution for grid stability and frequency control.

**Part 2: Semiconductors, Circuits, and Connectivity** shifts the focus to the micro-physics that underpin the information age. The narrative opens at the manufacturing edge with **TSMC**'s 3nm process, before exploring the application of efficient circuit design in **Nordic Semiconductor**'s IoT solutions. It then dives into processor architecture, contrasting **Intel**'s 3D transistor stacking with the parallel processing dominance of **AMD** and **NVIDIA**, followed by STMicroelectronics' use of **Silicon Carbide** for electric vehicle power management. The section concludes by connecting these distributed systems, tracing the evolution of global networks from **Ericsson**'s 5G infrastructure and **Huawei**'s autonomous network management to the orbital mechanics of **SpaceX**'s satellite constellations.

The dynamic nature of engineering means the need for constant adaptation and flexibility to challenge new solutions. This book is meant to be more than a reading experience; it is an active engagement with the challenges, successes, and unexpected outcomes inherent in major global engineering projects. By linking theoretical studies with operational realities, these case studies aim to sharpen readers' analytical skills and offer a unique view into the complex landscape of contemporary engineering.

## References


1 Prince, M. "Does active learning work? A review of the research," *Journal of Engineering Education*, 93(3), 223–231, 2004.

2 Herreid, C.F. and Schiller, N.A. "Case study: case studies and the flipped classroom," *Journal of College Science Teaching*, 42(5), 62–66, 2013.

3 Mesny, A. "Taking stock of the century-long utilization of the case method in management education," *Canadian Journal of Administrative Sciences/Revue Canadienne Des Sciences De l'Administration*, 30(1), 56–66, 2013.

4 Pfeffer, J. and Sutton, R.I. *The Knowing-Doing Gap: How Smart Companies Turn Knowledge into Action*, Boston, MA: Harvard Business School Press, 1999.

5 Lave, J. and Wenger, E. *Situated Learning: Legitimate Peripheral Participation*," Cambridge: Cambridge University Press, 1991.

6 Johri, A. and Olds, B.M. "Situated engineering learning: bridging engineering education and practice," *Journal of Engineering Education*, 100(1), 151–185, 2011.

7 Dewey, J., *How We Think*, Boston, MA: D.C. Heath & Co, 1910.

8 Bloom, B.S., Engelhart, M.D., Furst, E.J., Hill, W.H. and Krathwohl, D.R. *Taxonomy of Educational Objectives: The Classification of Educational Goals.* Handbook I: *Cognitive Domain*, New York: David McKay Company, 1956.

9 Anderson, L.W. and Krathwohl, D.R. *A Taxonomy for Learning, Teaching, and Assessing: A Revision of Bloom's Taxonomy of Educational Objectives*, New York: Longman, 2001.

10 Jonassen, D.H. "Toward a design theory of problem solving," *Educational Technology Research and Development*, 48(4), 63–85, 2000.

11 Hatano, G. and Inagaki, K. "Two courses of expertise," in *Child Development and Education in Japan*, ed. H. Stevenson, H. Azuma, and K. Hakuta, 262–272. New York: W. H. Freeman and Company, 1986.

12 Passow, H.J. and Passow, C.H. "What competencies should undergraduate engineering programs emphasize? A systematic review," *Journal of Engineering Education*, 106(3), 475–526, 2017.

13 Forum, W.E. *The future of jobs report, 2023*, World Economic Forum, Geneva, 2023.

14 Bucciarelli, L.L. *Designing Engineers*, Cambridge, MA: MIT Press, 1994.

15 Sovacool, B.K., Hess, D.J., Amir, S., Geels, F.W., Hirsh, R., Medina, L.R., Miller, C., Palavicino, C.A., Phadke, R., Ryghaug, M. and Schot, J. "Sociotechnical agendas: reviewing future directions for energy and climate research," *Energy Research & Social Science*, 70, 101617, 2020.

16 Kolodner, J.L. "An introduction to case-based reasoning," *Artificial Intelligence Review*, 6(1), 3–34, 1992.

17 Kolb, D.A. *Experiential Learning: Experience as the Source of Learning and Development*, FT Press, 2014.

18 Petroski, H. *To Engineer is Human: The Role of Failure in Successful Design*, St. Martin's Press, 1985.

19 Argyris, C. and Schön, D.A. *Organizational Learning: A Theory of Action Perspective, Reading*, MA: Addison-Wesley, 1978.

20 Love, P.E., Smith, J., Ackermann, F. and Irani, Z. "Making sense of rework and its unintended consequence in projects: the emergence of uncomfortable knowledge," *International Journal of Project Management*, 37(3), 501–516, 2019.

21 Flyvbjerg, B. "Five misunderstandings about case-study research," *Qualitative Inquiry*, 12(2), 219–245, 2006.

22 Trist, E.L. *The Evolution of Socio-Technical Systems*," Toronto: Ontario Quality of Working Life Centre, 1981.

# Part 1

# Renewable Energies and Power Infrastructure

## Relevant Subjects

Power Systems Engineering, High Voltage Transmission (HVDC/HVAC), Renewable Energy Technologies

## Relevant Learning Objectives

Evaluate renewable energy generation and storage systems, analyze high-voltage transmission infrastructure and grid stability strategies.

# 1

# BorWin2: HVDC Breakthrough for Offshore Wind

Siemens' Evolution from HVDC Light® to HVDC Plus
in Germany's Energy Transformation

## 1.1 Bridging Offshore Power

Visualize the challenge of transporting a thousand fragile wine glasses across a turbulent ocean. Now, amplify that difficulty by a million, and you'll gain an understanding of the monumental task that Siemens faced when commissioned to transmit substantial quantities of electrical power from offshore wind farms to the German mainland. Instead of wine glasses, the objective was to transfer megawatts of clean energy through the unpredictable North Sea without losing much power. In this case study, we delve into the innovative solutions Siemens employed to overcome the BorWin2 HVDC project's unique obstacles, establishing a blueprint for future offshore wind power transmission.

In 2010, Siemens embarked on one of the most ambitious power transmission projects in history – the BorWin2 HVDC link [1]. This project was not just another day at the electrical engineering office. It involved connecting massive wind farms capable of powering approximately one million German households, equivalent to the entire population of Munich deciding to watch their football team play while running their dishwashers simultaneously. This was double the BorWin1 project's capacity and a true test of the capabilities of high-voltage direct current technology [2].

What is the challenge here? Wind turbines produce fluctuating AC power. The system converts it to DC for efficient long-distance travel. Then back to AC for the grid. Sounds straightforward? Not in the North Sea. Its infamous mood swings create waves as tall as 14 meters. Imagine a five-story building of water crashing against the equipment. That's the daily reality offshore.

The BorWin2 platforms were subject to extreme environmental conditions because it is located approximately 100 kilometers northwest of the German coast, near the North Sea Island of Borkum. As illustrated in Figure 1.1, the platform

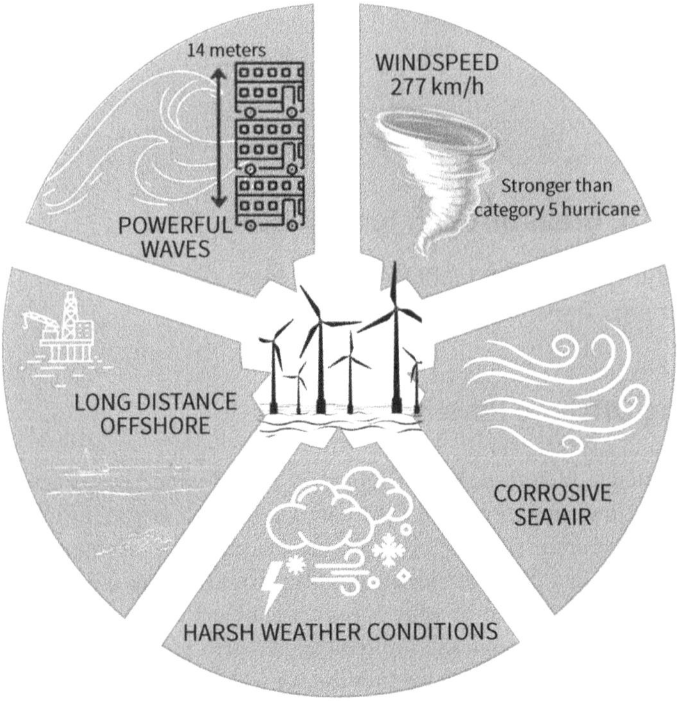


**Figure 1.1** Harsh environmental conditions around an offshore wind farm.
*Source:* Author's own work.

must endure a hostile marine environment characterized by high waves and gale-force winds. The system had to withstand wind speeds of up to 277km per hour, waves as high as 14 meters, and corrosive salty sea air. All these conditions, while transmitting 800 megawatts of electricity over 125 kilometers of open ocean and 75 kms long land route to the German grid [2]. Figure 1.2 provides a geographical overview of this extensive route, highlighting the distance between the offshore generation site and the onshore grid connection points.

Accomplishing this feat required Siemens to push the boundaries of HVDC transmission technology. The team pursued exceptional engineering solutions beyond standard industry practice. The project featured a new modular converter platform design, state-of-the-art cable router, and extensive system studies to ensure optimal performance [3].

Powerful winds, towering waves, and unpredictable currents presented immense challenges from the onset. The high-voltage direct current technology used was essential for efficiently transmitting the power generated over such a vast distance,

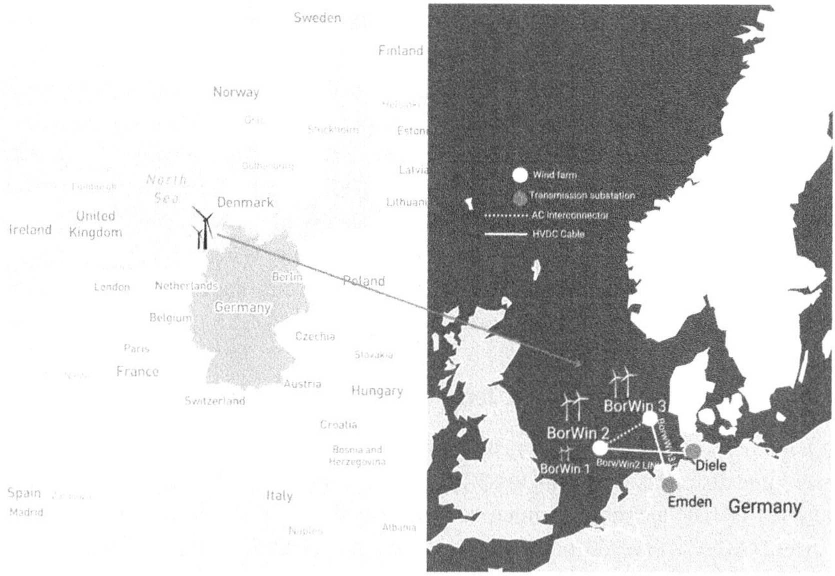


**Figure 1.2** BorWin2 geographical project overview. *Source:* Adapted from [5].

but it also came with its own set of complexities [4]. Developing the right control strategies to ensure grid stability and power quality was paramount.

This case study explores how Siemens navigated these challenges using a multi-pronged approach, leveraging cutting-edge technology, rigorous system modeling, and a deep understanding of the environment [6]. BorWin2 adopted Siemens' HVDC Plus with modular multilevel converter (MMC) based voltage source converter (VSC) technology, instead of ABB's HVDC Light® (as BorWin1 used). HVDC Plus offered greater transmission capacity and better grid compatibility [2]. The result was a groundbreaking offshore power link that set new standards for industry and paved the way for even more ambitious projects in the future. This required a different set of engineering expertise and system adjustments. Proving that sometimes the best way to solve a problem is to think differently about electricity transmission. This case highlighted how reconceptualizing power transmission fundamentals can lead to breakthrough results.

The core challenge of the project is clear here. How to efficiently transmit gigawatts of wind power over hundreds of kilometers of hostile marine environment. The solution was needed to maintain grid stability, minimize transmission losses, and ensure system reliability while balancing marine ecosystem protection with Germany's energy requirements.

## 1.2 Transmission Technology Considerations

Siemens is a German international company. They are responsible for several critical high-voltage direct current (HVDC) projects in the North Sea, including the BorWin2 project. Over the course of this case study, we explore the unique challenges associated with the BorWin2 HVDC transmission system and how Siemens overcame them to deliver reliable, sustainable power. The first offshore wind farm to use HVDC Light® technology was the BorWin1 project in Germany and was commissioned in 2010. Impressed by the technology's performance, the BorWin2 project was initiated shortly after.

### 1.2.1 The Evolution of HVDC Technology

What if early electrical engineers had come up with a technology to carry electricity over large distances with little loss? The entire modern world probably would have adopted HVDC technology much sooner than it did. The battle over alternating current (AC) vs. direct current (DC) transmission, in fact, was long-standing. Dating back to the late nineteenth century when Thomas Edison championed DC, while Nikola Tesla and George Westinghouse advocated for AC. Early DC systems suffered from significant voltage conversion inefficiencies and restricted transmission distances. These limitations ultimately led to AC becoming the dominant electricity transmission standard [7]. Yet with the advent of ever-increasing electrical power transmission needs over longer distances, engineers looked back at DC transmission and the result brought forth today's modern HVDC technology.

Before diving into BorWin2, let's understand why traditional AC transmission was not suitable for long-distance power transmission. Imagine trying to push water through a very long garden hose. The further it travels, the less pressure you have at the end. AC transmission over long distances faces similar challenges [2]. Although High Voltage Alternating Current (HVAC) is the norm for onshore grids, its efficiency drops significantly in long subsea cables due to the capacitive charging current (reactive power) generated by the cable insulation. HVAC transmission losses become unsustainable beyond 80–100 km, often requiring mid-point compensation stations, which are difficult to install offshore. In contrast, HVDC systems can remain highly efficient, with losses of around 3.5–5% per 1,000 km, and therefore represent the only viable engineering solution that makes sense for far-shore projects such as BorWin2 [8]. With HVDC transmission, it is like the superhighway of power transmission that literally upgraded the limitations of sustainable energy.

The first commercially successful HVDC system was constructed in the 1930s in Sweden. The Gotland link was built by a Swedish company ASEA: a 100 kV, 20 MW connection between the Swedish mainland and island of Gotland [8].

The development of this pioneering project indicated how HVDC could reliably transmit power over long distances with low losses. Over the decades thyristor converter technology evolved significantly. These advancements transformed high-voltage direct current (HVDC) transmission into a scalable solution for modern power systems and able to lock power-grid-neighboring grid links to the ±450 kV Pacific DC Intertie (United States).

In the 1990s, voltage source converters (VSC) burst onto the scene like the rockstars of power transmission, making HVDC systems smaller, more efficient, and flexible enough to handle offshore wind farms and weak AC grids [9]. Fast forward to today, modular multilevel converters (MMC) are the new headliners, with ABB's HVDC Light and Siemens' HVDC Plus stealing the show. These innovations are taking power transmission to a more insane level, and renewable energy is just walking the line of eligibility for it to appear in any superhero movie. From the first-generation debates of AC vs. DC to today's new applications like offshore wind, HVDC has grown into a stronghold technology for the sustainable energy supply.

One of the key advantages of HVDC transmission is to transmit power over long distances with low losses. Offshore wind farms are typically situated far from the shore, so this particularly applies. HVDC provides more control over power flow versus traditional AC transmission, as well as grid stability benefits and less of an environmental footprint.

The connection between the offshore wind farms and the onshore AC grid requires converting the variable-frequency AC power from the wind turbines into DC for transmission, then back to fixed-frequency AC power on the receiving end. This conversion is managed by converter stations located on offshore and onshore platforms. "When we first proposed transmitting gigawatts of power under the North Sea, some thought we were as crazy as Jules Verne writing about submarines," recalls a Siemens project manager. Now it's as routine as sending an email – well, almost [10]. Figure 1.3 outlines this sophisticated system architecture, tracing the energy flow from the wind turbines through offshore converter, along subsea cable, to the onshore station.

ABB's HVDC Light®, is a HVDC based voltage source converter known for its minimal size, efficient power loss management, and black start features such as being capable of restarting independently. In contrast, Siemens' HVDC Plus, is a competing HVDC technology with higher transmission capacity (up to 1,000 MW) and better grid integration. This means that it is well-suited for large-scale offshore wind power projects like BorWin2, which requires cross-country power transmission of significant amounts. Each HVDC technology has pros and cons. Therefore, all parameters must be analyzed by engineers before choosing any of them based on the project requirement. Considerations include power capacity, grid needs, reliability and cost-effectiveness.

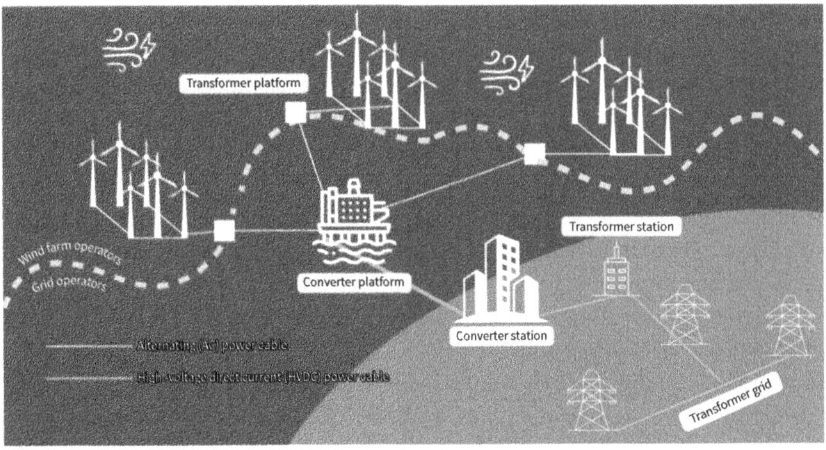


**Figure 1.3** BorWin2 HVDC system architecture – from wind to grid. *Source:* Adapted from [5].

| BorWin1 | | BorWin2 |
|---|---|---|
| 400 MW | Capacity | 400 MW supports 800 MW |
| 80 × Bard 5.0 MW | Turbines | 80 × Areva Wind M5000-116 (5.0 MW) |
| ~ 130 km | Distance to shore | ~ 140 km |
| Bard engineering | Developer | Global Tech I Offshore Wind GmbH |
| 2013 | Commissioned | 2015 |

**Figure 1.4** Technology comparison of BorWin1 and BorWin2. *Source:* Author's own work.

As Figure 1.4 illustrates, engineers must balance the project-unique needs and constraints when selecting the right HVDC technology for a project. HVDC Plus would be beneficial, for example, for HV applications with higher transmission capacity (3,000 MW per system) as it could be used for large-scale offshore wind farms such as BorWin2. It enables the transmission of larger amounts of renewable energy over 200 km links. Conversely, the compact design and black start capability of HVDC Light, which generally supports capacities up to 1,200 MW, can be more appropriate for smaller-scale projects with alternative power transmission requirements.

Beyond transmission capability, resolution of reliability, grid matching and cost-effectiveness are other things that should be considered in detail before choosing the most suitable HVDC solution that fits the project requirements and operational constraints [11]. Thoroughly evaluating these factors is essential to ensuring the successful implementation and long-term performance of the HVDC transmission system.

### 1.2.2 The Germans' HVDC Challenge

For many years, Europe has been a leader in the worldwide transition toward sustainable renewable energy with targets for carbon free renewables in their energy mix and overall energy efficiency. The European Green Deal, the aim of which is to achieve a climate-neutral Europe by 2050. Its interim targets of least 55% greenhouse gas (GHG) emission cuts on the way to 1990 levels by 2030 [12]. Key projects are under development to accelerate the phase out of fossil fuels notably due to the recent rise in energy security concerns. This plan rests largely on wind and solar as capacity, projected to total 300GW of offshore wind capacity in 2050. Germany has long been a frontrunner in this transformation, creating an example to the world with its very own Energiewende vision of the future or energy transition.

Germany's Energiewende is a long-term vision to phase out nuclear power and fossil fuels while significantly increasing the share of renewable energy sources. Germany's Energiewende (literally "energy transition") is the country's planned transition to a low-carbon, reliable, and affordable energy supply. It was launched in the early 2000s and accelerated after the 2011 Fukushima disaster. The country's goal is to achieve 80% of electricity from renewables by 2030 and climate neutrality by 2045 [13]. In the last two decades Germany has advanced tremendously; In 2023 renewable energy sources reached almost half of their overall consumption with 52%, from a mere 6% [14]. The dramatic growth of onshore and offshore wind, the broad promotion of solar photovoltaics (PV), and bioenergy has put wind and solar on a par with conventional power sources. It is in line with policy reforms including the Renewable Energy Act (EEG), such as feed-in tariffs that stimulated significant green energy investment. However, the transition has presented considerable challenges as well. The grid has been challenged during the withdrawal of nuclear power by 2023 and coal by phase out in favor of renewable 2038. It would require wide scale grid modernization of energy storage, more flexible backup systems and new grid modernization. Moreover, public opposition to onshore wind, grid delays and changing energy prices have made crossing Germany´s ambitious target much harder [15, 16]. Despite its challenges, in the arena of clean energy, Germany is still one of the world leaders with innovative solutions in hydrogen, offshore wind as well as smart grid technologies among others making way for the fully decarbonized energy system.

Germany hopes to be a leader in energy transformation as part of their "Energiewende" movement. One of their critical links is the 200 km long BorWin2 HVDC project which needs to transmit 800 MW offshore wind power from the North Sea to the German mainland grid. Siemens faced enormous challenges in this project, maintaining performance of equipment to withstand North Sea harsh environments and to deliver reliable power. BorWin2 is an order of magnitude and much bigger and more complex than its predecessor BorWin1. This increase in scale necessitated a technological shift from the HVDC Light® system used in the earlier project to Siemens' proprietary HVDC Plus solution. To understand the engineering logic behind this switch. Table 1.1 shows a comparison of their technical specifications.

The BorWin projects play a crucial role in Germany's renewable energy strategy. These HVDC transmission systems in the North Sea are designed to transport electricity from offshore wind farms to the German grid with minimal energy losses. With offshore wind farms increasingly being sited further from the shore, suitable long-distance transmission technologies like HVDC are needed to bring this power into the population. BorWin cluster is several projects – BorWin1 and its equivalents, BorWin2, BorWin3, BorWin5 as well as BorWin6 (projected) contribute to the maturing of German offshore wind.

Once fully completed, the BorWin projects will have a combined transmission capacity of approximately 5.5 GW, with BorWin1 (400 MW), BorWin2 (800 MW), BorWin3 (900 MW), BorWin5 (900 MW), and BorWin6 (980 MW) [17]. Over the

**Table 1.1** Comparison of the technical specifications of HVDC Light and HVDC Plus.

| Feature | HVDC Light® | HVDC Plus |
|---|---|---|
| Company | ABB/Hitachi | Siemens |
| Capacity | Up to 1,200 MW | Up to 3,000 MW |
| Design | Compact, modular | Standard, higher capacity |
| Black Start Capability | Yes | No |
| Grid Integration | Good | Enhanced |
| Typical Applications | Small-medium projects | Large-scale projects |
| Transmission Distance | Up to 200 km | 200 km+ |
| **Common Benefits:**<br>High efficiency<br>Low losses<br>VSC technology<br>Offshore capability | | |

*Source:* Author's own work.

next few years, the country aims to have its offshore wind capacity to reach 30 GW by 2030 and 70 GW in what they term as "first stabilization phase" no later than 2045 [18]. The 5.5 GW capacity of BorWin itself will correspond to something around 18% of Germany's planned offshore wind (2030) and around 8% of the country's entire renewable electricity demand. So, the BorWin projects are of key importance in Germany's offshore wind integration research roadmap. Its objective is to decrease the dependence on fossil fuels and foster a reliable and sustainable energy supply.

Although they have greater importance, these projects have also struggled in the realm of engineering and logistical challenges, as well as offshore platform construction risks. Nevertheless, advances in HVDC, modular offshore grid solutions plus wind turbine improvements to come allows BorWin to become the template for future German and international offshore projects.

To overcome the long-distance power transmission and harsh environmental conditions was no easy feat and the challenges did not end there. Integrating the BorWin2 HVDC system with the onshore grid necessitated overcoming a multitude of complex engineering and logistical obstacles. Transporting and installing the gigantic offshore converter platform was not exactly a weekend DIY project; it required specialized vessels, heavy-lift cranes, and a team of marine operators who probably deserve superhero capes for their precision. And if that wasn't enough, connecting the system to the existing AC grid felt like trying to peg a square plug into a round hole while also ensuring the lights didn't flicker like an old bulb.

800 MW of power over a 200-km subsea cable, introduced problems coming from voltage stability, power losses and electromagnetic compatibility aspects. In order to overcome these, optimization of cable design, converter station architecture and real-time control strategies are absolutely needed for grid stability and non-interruption of power supply. By testing various HVDC technologies, Siemens chose Siemens' HVDC Plus over ABB HVDC Light because of its higher transmission capacity. Its grid matching capability and modular multilevel converter (MMC) technology provides better voltage control and fault ride-through. The HVDC Plus converters had optimized reactive power compensation and higher order harmonic filtering, to facilitate seamless transition to the existing onshore AC network.

To mitigate operational risks, Siemens employed advanced system modeling and real-time simulations to assess the performance of the HVDC link under various grid conditions. These were used in simulations to tune control algorithms and protection schemes. It allows the system to dynamically react to changes in wind generation as well as grid demand. Following thorough planning, advanced engineering and use of cutting-edge power electronics Siemens has managed to build the BorWin2 HVDC transmission system, the new world best for offshore wind farming integration.

## 1.3 Think and Solve

The BorWin2 project was a difficult engineering challenge with novel approaches on power transfer, offshore infrastructure and linking it to the grid. Considering its size at the time of the commissioning, BorWin2 was one of the largest offshore HVDC systems. BorWin2 had to break new ground on several technical and organizational hurdles. The key challenges were transmitting offshore wind power efficiently over long distances, managing power quality and stability when connecting to the onshore grid. In parallel, it has to ensure the reliability of subsea cables and offshore converter platforms in the harsh North Sea environment.

This section outlines the key problems encountered through some questions. By thinking about these points, readers can review the engineering principles used to address them, alternative solutions or approaches considered, and the lessons learned from implementing these and other methods.

The BorWin2 project was designed to transmit 800 MW of offshore wind power over a 200 km HVDC submarine cable to the German mainland. It works at ±300 kV bipolar HVDC transmission for minimum losses and higher reliability. Environmental conditions can be very unstable for the subsea cable. High saltwater corrosion, intense weather exposure and seabed vibrations can cause unpredictable mechanical load. Power losses had to be kept to a minimum so that minimum loss in the transmission can be successfully accomplished by the project.

To analyze the system's efficiency, consider the following engineering scenario:

- The total resistance per conductor of the HVDC cable is 0.02 Ω/km.
- The transmission voltage is ±300 kV, meaning the total voltage difference is 600 kV.
- Power transmitted (P) = 800 MW
- Cable length (L) = 200 km
- The HVDC system efficiency is **96%**.

**Q 1.1** Using the stated facts above, solve the following electrical power questions.

Small discrepancies may arise due to rounding.

- Calculate the total transmission resistance of the cable, considering both conductors. $R_{total} = R' \times L \times 2$
- Determine the DC current (I) flowing through the HVDC link. $I = P/V$
- Compute the power loss (MW) in the transmission cable due to resistance. $P_{loss} = I^2 \times R_{total}$

Based on a 96% efficiency, estimate the amount of power actually received at the onshore converter station. $P_{received} = P - P_{loss}$

This problem requires an understanding of power transmission losses, Ohm's Law, and system efficiency calculations, which are fundamental to electrical power engineering.

**Q 1.2** What was the primary engineering challenge in the BorWin2 project?

**A** Insufficient wind turbine efficiency.
**B** Difficulty in integrating HVDC with the onshore AC grid.
**C** Limited offshore wind potential in the North Sea.
**D** Lack of available submarine cable technology.

**Q 1.3** Which of the following engineering principles was critical in designing the HVDC transmission for BorWin2?

**A** Ohm's Law and Kirchhoff's Voltage Law.
**B** Electromagnetic Induction in Transmission Lines.
**C** Power Factor Correction and Reactive Power Management.
**D** Both B and C.

**Q 1.4** What alternative solution could be considered if HVDC transmission were not used for BorWin2?

**A** Using low-voltage AC transmission.
**B** Building an onshore wind farm instead.
**C** Implementing an HVAC (High Voltage Alternating Current) transmission system.
**D** Using hydrogen storage to transport wind energy.

## 1.4 Conclusion and Key Take-Away

The BorWin2 project is a big step forward for offshore wind power transmission. The project has proven that far distance high-voltage DC transmission can be done smoothly and efficiently by putting an 800 MW offshore wind farm onto the grid using HVDC. The challenges faced in transportation and installation of new platforms, the grid integration and system stability in such mega offshore DC projects have demonstrated that engineering solutions can be developed to solve them.

One of the key lessons learned from BorWin2 is the selection of proper transmission technology. The HVDC Plus solution of Siemens was essential for a stable grid, improved power quality and for the provision of reactive power compensation with quality and reliability overall of the complete system. VSC-based HVDC converters can provide reactive power compensation to the AC grid at both sending and receiving ends. These converters adjust reactive power output even when there is no active power transmission. It is one of the major advantages of modern VSC-HVDC systems over traditional Line Commutated Converter (LCC) HVDC systems. LCC based HVDC systems consume reactive power and also require additional compensation equipment. So, this project highlighted the requirement of careful planning, systematic modeling and dynamic research in power electronics to enable seamless integration with existing AC grids.

Furthermore, BorWin2 contributes to Germany's broader Energiewende (energy transition) goals by enabling a greater share of renewable energy in the national energy mix. As offshore wind capacity continues to expand, projects like BorWin2 serve as critical building blocks for achieving Europe's sustainability targets and reducing dependence on fossil fuels.

Looking ahead, the experiences from BorWin2 can be valuable for upcoming offshore HVDC projects regarding various areas from converter station design to operational efficiency gains and transmission loss reduction. With additional BorWin projects going live, they will provide an ever more crucial function in the construction of Europe's renewable energy backbone and move us a step closer toward a sustainable power system.

## 1.5 Discussion Points

BorWin2 is one of the most inspiring case studies on the transmission of offshore wind power, which brilliantly shows both the potential and problems of integrating large-scale renewable energy. The successful implementation of the HVDC solutions in this project demonstrates aspects of major engineering achievements, strategic decisions and offshore wind contribution for Germany to succeed with its transition to renewables.

The following discussion points aim to encourage further exploration of the key aspects of BorWin2 and its significance in the evolving landscape of renewable energy:

1. **HVDC vs. HVAC Transmission:** Why is HVDC preferred over traditional HVAC for offshore wind power transmission? What are the benefits and potential drawbacks?
2. **Technology Selection:** How did Siemens' HVDC Plus technology improve grid stability and efficiency in comparison to other available options?
3. **Global Impact:** How do large-scale offshore wind projects like BorWin2 influence renewable energy policies and market trends worldwide?
4. **Optimization and Future Improvements:** What advancements could be made to enhance the efficiency and cost-effectiveness of future HVDC transmission projects?

## 1.6 Limitations of the Case Study

While the BorWin2 project provides valuable insights into the complexities of offshore HVDC transmission, certain limitations must be acknowledged when analyzing its impact and broader applicability. There are technological constraints, economic, grid compatibility and expansion difficulties, environmental and regulatory barriers working on top of the system scalability and future adaptability in addition to operational maintenance issues are some of the limitations.

1. **Technological Constraints:** Though HVDC technology has evolved for BorWin2, there were still some problems with system losses, converter efficiency and reactive power handling in this system. Further optimization of the power transmission capability might be enabled by the concurrent advancements in converter station technology and cable technology.
2. **Economic Considerations:** Offshore HVDC platforms are costly to deploy initially, which acts as a barrier to full introduction of such systems. The BorWin2 project required substantial infrastructure, dedicated vessels, and maintenance costs sparked questions of BorWin2 cost-effectiveness in comparison with other renewable energy integration methods.
3. **Grid Compatibility and Expansion:** Integrating offshore wind farms into existing AC grids can pose stability concerns. While BorWin2 addressed these grids integration issues, different regional and national grid configurations may require tailored solutions to ensure compatibility and stability, which could limit the universal application of the specific approaches used in the BorWin2 project. The unique characteristics and requirements of each grid system must be carefully considered when planning and designing offshore HVDC transmission projects to ensure optimal performance and seamless integration.
4. **Environmental and Regulatory Challenges:** Offshore wind projects must comply with strict environmental regulations and permitting processes. The BorWin2 project overcomes these challenges, but other projects in different regions may encounter greater delays and restrictions.
5. **Operational and Maintenance Complexities:** The harsh offshore environment introduces formidable maintenance hurdles, demanding regular inspections and repairs. Adverse weather conditions can impede accessibility and drive-up operational expenditures over the lifespan of the project.
6. **Scalability and Future Adaptability:** As BorWin2 demonstrated, future offshore HVDC projects may need to handle even higher transmission capacities and build more resilient infrastructure. The lessons from BorWin2 will need to be applied to accommodate growing offshore wind development.

## References


1 Misyris, G., Van Cutsem, T., Møller, J., Dijokas, M., Estragués, O.R., Bastin, B., Chatzivasileiadis, S., Nielsen, A., Weckesser, T., Østergaard, J. and Kryezi, F. "North Sea wind power hub: system configurations, grid implementation and techno-economic assessment," arXiv preprint arXiv:2006.05829, 2020.

2 Xu, L. and Andersen, B.R. "Grid connection of large offshore wind farms using HVDC," *Wind Energy*, 9(4), 371–382, 2006.

3 Papathanassiou, S. and Papadopoulos, M. "A study case on harmonic distortion created by wind turbines," in *18th International Conference and Exhibition on Electricity Distribution (CIRED 2005)*, v2–21. Stevenage UK: IET, 2005.

4 Jiao, Y. and Nian, H. "Grid-forming control for DFIG based wind farms to enhance the stability of LCC-HVDC," *IEEE Access*, 8, 156752–156762, 2020.

5 TenneT. "BorWin2 project description," https://www.tennet.eu/de-en/projects/borwin2 (accessed February 21, 2026].

6 Stadtmann, F., Rasheed, A., Kvamsdal, T., Johannessen, K.A., San, O., Kölle, K., Tande, J.O., Barstad, I., Benhamou, A., Brathaug, T. and Christiansen, T. "Digital twins in wind energy: emerging technologies and industry-informed future directions," *IEEE Access*, 11, 110762–110795, 2023.

7 McNichol, T. *AC/DC: The Savage Tale of the First Standards War*, John Wiley & Sons, 2011.

8 Ardelean, M. and Minnebo, P. *HVDC Submarine Power Cables in the World: State-of-the-Art Knowledge*, Publications Office of the European Union, 2015.

9 Mitra, P., Wikström, C., Johannesson, N. and Larsson, T. "First real-time implementation of DC grid protection strategy," in *11th IET International Conference on AC and DC Power Transmission*, 060. Stevenage UK: IET, 2015.

10 Lesnicar, A. and Marquardt, R. "An innovative modular multilevel converter topology suitable for a wide power range," in *2003 IEEE Bologna Power Tech Conference Proceedings*, vol. 3, 6. IEEE, 2003.

11 Thompson, M. "Interview with Klaus Schmidt, Siemens HVDC project manager," *Power Engineering International*, 28(4), 45–48, 2020.

12 Ryndzionek, R. and Sienkiewicz, Ł. "Evolution of the HVDC link connecting offshore wind farms to onshore power systems," *Energies*, 13(8), 1914, 2020.

13 Fetting, C. "The European green deal," *ESDN Report*, 2(9), December 2020.

14 Schwäbe, C., Dreher, C., Voss, F. and Kovac, M. "How to govern mission-oriented innovation policy–the case of the German Energy Research Program," Conference: Eu-SPRI, University of TwenteAt: Twente, Enschede, The Netherlands, 2024.

15 Koehler, H., Halfmann, D., Hercegfi, P., Kuhlenkampff, F., Armbruster, C. and Rogalla, S. "Material demand analysis for power electronics in the German energy transition across generation, storage, and electric vehicles," in *2024 Energy Conversion Congress & Expo Europe (ECCE Europe)*, 1–7. IEEE, 2024.

**16** Reichardt, K., Negro, S.O., Rogge, K.S. and Hekkert, M.P. "Analyzing interdependencies between policy mixes and technological innovation systems: the case of offshore wind in Germany," *Technological Forecasting and Social Change*, 106, 11–21, 2016.

**17** TenneT, "TenneT Germany," https://www.tennet.eu (accessed February 21, 2026).

**18** Hinrichs-Rahlwes, R., "Massive growth of PV capacity as a major cornerstone of Germany's energy security and climate policies," in *Reducing the Effects of Climate Change Using Building-Integrated and Building-Applied Photovoltaics in the Power Supply*, 149–161. Cham: Springer Nature Switzerland, 2024.

# 2

# GreenHyScale Denmark: Large-Scale Green Hydrogen Engineering

From Fossil-Based to Renewable Hydrogen Through Advanced Electrolysis Technologies

## 2.1 Synopsis

How do we transform abundant renewable resources into versatile, clean fuel? Imagine converting the power of wind and sun into an electron flow that can fuel industries, heat homes, and power vehicles. Green Hydrogen is a valuable and naturally available resource. This element is vital for reducing human-made carbon dioxide emissions and transitioning to a renewable energy system. However, the current methods of hydrogen production are heavily reliant on fossil fuels, undermining their environmental benefits [1]. However, the less talked about green hydrogen is now being pushed to the center of scientific and industrial debates. This case study explores the frontier of this renewable energy engineering, where electricity meets water to produce green hydrogen – potentially the most significant energy carrier of our sustainable future.

However, the scale-up and cost-competitiveness of green hydrogen remain major hurdles. This case study examines the latest advancements, obstacles, and opportunities in scaling up green hydrogen production to an industrial level, with a focus on the "GreenHyScale" project in Denmark. The goal of GreenHyScale is to create a replicable solution for companies in both Denmark and the rest of Europe to accelerate the development of large-scale green hydrogen infrastructure [2].

This case study will unpack the various elements that will shape the future of industrial-scale green hydrogen in the energy transition. The case study examines how engineering solutions in electrolysis technology are reshaping our approach to renewable energy storage and utilization, marking a crucial step toward industrial decarbonization and energy system integration. As we stand at the threshold of the hydrogen economy, the technical decisions made today will shape the energy landscape for decades to come.

## 2.2 Background and Case Scenario

Hydrogen is the most abundant element in the universe. It has an energy density about three times higher than that of conventional fossil fuels on a weight basis. When combusted, hydrogen reacts with oxygen to produce water vapor, without emitting any greenhouse gases [3]. This already makes hydrogen an interesting clean energy carrier with the possibility to be a large stake of the sustainable energy system. To be precise, a kilogram of hydrogen has an energy equivalent to 7–8 kg of coal or 2.8 liters of gasoline [4]. Therefore, hydrogen has been classified as the highest carbon free energy source compared to any other energy resources [5].

In 1800, William Nicholson and Anthony Carlisle first demonstrated water electrolysis by splitting water into hydrogen and oxygen. Two centuries later, this fundamental principle evolved into a cornerstone of the energy transition. Figure 2.1 shows the fundamentals of hydrogen production from its discovery to a green transition today. While hydrogen production has traditionally been dominated by fossil fuels – with 95% of current hydrogen production emitting 900 million tons of $CO_2$ annually. Thus, the engineering challenge of our time lies in scaling up clean production methods without fossil fuels.

Green hydrogen is produced via electrolysis of water using renewable energy. It significantly reduces greenhouse gas emissions compared with conventional hydrogen production methods. The growth of green hydrogen sector has catalyzed interest in electrolyzer technology, which also helps in smart grid management with respect to supply and demand and means of utilizing excessive renewable energy generation. In addition, biological hydrogen production processes utilizing microorganisms or

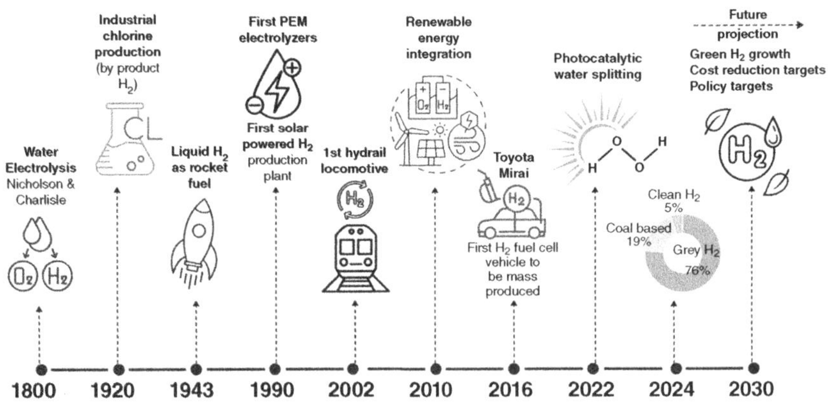


**Figure 2.1** Major milestones of hydrogen production – from discovery to green transition. *Source:* Author's own work.

enzymes to split water molecules are also being explored as viable green options [6]. The key lies in using renewable electricity to power the electrolysis of water, creating "green" hydrogen with zero direct emissions.

The global push for sustainable energy solutions has brought the hydrogen economy to the forefront of scientific and industrial discussions. Hydrogen is an important chemical feedstock for many large-scale industrial applications. Today, the majority of this feedstock is generated from fossil fuel sources such as reforming natural gas [7]. However, this production process results in significant greenhouse gas emissions, contributing to climate change.

The challenge pulls in different players: renewable energy developers, electrolyzer suppliers, industrial users, energy utilities, and policymakers. The European long-term decarbonization strategy "A Clean Planet for all" emphasizes the importance of hydrogen by considering it a key European strategy. The scale-up of current technologies to gigawatt-scale is a monumental challenge. The European Union's ambitious target of producing 10 million tons of renewable hydrogen by 2030 has set the stage for unprecedented technological innovation in this field.

Currently green hydrogen serves as a crucial link for renewable electricity generation and re-innovation of the traditional energy sector. It bridges the gap as an energy carrier and storage medium, therefore potentially solving the problem of intermittency faced by wind and solar power. Companies like Siemens Energy (with their SILYZER technology), ITM Power, and ThyssenKrupp have been instrumental in scaling up the technology. In this direction, the green initiative like the GreenHyScale project is the potential large-scale implementation. It pioneering to make smooth transition of renewable energy production to industrial energy consumption.

The GreenHyScale project represents a critical milestone in renewable energy engineering because it potentially addresses the complex challenges of scaling up green hydrogen production. With global electrolyzer capacity projected to grow from 0.3 GW in 2020 to potentially 850 GW by 2050, engineers face intricate challenges in system design, efficiency optimization, and integration with renewable power sources. Sustainable scale up involves the need for technology innovations in electrolyzer, and power electronics while maintaining economic stability.

### 2.2.1 The Evolution of Renewable Energies

Wind as well as solar energy have over the last decades changed the renewable energy landscape on a global scale [8]. These clean energy technologies have become increasingly cost-competitive and starting to disrupt fossil fuels based on traditional energy systems. With the increase in renewable energy generation come challenges to handle its intermittency and variability.

The journey of renewable energy technologies spans nearly two centuries. The commercialization of hydropower began in 1882, which harness water into a source for electricity. In 1954 the start of the commercialization of solar photovoltaics (PV) occurred. This was followed by the emergence of wind turbines in the 1970s. The late 1980s brought with its biomass power plants and the 1990s was the expansion in geothermal power applications. As we entered the twenty-first century, Concentrated Solar Power (CSP) gained traction, followed by offshore wind farms in the 2010s. Figure 2.2 shows the timeline of key renewable energy milestones. These renewable energy technologies have experienced significant cost reductions and efficiency improvements over time. It is becoming increasingly competitive with traditional fossil fuel-based power generation.

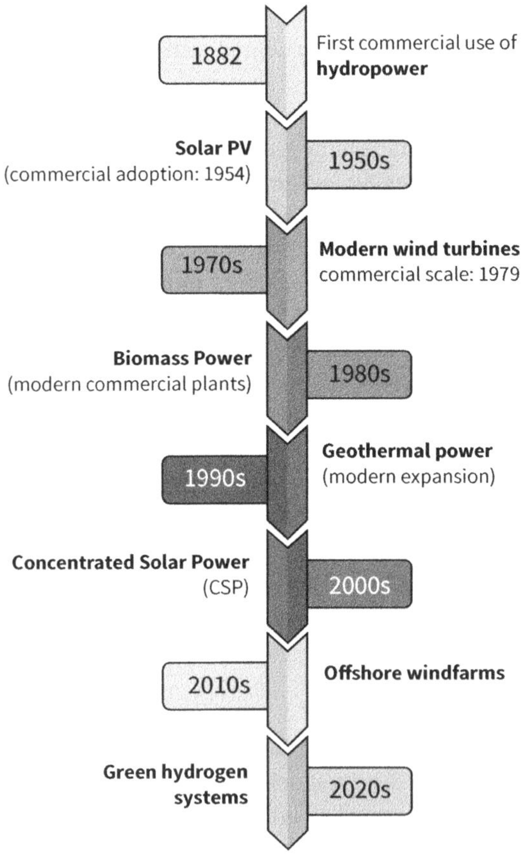


**Figure 2.2** Timeline of renewable energy technologies. *Source:* Author's own work.

Green hydrogen systems, emerging in the 2020s, represent the newest addition to the renewable energy portfolio. Current worldwide production of green hydrogen in small scale should grow to a total capacity of 850GW by 2050 [9]. The relentless exponential increase includes decreasing electrolyzer costs (expected to be near 40–80% reduction by 2030), with renewable energy capacity ramp up as well as supportive policy worldwide. Particularly, the EU aims to achieve 10 MTOEP of renewable hydrogen by 2030 itself. While the transition to renewable energy is well underway, significant hurdles remain with respect to energy storage, grid integration, and decarbonizing sectors like heavy industry and long-distance transportation.

### 2.2.2 The Road to Green Hydrogen

Hydrogen can be produced from various sources, including fossil fuels as mentioned in section 2.1, as well as nuclear power, and renewable energy. Green hydrogen's journey in the renewable energy sector began during the 1970s oil crisis. This situation provided researchers with the drive to discover new ways of producing energy other than fossil fuels [10]. But the first real commercial attempt was made in 1991, when the German Solar Wasserstoff Bayern (SWB) built a solar powered hydrogen production facility [11]. The early initiative may be limited, but it has already proved that renewable energy can be used to generate hydrogen.

The modern green hydrogen movement really hit the gas in 2017 when NEL ASA, a Norwegian company, decided to flex its innovation muscles and launched the first commercially viable alkaline electrolyzer production line. This wasn't just any turning point but it was the kind of turning point that makes you wonder why it took so long. Suddenly, green hydrogen production became scalable and cost-effective. Since then, green hydrogen technology has been riding rapid growth of renewable energy sources, especially with solar powered plants.

Today, several demonstration projects and pilot plants are being constructed around the world, showcasing the potential of green hydrogen to support the global energy transition. But before discussing one of these projects, it's crucial to explain the different hydrogen production technologies and their advantages and disadvantages [2].

Most of the green hydrogen production takes electrolysis, which is used to separate water ($H_2O$) into hydrogen ($H_2$) and oxygen ($O_2$) by passing electricity. Electrolysis technology has evolved through three major advancements over the years, each with its distinct advantages and challenges; these are alkaline electrolysis, proton exchange membrane electrolysis and solid oxide electrolysis [1].

The most mature and commercially deployed technology of hydrogen production is alkaline electrolysis (AEL). It accounts for about 70% of global electrolytic hydrogen production [12]. The oldest method uses a liquid alkaline electrolyte that operates at relatively low temperatures of 60–80 °C. Advantages include lower capital cost, a relatively robust solution and the ability to use low-purity water. While AEL systems are very reliable and inexpensive in producing hydrogen, they are often larger in size so could respond slowly to variations of electricity supply. This makes them less appropriate for integration with renewable energy. Also, alkaline electrolyzers have lower current densities and hydrogen purity compared to other technologies [13].

On the other hand, proton exchange membrane (PEM) electrolysis employs a solid polymer electrolyte and operates at much higher current densities for larger and more efficient systems [14]. PEM electrolyzer systems were developed in the 1960s by General Electric, and are inherently more suitable to handle the intermittency of renewable energy sources due to their higher dynamic response compared with alkaline electrolyzers [15]. They are ideal for matching solar or wind power. However, PEM electrolyzers need a strictly high-purity feed water, and require expensive metal catalysts leading to higher capital and operating costs than alkaline systems.

Solid oxide electrolysis (SOEC), the third technology operates at extreme high temperatures of 700–850 °C, enabling better water splitting and the option for co-electrolysis of steam with $CO_2$ to produce syngas [16]. Though this is the most modern method and promises higher efficiencies it faces problems related to materials, durability, and maintenance costs. Unlike steam reformation, it can harness industrial waste heat or nuclear reactor thermal output and is thus a strong contender for large-scale hydrogen production [17]. Unfortunately, the lifetime of SOEC systems is still a challenge under extreme operating conditions so that materials science have to be improved further.

Each of these electrolysis pathways has its own merits and drawbacks, and the optimal choice depends on factors such as scale, intermittency of power supply, water quality, and capital/operating costs [18]. Table 2.1 summarizes the basic characteristics of each technology with their advantages and disadvantages.

Aside from these well-known hydrogen electrolysis production methods, some other emerging approaches and cutting-edge research works are ongoing. For instance HydroGEN Consortium, one of the early-stage research programs supported by the US Department of Energy, is at the frontier of these emerging methods [19]. Their work explores multiple advanced water-splitting processes, including thermochemical, photoelectrochemical, and high-temperature electrolysis accomplished only using renewable energy [20]. These new efforts promise better hydrogen production than the conventional alkaline or PEM

**Table 2.1** Comparison of hydrogen electrolysis production technologies.

| | Alkaline (AEL) | Proton Exchange Membrane (PEM) | Solid Oxide (SOEC) |
|---|---|---|---|
| **Development Period** | 1920s | 1960s | 2000s |
| **Operating Temperature** | 60–80 °C | 50–80 °C | 700–900 °C |
| **Electrolyte** | Liquid KOH/ NaOH | Solid polymer membrane | Ceramic |
| | | **ADVANTAGES** | |
| **Cost** | Low-cost materials | Moderate | High material costs |
| **Maturity** | Well-established | Commercial | Emerging |
| **System Size** | Large | Compact | Moderate |
| **Response Time** | Slow | Fast | Moderate |
| **Efficiency** | 65–82% | 65–83% | Up to 95% |
| | | **DISADVANTAGES** | |
| **Technical** | • Low current density<br>• Low pressure operation<br>• Corrosive electrolyte | • Membrane degradation<br>• Complex water management<br>• Sensitive to impurities<br>• Limited stack size | • Material degradation<br>• High temperature requirements<br>• Brittle ceramic components<br>• Sealing challenges |
| **Economic** | • Large space requirement | • High capital costs<br>• Expensive catalysts<br>• Higher O&M costs<br>• High maintenance costs | • High manufacturing costs<br>• Limited commercial usage |
| **Operational** | • Slow response time<br>• Poor dynamic | • Limited durability<br>• Shorter lifetime than AEL | • Limited long-term stability<br>• Thermal cycling issues<br>• Long start up time<br>• Complex heat management |

*Source:* Author's own work.

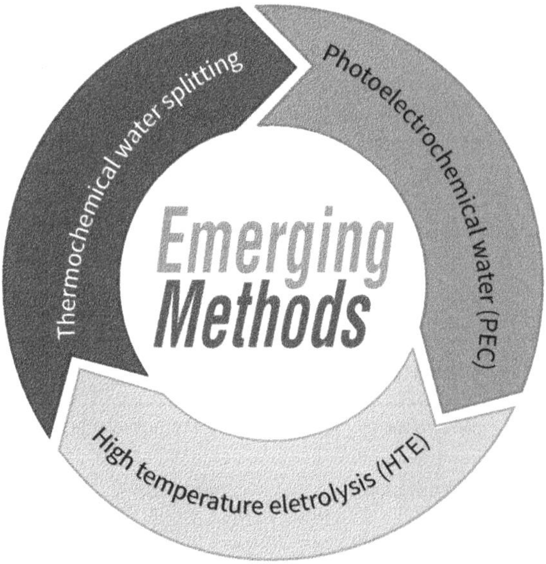


**Figure 2.3** Emerging methods of hydrogen electrolysis production. *Source:* Author's own work.

electrolysis. This area is being actively researched and pursued to prepare us for the transition to a more sustainable hydrogen economy powered by renewables (Figure 2.3).

Thermochemical water splitting is a newer approach in which extreme heat is used for water splitting. This heat is achieved through concentrated solar power or nuclear reactors to produce the electrical energy required for hydrogen production. By utilizing a high temperature environment, this method breaks down water molecules and obtain hydrogen without electricity-driven electrolysis. Instead, the water splitting reactions are driven by the heat, potentially reaching impressive hydrogen production efficiencies of 50% or more [21].

On the other hand, photoelectrochemical water (PEC) splitting capitalizes on the use of solar and special semiconductor materials to dissociate water molecules directly [22]. This was inspired by photosynthesis in nature. The direct conversion of solar energy into hydrogen is a promising route for the efficient and low-cost production of $H_2$.

Additionally, high-temperature electrolysis (HTE) is the process that super-optimizes hydrogen production by powering both heat and electricity simultaneously [23]. This general approach is usually integrated with solid oxide electrolysis cell technology which is being able to operate at extreme high temperatures of 700–850 °C. This enables better water splitting and add the option for co-electrolysis of steam and carbon dioxide to produce syngas. Syngas is a versatile feedstock that can be useful for further processing.

These new hydrogen technologies have their own pros and cons, and further development of them is key to growing the hydrogen industry and supporting global clean energy transition. They are still not ready for the market, but the constant research in this area has immense potential that could open new ways to make hydrogen cost-effectively on a large scale and with zero carbon emissions. When you look at the conventional hydrogen production methods, there are some clear connections between the old and new approaches.

The key overlap between conventional hydrogen technologies and emerging approaches lies in their shared goal of efficient hydrogen production while minimizing environmental impact. High-Temperature Electrolysis (HTE) is closely related to SOEC [24]. SOEC itself is a high temperature electrolyzer, meaning SOEC systems can incorporate HTE principles to increase efficiency. The other two (Thermochemical and PEC) are separate from electrolysis. They use different scientific principles, relying more on heat or direct solar reactions rather than electricity [25].

Figure 2.4 shows the comparison between mainstream and emerging hydrogen production methods. The overlap is evident in the shared interest in lowering carbon footprints, increasing efficiencies and integrating hydrogen production with renewable resources [26]. Both conventional and emerging methods feed into a diversification of technology routes toward a sustainable hydrogen economy. In both methods, hybrid and water-splitting approaches offer more long-term innovative potential as electrolysis is commercially well advanced. TRL stands for Technology Readiness Level. As shown in Figure 2.4, TRL represents a standardized scale from 1 to 9 which measures the maturity level of a particular technology [27]. Below is what each level means and refers to the figure for the context in hydrogen technologies.

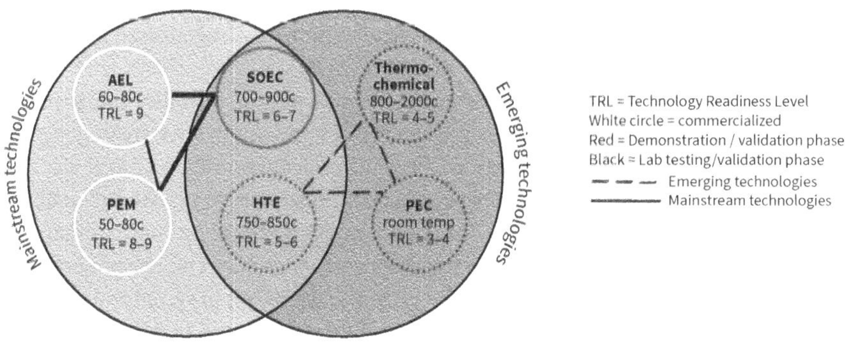


**Figure 2.4** Mainstream electrolysis technologies vs. emerging advanced water splitting techniques. *Source:* Author's own work.

TRL 1: Basic principles observed.
TRL 2: Technology concept formulated.
TRL 3: Experimental proof of concept.
TRL 4: Technology validated in lab.
TRL 5: Technology validated in relevant environment.
TRL 6: Technology demonstrated in relevant environment.
TRL 7: System prototype demonstration in operational environment.
TRL 8: System complete and qualified.
TRL 9: Actual system proven in operational environment.

In the context of hydrogen production technologies, so far only AEL and PEM are commercialized while SOEC is in the demonstration phase. Out of the three emerging techniques, only HTE is close to commercialization and is currently in validation and nearly demonstration phase. While each technology offers unique advantages, the choice for large-scale implementation depends on factors like maturity, efficiency, and cost-effectiveness. The GreenHyScale project, which is the focus of this case study, has chosen alkaline electrolysis technology for its proven reliability and scalability in industrial applications.

### 2.2.3 The Ambition of the GreenHyScale Project

What if a single project could transform not just a region but the entire continent's approach to renewable energy? Much like seeding the forest that could grow into thousands of trees, the GreenHyScale project in Skive, Denmark is laying the foundation for large-sized green hydrogen production. The ambitious five-year project aims to build a 100MW electrolysis plant [28]. This will harness Denmark's offshore wind resources to power this state-of-the-art high efficiency electrolyzer system. This system will be capable of 20,000 tones production of green hydrogen annually. By creating a scalable and adaptable model for both onshore and offshore applications, GreenHyScale seeks to act as a blueprint for global green hydrogen initiatives, aligning perfectly with the European Union's hydrogen strategy and offshore renewable energy goals [29].

GreenHyScale is already unique in terms of scale and replicability compared to other hydrogen projects. Although Europe has a lot of innovative hydrogen projects on its hands (e.g., the German REFHYNE project running a 10MW electrolyzer in the Shell Rheinland refinery and the H2Future project in Austria using a 6MW PEM electrolyzer for steelmaking by hydrogen) these are not comparable with GreenHyScale's goal of being at 100MW capacity [30]. In UK, the projects like HyNet focuses on blue hydrogen – using carbon capture but from natural gas [31]. In contrast to that, GreenHyScale's commitment to 100% green hydrogen powered by renewable energy is a league of its own and a true sustainable solution.

GreenHyScale is currently in advanced development phase with ambitions to create a reference for big-scale green hydrogen projects. The progress of the project is being closely scrutinized by stakeholders in both industry and political world. If it is successful, the most important impact will be reducing the cost of hydrogen while jump-starting a pan-European green hydrogen economy. In contrast to small pilot projects, GreenHyScale opens a replicable path that could speed up the commercial deployment of green hydrogen by providing a durable framework. It paves the road toward large scale but cost-competitive green hydrogen solution while achieving zero carbon emissions [32].

But how did GreenHyScale begin and who are the initiators? A historical analysis revealed that the GreenHyScale project emerged from a strategic collaboration initiated in late 2021, bringing together some of Europe's leading energy and hydrogen technology companies. The consortium structure reflects a carefully balanced mix of technology providers, research institutions, and industrial end-users, each bringing specific expertise to the project. Figure 2.5 contains the main collaborators that were brought together by the driving factors of the project.

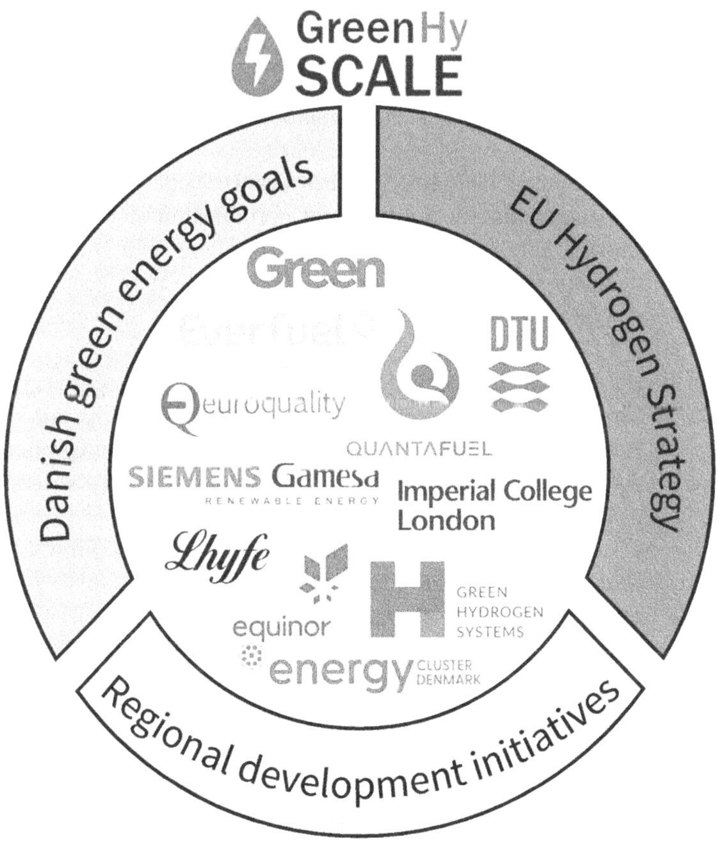


**Figure 2.5** GreenHyScale project ecosystem and driving factors. *Source:* Author's own work.

The consortium is led by Green Hydrogen Systems as the project coordinator, a Danish technology company responsible for overall project management [33]. They bring their expertise from previous successful smaller-scale installations and contribute with their core pressurized alkaline electrolysis technology. Their leadership role ensures technological innovation remains at the forefront of the project's development.

Siemens Gamesa Renewable Energy is a key technology partner with deep wind power integration heritage [34]. It is critical to both power management and grid balancing solutions, as well as driving offshore applications advancement of the project. Their involvement ensures smooth operational integration of renewable energy generation and hydrogen production. For the operation of hydrogen distribution infrastructure and developing systemic safety procedures as well as emergency handling, Everfuel is responsible. They bring relevant experience in hydrogen logistics to the project with substantial value to its operational scope.

As for the site provider, Greenlab Skive also serves as the industrial symbiosis partner. They manage local stakeholder relations and provide existing infrastructure and utilities [35]. Their established presence in the region and experience in industrial park management provides a solid foundation for the project's physical implementation.

For research and development activities, the Technical University of Denmark (DTU) leads crucial performance in analysis and optimization studies. Their participation in technical validation and certification, along with their dedication to training and knowledge transfer, enables the project to keep high academic, technical standards and at the same time contribute to workforce development [36].

Energinet as the Danish transmission system operator who has an essential grid integration and generation management expertize in Denmark. Participation of such partners guarantees compliance with regulation and seamless integration into the current power backbone – an essential pillar to the project's success partner.

The inception of GreenHyScale was driven by three main factors. First, the policy context aligned perfectly with the EU Hydrogen Strategy, Danish government's green energy transition goals, and regional development initiatives in the Central Denmark Region. Secondly, market drivers were the rise of green hydrogen solutions for big commercial applications; especially reductions in hydrogen production cost and industrial sector decarbonization. Third, technological readiness via matured alkaline electrolysis technology integration with renewable energy, and system efficiency following scalability related improvements.

The project's financial structure comprises €30 million from EU Horizon 2020 (Grant agreement: 101036935), complemented by €45 million in private

**Figure 2.6** Main financial support for GreenHyScale project. *Source:* Author's own work.

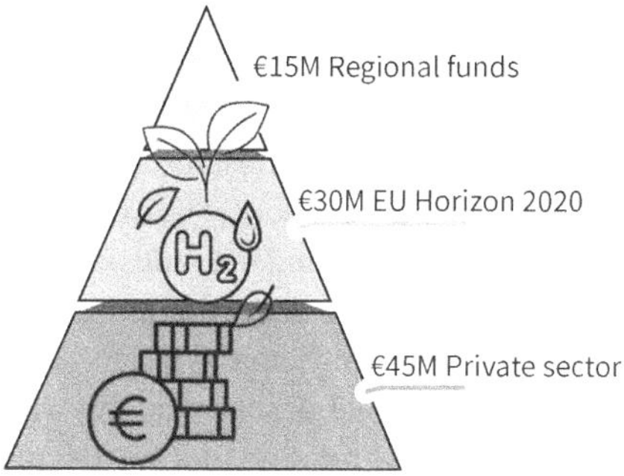


sector investment and €15 million from regional development funds. Additional support comes from the Danish Energy Agency while creating a robust financial foundation for the project's execution [37]. The GreenHyScale project is in very ambitious terms and with a compact, carefully built consortium and solid financial support. It can act as a systemic demonstration and blueprint for green hydrogen at industrial scale worldwide [38] (Figure 2.6).

With an established development timeline, the project started with initial concept development in 2020. Then in 2021, the project gained EU funding appraisal, achieved advanced planning of the project and signed detailed site selection and contractual milestone. In the following year, the official launch took place which signified the start of detailed engineering design, environmental impact assessment and stakeholder engagement programs.

## 2.3 Think and Solve

Scaling up hydrogen production for the practical project faces substantial technical obstacles. The 100 MW system represents an unprecedented engineering challenge, far exceeding the scale of existing installations. A primary hurdle is managing the electrolyzer's operation with wind power's intermittent nature. The system must respond rapidly to power fluctuations while maintaining optimal efficiency and preventing accelerated equipment degradation. The integration of onshore hydrogen production infrastructure with offshore wind generation presents additional logistical complexities. The project site's distance from offshore wind farms necessitates carefully planned energy transmission and hydrogen distribution networks. This geographical separation requires robust infrastructure solutions to ensure efficient energy transfer and hydrogen transport throughout the system [39].

Besides technological hurdles, there are also limited infrastructure and economic concerns. The integration of onshore hydrogen production with offshore wind power generation will require vast transmission systems, which can be convoluted logistically and capital intensive. One of the key considerations when looking at this scale of project, and the heavy front-end investment that must accompany it, is the wider economic viability of the overall facility [8].

The existing infrastructure needs major overhaul to support hydrogen transport and storage. Without substantial new investment in hydrogen pipelines there is no dedicated supply network. The existing compression and storage facilities could also run the risk of needing upgrades or need to be completely rebuilt for the hydrogen that is being projected. Furthermore, the high capital cost of electrolyzer and renewable energy systems remains a significant barrier, which the project aims to address through economies of scale and innovations in production processes [40].

Despite public funding support, the project faces economic uncertainties. Current production costs of 3–4 €/kg $H_2$ remain significantly higher than fossil fuel alternatives. However, high capital expenditure on electrolysis equipment and infrastructure coupled with high operational costs and mostly associated with the price of electricity. It raises the question of the long-term economic viability of the project. Uncertainty about the market demand and competition from other energy vectors also weigh on this aspect.

Apart from the technical and economic hurdles, GreenHyScale is surrounded by a web of environmental and social issues. While the environmental footprint is not limited to its profile of greenness with large amounts of water required. For instance, nine liters of water can produce 1 kg of hydrogen, and large land use implications for both the site and some parts of its conversion infrastructure. Due to hydrogen's highly flammable nature, safety is a paramount concern. This requires extensive safety protocols and specialized training programs.

To assess understanding of the GreenHyScale project's technical challenges and system integration complexities, below is a set of evaluation questions. These questions focus on critical aspects of large-scale hydrogen production systems, particularly emphasizing the interaction between renewable energy sources and electrolyzer operations. The following assessment combines multiple-choice questions testing conceptual understanding and a comprehensive problem-solving exercise that requires practical application of engineering principles. These questions are designed to evaluate both theoretical knowledge and practical problem-solving abilities at a bachelor engineering level, with emphasis on energy systems integration and process control fundamentals.

**Q 2.1** Which of the following represents the primary engineering challenge in the GreenHyScale project?

**A** Managing water supply for electrolysis.
**B** Stabilizing electrolyzer operation with intermittent wind power.
**C** Securing project funding.
**D** Training operational staff.

**Q 2.2** In the context of system integration, what is the main concern when scaling up to 100 MW?

**A** Environmental permits.
**B** Marketing the hydrogen product.
**C** Power quality and grid stability.
**D** Community acceptance.

**Q 2.3** When considering the hydrogen production system's response to wind power fluctuations, which parameter is most critical?

**A** Storage capacity.
**B** Response time.
**C** Water pressure.
**D** Ambient temperature.

**Q 2.4** An electrolyzer system in the GreenHyScale project needs to handle wind power fluctuations while maintaining stable hydrogen production. Consider the following facts and answer the questions below:

Nominal power: 100 MW
Wind power variation: 40% to 100% of nominal power
Required response time: <30 seconds
Minimum hydrogen production efficiency: 70%
Water consumption: 9 L/kg $H_2$
Target hydrogen production: 50 kg/hour at nominal power

**Calculate:**

**A** The minimum and maximum hydrogen production rates (kg/h).
**B** The required water flow rate range (L/h).
**C** The energy consumption per kg of hydrogen at both minimum and maximum power.
**D** Suggest two technical solutions to maintain stable hydrogen production despite power fluctuations.

## 2.4 From Vision to Reality

The GreenHyScale project is like the overachiever of green hydrogen production, setting the bar high and making the rest of us wonder why we are still struggling with basic recycling. By diving into the project's twists, turns, and inevitable hiccups, this case study uncovers some key takeaways – because we all should learn from someone else's hard work.

From a technical feasibility and innovation perspective, the successful integration of 100 MW electrolyzer systems with renewable energy sources has convincingly demonstrated the technical viability of large-scale green hydrogen production. Advanced control systems and power management solutions have proven essential for managing intermittent renewable energy inputs, while the project establishes new benchmarks for system efficiency and operational stability in industrial-scale hydrogen production.

The project also carries substantial economic and market implications. The multi-level funding model – from the EU, private sector, and with regional funds will offer replicable results also for future projects. The economies of scale achieved through this project contribute substantially to reducing green hydrogen production costs, while the demonstration of commercial viability at this scale helps de-risk future investments in the sector.

When it comes to infrastructure development, combining old infrastructure with new hydrogen production facilities is like trying to teach your grandparents how to use a smartphone – it's all about strategic placement and a lot of patience. This project not only sets the gold standard for hydrogen distribution networks but also hand over grid integration solutions that future renewable energy projects can easily copy and paste.

The GreenHyScale project has brought forward several important realizations that will direct how green hydrogen is produced in future. Extracted from lessons learned, all in combination with successes and problems uncovered during implementation these frameworks assure a coherent blueprint for future projects in this area. The following key takeaways represent the most significant learning outcomes from this groundbreaking initiative:

1. Scalability: Transitioning from small-scale demonstration projects to 100 MW requires an engineered solution of massive scale, as well as institutional safeguards.
2. Alliance Model: The formation of the consortium shows that technical and operational experience, together with research capabilities, can address key factors of the project.
3. Risk Management: It is very important that such technical challenges are identified early, for the objectives to be hit.
4. Transfer of Knowledge: Correct documentation and sharing of experience is the lifeline to develop the industry.
5. Regulatory Framework: Joint work with authorities, up-to-date regulatory compliance are necessary for implementation of project.

The GreenHyScale project thus serves as a roadmap for future green hydrogen flagship initiatives underway. The case of success shows that with accurate planning, technical know-how, and stakeholder engagement, large-scale industrial green hydrogen production is feasible, and it is essential for decarbonization efforts. The project not only proves that large-scale green hydrogen production is as feasible as finding socks that match on the first try, but also lays out a roadmap for future implementations in the ever-expanding hydrogen economy – because who doesn't love a good roadmap to success?

## 2.5 Discussion Points

What if scaling up green hydrogen production was like conducting a symphony orchestra, where each instrument must play perfectly in harmony to create a masterpiece? Just like the conductor needs to harmonize all musical parts in different well-balanced manners, the GreenHyScale project gracefully handles the intricate ecosystem of technical, economic, and operation elements for large-scale hydrogen production.

This analogy becomes even more feasible when considering the web of challenges and outcomes that emerged during the project journey to be examined.

**Discussion Points:**

1. **Technical Integration Challenges:** How does the intermittency of wind power truly impact system stability at this scale?
   What are the limitations of current electrolyzer technology in handling rapid power fluctuations?
2. **Economic Viability and Market Development:** Is the current cost structure sustainable without government support?
   How do economies of scale manifest in real-world operations?
   What role do carbon prices play in project economics?
   Life cycle analysis (LCA) of equipment and infrastructure.
3. **Social and Policy Implications:** Policy framework requirements for project replication.
   Workforce development and skills requirements.
4. **Future Scalability and Replication:** What are the realistic limitations for further scaling?
   How transferable are the lessons learned to different geographical contexts?
   Potential for standardization across future projects.
5. **Risk Management and Safety Considerations:** Emergency response planning for large-scale operations.
   Cybersecurity implications for automated systems.

Each of these discussion points merits deeper exploration and could benefit from input across multiple stakeholder groups. The complexity of these issues highlights the importance of continued dialogue and knowledge sharing within the industry.

## 2.6 Limitations of the Case Study

The GreenHyScale project gives us a lot of useful information about large-scale green hydrogen production, most important key limitation should be kept in mind to keep the scientific rigor and provide proper context for future implementation. This section discusses important constraints and boundaries for the understanding of the outcomes of the project.

1. **Geographic and Climate-Specific Constraints:** The project's location in Denmark presents specific geographic and climatic conditions that may not be representative of other potential deployment sites. Wind patterns, grid infrastructure, and ambient temperature variations encountered during the study period are regional in nature. As such, the performance metrics and operational strategies created may be drastically adjusted for projects located in other regions or with distinct climatic zones of renewable energy.
2. **Temporal Limitations:** The provisional set up that was in place for a rather short operational period hinders our knowledge of the long-term attenuation trends and maintenance frequency required by the full-scale system. It tells us about experiments through accelerated and realistic testing, but not only in laboratory environments. For example, how the components degrade under real-life use is only partially known. In addition, the study period might not have captured the complete spectrum of seasonal variations and extreme weather events that affect system performance.
3. **Technology Evolution Considerations:** The rapid pace of technological advancement in both electrolyzer and renewable energy systems means that some of the technical solutions implemented in GreenHyScale may become outdated relatively quickly. The project's findings are inherently tied to the state of technology at the time of implementation, and future projects may have access to more advanced or more cost-effective solutions.
4. **Economic Context Dependencies:** The economic analysis is case-specific for the prevailing market conditions, energy prices, and policy settings that existed in the project period. The scope of economic conclusions is constrained by the fact that energy markets are dynamic, regions have different incentivized structures and burdens shift over time through carbon pricing. The project's cost recovery structure, assuming substantial public funding may not be realistic in other contexts.
5. **Regulatory Framework Specificity:** The regulatory compliance strategies created throughout the project are fully oriented toward European Union and Danish regulations. Approaches, however, may not be as universally applicable given differing regulatory regimes, safety standards and permitting processes in other jurisdictions.

## References


1 Alam, S., Khan, M.Y.N., Das, A., Koushik, M.R.H., Shuvo, M.R.A. and Das, T. "Impact of green technologies on the production of hydrogen in a sustainable manner," *International Journal of Science and Research Archive*, 10(1), 106–112, 2023.

2 Cassol, G.S., Shang, C., An, A.K., Khanzada, N.K., Ciucci, F., Manzotti, A., Westerhoff, P., Song, Y. and Ling, L. "Ultra-fast green hydrogen production from municipal wastewater by an integrated forward osmosis-alkaline water electrolysis system," *Nature Communications*, 15(1), 2617, 2024.

3 Ogden, J.M., "Hydrogen: the fuel of the future?" *Physics Today*, 55(4), 69–75, 2002.

4 Guarieiro, L.L., Anjos, J.P.D., Silva, L.A.D., Santos, A.Á., Calixto, E.E., Pessoa, F.L., Almeida, J.L.D., Andrade Filho, M., Marinho, F.S., Rocha, G.O.D. and Andrade, J.B.D. "Technological perspectives and economic aspects of green hydrogen in the energetic transition: challenges for chemistry," *Journal of the Brazilian Chemical Society*, 33(8), 844–869, 2022.

5 Aboelela, D. and Soliman, M.A. "Hydrogen production from microbial electrolysis cells powered with microbial fuel cells," *Journal of King Saud University-Engineering Sciences*, 36(6), 369–374, 2024.

6 Demirci, U.B. and Miele, P. "Overview of the relative greenness of the main hydrogen production processes," *Journal of Cleaner Production*, 52, 1–10, 2013.

7 Ayers, K., "The potential of proton exchange membrane–based electrolysis technology," *Current Opinion in Electrochemistry*, 18, 9–15, 2019.

8 Gederø, T., "Tildeler havvindområder på Utsira Nord," Offshore Norge, 2025.

9 Mneimneh, F., Ghazzawi, H., Abu Hejjeh, M., Manganelli, M. and Ramakrishna, S. "Roadmap to achieving sustainable development via green hydrogen," *Energies*, 16(3), 1368, 2023.

10 Noussan, M., Raimondi, P.P., Scita, R. and Hafner, M. "The role of green and blue hydrogen in the energy transition—a technological and geopolitical perspective," *Sustainability*, 13(1), 298, 2020.

11 Hoffmann, P. *Tomorrow's Energy*, Revised and Expanded Edition: *Hydrogen, Fuel Cells, and the Prospects for a Cleaner Planet*, MIT Press, 2012.

12 Balcombe, P., Chatenet, M., Deseure, J., Schäfer, H. and Staffell, I. "Markets and costs for hydrogen electrolysis," in *Powerfuels: Status and Prospects*, ed. N. Bullerdiek, U. Neuling and M. Kaltschmitt, 235–256. Cham: Springer Nature Switzerland, 2024.

13 Sharshir, S.W., Joseph, A., Elsayad, M.M., Tareemi, A.A., Kandeal, A.W. and Elkadeem, M.R. "A review of recent advances in alkaline electrolyzer for green hydrogen production: performance improvement and applications," *International Journal of Hydrogen Energy*, 49, 458–488, 2024.

**14** Sebbahi, S., Nabil, N., Alaoui-Belghiti, A., Laasri, S., Rachidi, S. and Hajjaji, A. "Assessment of the three most developed water electrolysis technologies: alkaline water electrolysis, proton exchange membrane and solid-oxide electrolysis," *Materials Today: Proceedings*, 66, 140–145, 2022.

**15** Pasture, M. "Study of the dynamic behaviour of large scale PEM electrolysers in a fast demand response perspective," https://thesis.dial.uclouvain.be/entities/masterthesis/6f5e213e-1e5f-4e31-a551-949a6116a6ac (accessed February 21, 2026).

**16** Laguna-Bercero, M.A. and Orera, A. "High temperature co-electrolysis–a route to syngas," in *Energy Storage and Conversion Materials*, ed. S. Skinner, 42–99, 2019.

**17** Milewski, J., Kupecki, J., Szczęśniak, A. and Uzunow, N. "Hydrogen production in solid oxide electrolyzers coupled with nuclear reactors," *International Journal of Hydrogen Energy*, 46(72), 35765–35776, 2021.

**18** Azig, D. "Hydrogen mini-factory for domestic purposes (wind version)," *Scientific Reports*, 13(1), 13154, 2023.

**19** Miller, E.L., Thompson, S.T., Randolph, K., Hulvey, Z., Rustagi, N. and Satyapal, S. "US Department of Energy hydrogen and fuel cell technologies perspectives," *MRS Bulletin*, 45(1), 57–64, 2020.

**20** Budama, V.K., Duarte, J.P.R., Roeb, M. and Sattler, C. "Potential of solar thermochemical water-splitting cycles: a review," *Solar Energy*, 249, 353–366, 2023.

**21** Zhang, X.L., Yu, P.C., Su, X.Z., Hu, S.J., Shi, L., Wang, Y.H., Yang, P.P., Gao, F.Y., Wu, Z.Z., Chi, L.P. and Zheng, Y.R. "Efficient acidic hydrogen evolution in proton exchange membrane electrolyzers over a sulfur-doped marcasite-type electrocatalyst," *Science Advances*, 9(27), eadh2885, 2023.

**22** Wei, S., Xia, X., Bi, S., Hu, S., Wu, X., Hsu, H.Y., Zou, X., Huang, K., Zhang, D.W., Sun, Q. and Bard, A.J., "Metal–insulator–semiconductor photoelectrodes for enhanced photoelectrochemical water splitting," *Chemical Society Reviews*, 53, 6860–6916, 2024.

**23** Posdziech, O., Schwarze, K. and Brabandt, J. "Efficient hydrogen production for industry and electricity storage via high-temperature electrolysis," *International Journal of Hydrogen Energy*, 44(35), 19089–19101, 2019.

**24** Buttler, A., Koltun, R., Wolf, R. and Spliethoff, H. "A detailed techno-economic analysis of heat integration in high temperature electrolysis for efficient hydrogen production," *International Journal of Hydrogen Energy*, 40(1), 38–50, 2015.

**25** Cheng, W.H., De La Calle, A., Atwater, H.A., Stechel, E.B. and Xiang, C. "Hydrogen from sunlight and water: a side-by-side comparison between photoelectrochemical and solar thermochemical water-splitting," *ACS Energy Letters*, 6(9), 3096–3113, 2021.

**26** Franco, A. and Giovannini, C. "Recent and future advances in water electrolysis for green hydrogen generation: critical analysis and perspectives," *Sustainability*, 15(24), 16917, 2023.

**27** International Energy Agency (IEA), "The future of hydrogen report," 2019.

**28** Energy, C., "Water electrolysis and hydrogen in the European Union," 2024.

**29** "Greenhyscale official website," https://www.greenhydrogensystems.com/projects/greenhyscale (accessed March 23, 2026).

**30** Biabani, H., Aminlou, A., Hayati, M.M., Majidi-Gharehnaz, H. and Abapour, M. "Green hydrogen research and development projects in the European Union," in *Green Hydrogen in Power Systems*, ed. V. Vahidinasab, B. Mohammadi-Ivatloo, and J. Shiun Lim, 301–320. Cham: Springer International Publishing, 2024.

**31** Badejo, A.O., Aminu, N.I., Kumar Sekar, L. and Rita Okoroafor, E. "Assessment of the environmental risks and policies associated with carbon capture and storage in the United Kingdom–a case study of the HyNet North West Project," *In Abu Dhabi International Petroleum Exhibition and Conference*. SPE, 2024.

**32** He, W., Zheng, Y., You, S., Strbac, G., Therkildsen, K.T., Olsen, G.P., Howie, A., Byklum, E. and Sharifabadi, K.T. "Case study on the benefits and risks of green hydrogen production co-location at offshore wind farms," *In Journal of Physics: Conference Series*, 2265(4), 042035. IOP Publishing, 2022.

**33** CTCN, and Green, E.I. "Green hydrogen technologies for systems transformation, national strategies, plans and projects," 2023. https://www.ctc-n.org/resources/green-hydrogen-technologies-systems-transformation-national-strategies-plans-and-projects (accessed February 21, 2026).

**34** Afridi, S.K., Koondhar, M.A., Jamali, M.I., Alaas, Z.M., Alsharif, M.H., Kim, M.K., Mahariq, I., Touti, E., Aoudia, M. and Ahmed, M.M.R. "Winds of Progress: An In-depth Exploration of Offshore, Floating, and Onshore Wind Turbines as Cornerstones for sustainable energy generation and environmental stewardship," *IEEE Access*, 12, 66147–66166, 2024.

**35** "Greenlab official website," www.greenlab.dk (accessed February 21, 2026).

**36** Tækker, C. "Electrolysis Plant to Accelerate Hydrogen Production," DTU Electrotechnology, 2021.

**37** Schreiber, L. "Contribution of the EU Research & Innovation programmes to the uptake of hydrogen in Europe: Monitoring flash," 2024.

**38** Pio, D.T., Vilas-Boas, A.C.M., Araújo, V.D., Rodrigues, N.F.C. and Mendes, A. "Decarbonizing the aviation sector with Electro Sustainable Aviation Fuel (eSAF) from biogenic $CO_2$ captured at pulp mills," *Chemical Engineering Journal*, 463, 142317, 2023.

**39** Shields, M., Stefek, J., Oteri, F., Kreider, M., Gill, E., Maniak, S., Gould, R., Malvik, C., Tirone, S. and Hines, E. "A supply chain road map for offshore wind energy in the United States (No. NREL/TP-5000-84710)," *National Renewable Energy Laboratory (NREL)*, Golden, Colorado, United States, 2023.

**40** Tebibel, H. "Methodology for multi-objective optimization of wind turbine/battery/electrolyzer system for decentralized clean hydrogen production using an adapted power management strategy for low wind speed conditions," *Energy Conversion and Management*, 238, 114125, 2021.

# 3

# DONG to Ørsted: Viking Longship Green Hydrogen Strategy

Navigating Corporate Transformation and Large-Scale Hydrogen Project Integration

## 3.1 From Black to Green

How did a company that dominated fossil fuel production become a pioneer in green hydrogen innovation? Like a master chess player who sees several moves ahead, Ørsted (formerly DONG Energy) transformed itself from Denmark's largest oil and gas company into a global renewable energy leader. They demonstrated that strategic foresight and bold transformation are possible in the energy sector.

The Ørsted story is a modern-day paragon of transformation into energy corporate history. Having completely divested its oil and gas business in 2017, this company has gone above and beyond in its campaign to shrink its carbon footprint. It had decreased emissions by 87% since 2006 and no other major energy company has come close to achieving comparable reductions just by pulling out of a business. The 360-degree turn demonstrates Ørsted's dedication to a sustainable future as it aims to be the world's first carbon-neutral utility company by 2025. It may be telling to give the name of the discoverer of electromagnetism in 1820, Hans Christian Ørsted, to a company looking to lead another revolution of energy innovation in the twenty-first century, as it continues.

The global push toward a decarbonized energy landscape has led to a surge in offshore wind development, with Denmark's Ørsted emerging as a prominent player in this rapidly evolving sector. The case study investigates Ørsted's portfolio of green hydrogen projects with a view to three case studies highlighting the differences in scaling up production for hydrogen. The analysis covers the SeaH2Land project in the Dutch–Flemish North Sea Port cluster, the Gigastack project in the UK, and the Green Fuels for Denmark project, each representing unique approaches to integrating offshore wind power with hydrogen production.

How did Ørsted leverage its extensive offshore wind experience to overcome the challenges of large-scale green hydrogen production? What were the company's approaches to create an industrial symbiosis, where hydrogen production facilities are strategically located near both renewable energy sources and major industrial consumers?

## 3.2 Transforming Offshore Wind into Maritime Gold

Where oil rigs once stood like sentinel lighthouses across the horizon of the North Sea, a stunning reversal was underway. The state-owned oil and gas large-scale player of Denmark, DONG Energy was about to start something that many industry veterans saw as corporate self-destruction. Like a butterfly emerging from its chrysalis, the company was shedding its fossil fuel identity to become Ørsted, named after the Danish scientist who discovered electromagnetism – a fitting symbol for the company's electrifying transformation. Figure 3.1 summarizes the major milestones of Ørsted's transformation. The years 1972–2006 were focused on fossil fuel then renewable transition started from 2009 and hydrogen initiatives runs from 2019 to present.

It was 2009 when the leadership at DONG Energy, facing the realities of climate change, and rapidly depleting North Sea oil reserves, made a strong decision that rippled across the energy sector. This was the period when oil was selling for around USD$60 a barrel and renewable energy was not yet widely accepted by

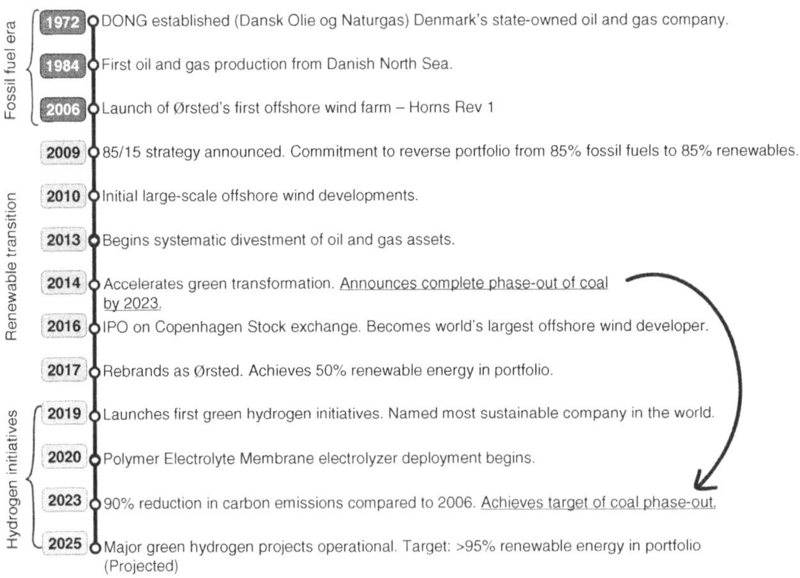


**Figure 3.1** Timeline chart for Ørsted's transformation. *Source:* Author's own work.

most investors. The company revealed that it is breaking away completely from where it stood. The "85/15" strategy is a complete reversal of its energy portfolio from 85% fossil fuels and 15% renewables to the opposite within 10 years. It's like either they are going to be the last ones turning off the lights in the oil and gas industry, or the company could be the first ones to seriously build the energy system of tomorrow [1].

Ørsted's strategic positioning and project selection criteria, dictates how the company masterfully integrates offshore wind with electrolyzer systems. This extends to their innovative approaches in maritime applications and industrial clusters, while valuing partnership models and stakeholder engagement strategies. The company especially navigated complex regulatory frameworks and influenced policy development, all while maintaining focus on economic viability and business model evolution.

By 2017, when DONG Energy rebranded as Ørsted, the company had already established itself as the global leader in offshore wind power. But the intermittent challenge of renewable energy remained – like trying to run a marathon while only being able to breathe every other minute. This is where hydrogen entered the picture, offering a solution as elegant as it was ambitious.

The technology behind Ørsted's green hydrogen projects builds on two centuries of scientific progress. In 1800, just twenty years before Hans Christian Ørsted discovered electromagnetism, William Nicholson and Anthony Carlisle first demonstrated water electrolysis. Fast forward to 2020, Ørsted was gearing to put Polymer Electrolyte Membrane electrolyzers on a scale that would have been impossible for early pioneers.

### 3.2.1 The Viking Longship Strategy

Unlike traditional single-site case studies, this case study looks into how a major energy company is developing multiple parallel projects, each tailored to specific market conditions and local requirements. What are the advantages of a portfolio approach to green hydrogen development? The answer demonstrates how lessons learned from one project inform the development of others. This timeline is from 2018 all the way up until now (focusing on project conception, development phases and very early operational insights) providing a higher resolution view of how Ørsted's hydrogen strategies have been ticking over thanks to market conditions and innovation.

Understanding that no company could single-handedly create a hydrogen economy, Ørsted adopted what they called the "Viking longship approach" – different partners rowing in the same direction. This means teaming up with industrial gas companies like Air Liquide and Linde, maritime groups Maersk/DFDS, technology developers like ITM Power/Siemens and a series of port

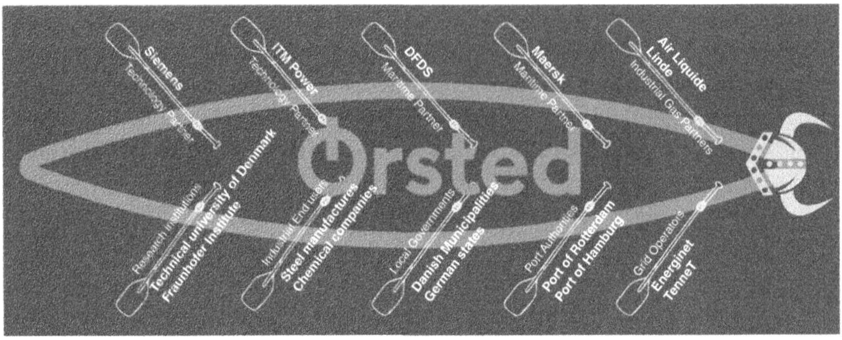


**Figure 3.2** Ørsted's Viking Longship Strategy: synchronized partnership approach (2018–2025). *Source:* Author's own work.

authorities and local governments. And by 2025 these would stand not only as technical breakthroughs, but also as evidence that the industrial ecosystems could be reinvented for profit and purpose. The hydrogen strategy was often described as not just building production facilities but creating bridges to a sustainable future – both metaphorically and literally through pipeline networks [2].

Figure 3.2 visualizes the combination of the Viking longship metaphor with modern business strategy. Ørsted's collaborative approach requires that the companions sail together in harmony. Much like Viking ships had to row in time to navigate rough waters, Ørsted's model has multiple partners coordinating their efforts toward a common goal. The timeline (2018–2025) is the hull of a ship, and oars of various stakeholders are on both sides. It maps out how industrial gas companies, maritime partners and technology developers need to row together to propel green hydrogen projects forward. Pointing of a starboard tack, condensing market forces with the waves below are regulatory cross currents in which the collaboration must battle. This synergistic approach has been essential in surmounting technical, financial and regulatory hurdles needed to create a functional hydrogen economy.

During the period from 2018 to 2025, Ørsted created three hydrogen flagship projects that covered various parts of the energy transition. Based on the North Sea Port cluster, SeaH2Land project seeks to provide 1 GW electrolyzer capacity by 2030 for over 20 industrial partners and cross-border infrastructure connecting Netherlands/Belgium. In the UK, Gigastack initiates with 100 MW at Hornsea offshore wind farm and is about industrial decarbonization from this project. Green Fuels for Denmark projects consisting in three phases, from 10 MW to 1.3 GW, mainly focused on maritime sector and Copenhagen infrastructure hub.

Just as a seasoned surfer catching the winning wave, Ørsted got the timing right. In 2019, the EU launched its ambitious 'Green Deal' doubling its ambition on green hydrogen – 40 GW of electrolyzer capacity by 2030. Hydrogen production costs, on the other hand, were falling about more than analysts had forecast with prices plummeting from €6/kg to €2/kg by 2025.

Finding its way through the labyrinth of regulation was just as difficult as choreographing a ballet where each dancer has his or her own sheet of music. Every project demanded delicate balancing of national and the EU level policies, port authorities, industrial partners, and grid operators. Ørsted's offshore wind development know-how was of paramount importance in this intricate regulatory landscape.

### 3.2.2 The Triple Helix of Innovation

The North Sea, once a theater of maritime trade and oil exploration, is where Ørsted is working on its most ambitious green hydrogen initiatives. While each project is a different way of doing the hydrogen economy, they all have in common the use offshore wind power to make clean hydrogen. Let us go through these game changing projects that are reshaping the face of European energy (Table 3.1).

**Table 3.1** General comparative analysis of Ørsted's flagship hydrogen projects.

| | Gigastack | Green Fuels for Denmark (GFDK) | SeaH2Land |
|---|---|---|---|
| **Location** | UK (Yorkshire) | Denmark (Copenhagen) | Netherlands/Belgium |
| **Total Capacity** | 100 MW (Phase 1) | 1.3 GW (Final Phase) | 1 GW |
| **Primary Focus** | Industrial Decarbonization | Maritime/Aviation | Industrial Clusters |
| **Wind Power Source** | Hornsea Wind Farm | Bornholm Offshore Wind | North Sea Wind Farms |
| **Key Partners** | Phillips 66, ITM Power | Maersk, Copenhagen Airport | 20+ Industrial Users |
| **Unique Feature** | Modular Scaling | Three-phase Development | Cross-border Pipeline |
| **Development Phase** | 2019–2021 | 2020–2023 | 2021–2022 |
| **Implementation Phase** | 2021–2025 | Phase 1: 2023–2025<br>Phase 2: 2025–2027<br>Phase 3: 2027–2030 | 2025–2030 |
| **Target Completion** | 2025 (Phase 1) | 2030 (All Phases) | 2030 |
| **$CO_2$ Reduction** | 0.8 Mt/year | 0.85 Mt/year | 1.5 Mt/year |

*Source:* Author's own work based on data from websites of Gigastack, GFDK and SeaH2Land.

Despite varied architectures and missions, these three projects represent Ørsted's general philosophy of combined energy solutions much like a tightly woven trinity, each project reinforces the other to form a robust hydrogen economy for Europe. The trio is like the strands of a rope, trying tighter and tighter, forming a collective framework for Europe's hydrogen economy. Illustrating how local context and needs drive the development of green hydrogen infrastructure before it can be mapped to a larger mosaic sustainable energy system.

What happens when you combine the might of the world's largest offshore wind farm with cutting-edge hydrogen technology? Imagine a modern-day Industrial Revolution, but instead of coal-fired furnaces darkening Yorkshire's skies, clean energy flows from wind and water, transforming the region once again.

#### 3.2.2.1 Gigastack

Located offshore from the coal powered industrial lands of Yorkshire is project Gigastack. It seeks to show green industrialization in Britain. The project was initiated all the way back to 2019 with Ørsted, leveraging its expertise from offshore Hornsea wind farm. The company partnered with ITM Power (innovator of electrolyzer technology), Phillips 66 (one of the leading industrial energy companies) and Element Energy (world market leading expert in the hydrogen and clean energy sectors) [3]. This strategic collaboration aimed to mobilize all the stakeholders of the green hydrogen value chain in order to advance the Gigastack project. This collaboration was not coincidental. It represented the convergence of generative renewable power generation, innovative electrolyzer manufacturing, and industrial applications.

Historically the Humber region is known as Britain's Energy Bay. It provides the ideal setting with a rich industrial past, deep-water ports and close proximity to major offshore wind developments. To generate the clean electricity that hydrogen production requires, the Hornsea wind farm complex will be utilized. The complex comprises Hornsea 1 (1.2 GW), Hornsea 2 (1.32 GW) and the upcoming Hornsea 3 (2.4 GW) will form the core of their renewables power portfolio (Figure 3.3).

The project's initial 100 MW phase focuses on decarbonizing the Phillips 66 Humber Refinery, one of the UK's most energy-intensive industrial facilities. This first step represents more than just industrial decarbonization – it's a proof of concept for the entire green hydrogen economy. The technology at the heart of this transformation comes from ITM Power's Sheffield facility, where PEM electrolyzers are manufactured with over 75% efficiency and remarkable responsiveness, capable of adjusting to wind power fluctuations in less than a second.

The project progresses through a series of well-choreographed phases, starting with an initial 100 MW built in place as a steppingstone for future expansion. By 2027, the project is set to reach 500 MW of capacity in conjunction with the larger

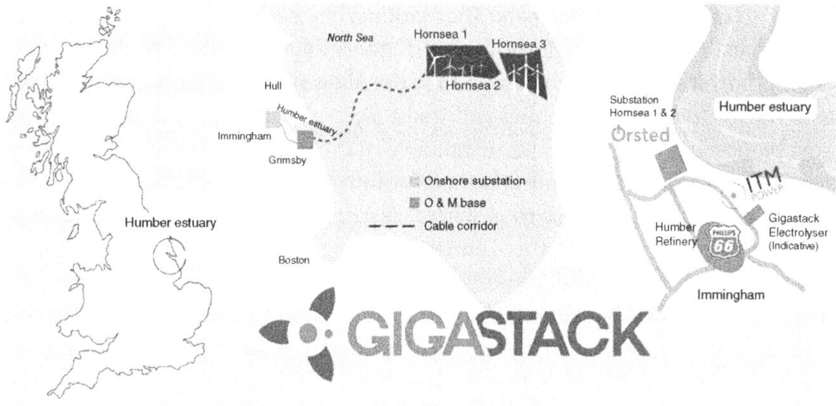


**Figure 3.3** Gigastack concept facility with indicative location at South Humber Bank, UK. *Source:* Adapted from [3].

industrial park and existing gas network. Phase three, aiming for 1 GW all within a regional hydrogen network in 2030. It can supply enough clean hydrogen to heat 350,000 homes or power up 1,200 buses annually.

The economic impact ripples through the region like waves from the North Sea itself. The project generates hundreds of direct construction jobs and permanent operational positions, while creating thousands more in the supply chain. The investment of £500 million, split between private and public funding, catalyzes a new industrial ecosystem. The project's learning curve promises to drive hydrogen production costs down from an initial £4–5/kg to £2/kg by 2030, making green hydrogen increasingly competitive with fossil fuel alternatives.

Innovation underpins all aspects of the project from advanced process control to improved electrolyzer efficiency through new business models combining multiple revenue streams. The modular design of the project enables rapid scaling and low development expense along with integration of existing infrastructure. This method-driven innovation comprises technology and goes deeper to include workforce development while taking up the challenge of educating the next generation of hydrogen economy workforce.

The Gigastack project is about industrial decarbonization. It has become the blueprint of regional transformation. With the combined knowledge and strengths of Ørsted on renewable energy, ITM Power's manufacturing infrastructure and Phillips 66 manufacturing heritage, the project illustrates how established industrial regions can re-focus on a sustainable future while preserving economic vibrancy. As this modern industrial revolution unfolds, the Humber region is at the vanguard of energy innovation once again.

#### 3.2.2.2 Green Fuels for Denmark: The Maritime Pioneer

What does it take to transform a historic maritime hub into a futuristic green energy powerhouse? Much like how Copenhagen's harbor once served as a gateway for traditional shipping and trade, Green Fuels for Denmark (GFDK) is now pioneering the gateway to sustainable maritime and aviation fuels, demonstrating how centuries-old ports can evolve into clean energy hubs.

In Copenhagen's South Harbor, where container ships have docked for generations, GFDK emerged in 2020 as a groundbreaking collaboration between industry giants like Ørsted's renewable energy, Maersk's maritime know-how, Copenhagen Airport and technology partners as Nel Hydrogen and Haldor Topsøe. The strategic alliance is just Denmark's aggressive attempt at tackling the challenges in decarbonizing its maritime sector by utilizing the country's strong wind prowess and shipping heritage.

The project's three-phase approach reflects careful on learning from earlier initiatives like Gigastack. While Gigastack focused on industrial decarbonization, GFDK instead used those insights from electrolyzer scaling and grid integration to focus on how this could be applied to aviation and maritime through efficient transportation of fuels. The first 10 MW phase is set for heavy road transport, with the objective of testing technology and operational procedures. This conservative start allows the project to refine its process while ramping before scaling to 250 MW by 2027. At this point, the main goal becomes producing sustainable aviation fuels.

The final phase, expanding to a total of 1.3 GW in 2030 highlights GFDK's grand strategy for both maritime and aviation applications. The project utilizes renewable electricity to power an electrolyzer through advanced Power-2-X technology. This then generates green hydrogen and is blended with captured $CO_2$ to make sustainable fuels. The complexity of integrating all these technologies (offshore wind power, carbon capture and fuel synthesis) into a single project goes beyond what previous endeavors have realized. Figure 3.4 shows the process flow on how offshore wind power drive the power to produce green hydrogen and sustainable fuels through the integration of $CO_2$ capture.

There are various ways in which GFDK's impact can be felt by projects that followed up like SeaH2Land. The multi-stakeholder approach, representing both public and private sectors was a milestone for SeaH2Land to be developed within the Dutch–Belgian North Sea Port cluster. GFDK's emphasis on end-user integration, specifically with shipping and aviation companies, have influenced the prospective SeaH2Land on industrial off-takers and port facilities.

The strategic location of the project in Copenhagen's South Harbor grants direct access to both maritime shipping lanes and airport infrastructure. That placement allows for seamless fuel routing; a demonstration of how existing industrial infrastructure can support green hydrogen projects. The annual reduction of 850,000

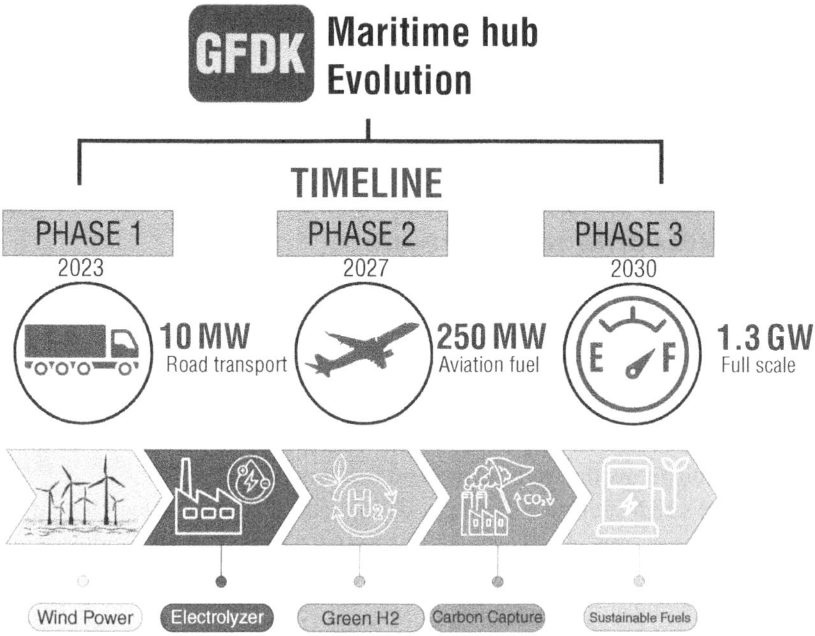


**Figure 3.4** GFDK's maritime hub evolution from phases 1–3. *Source:* Adapted from [4].

tons of $CO_2$ that this project attempts to deliver every year. It is about 2% of total emissions nationwide, indicating how hard a particular industrial sector can be pressed into on a national climate goal.

Learning from Gigastack's experience with grid integration and electrolyzer operation, GFDK incorporates advanced control systems and storage solutions to manage the variable output from offshore wind farms. The project's modular design allows for gradual scaling while maintaining operational efficiency, a lesson that directly influenced SeaH2Land's development strategy.

As GFDK matures, it serves as both a technical and organizational template for future green hydrogen projects. For a larger intention on behalf of Denmark, its impact runs into newer projects such as the SeaH2Land that incorporates innovative approach to stakeholders' engagement. It is an example of utilizing historical naval infrastructure for today's green energy transition, potentially providing a template for ports all over. SeaH2Land is in the process of developing its own equally ambitious plans, led by lessons learning from GFDK in areas of multi-stakeholder governance as well as maritime infrastructure to green hydrogen production coordination.

This connection between Gigastack, GFDK, and SeaH2Land showcases how each project builds upon the lessons of its predecessors, creating an increasingly sophisticated approach to green hydrogen deployment in maritime and industrial applications. As these projects continue to develop, they collectively contribute to the growing body of knowledge which is essential for scaling up green hydrogen production across Europe and beyond.

#### 3.2.2.3 SeaH2Land: The Cross-Border Pioneer

The SeaH2Land project rises from the historic routes of trade between the Netherlands and Belgium, reflecting best European cooperation. Positioned at the North Sea Port cluster, spanning through Vlissingen in the South and Ghent in the North, this bold project brings two countries together with a view of industrial decarbonization [5]. Figure 3.5 illustrates this cross-border ecosystem, showing how the planned hydrogen backbone connects offshore wind capacity to industrial hubs across the Netherlands and Belgium.

The region's rich industrial heritage, dating back to the seventeenth century Dutch golden age, now serves as the foundation for the sustainable industries of the future, steering in a new era of green economic development. The century of industrial heritage provides a robust base for establishing a Green economic model based on renewable energy, clean technologies and circular economy [6]. Regionally accelerating wind, solar and river power can help materially lower the carbon footprint of the region while also securing energy.

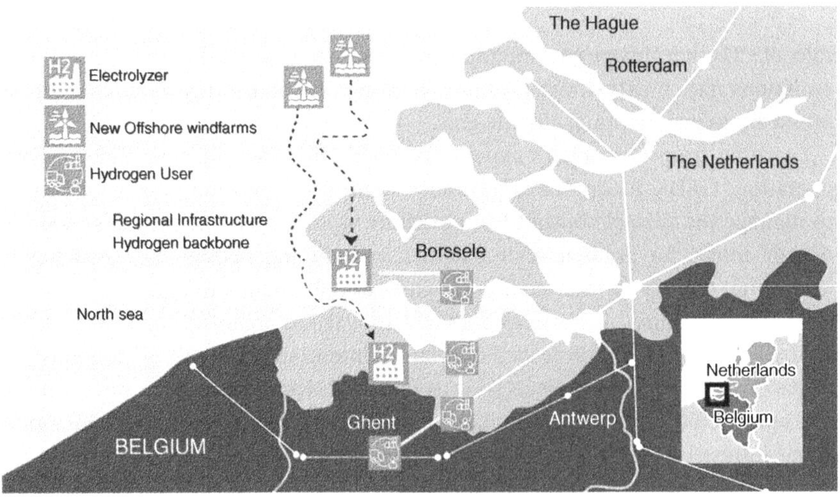


**Figure 3.5** Ecosystem of SeaH2Land project (to be completed by 2030). *Source:* Adapted from [7]. *Note:* Location of icons are figurative and does not indicate the exact location.

The change from relying on a fossil-fuel led past to green power is nothing short of miraculous. The industrial heritage of the region, human resources, and assets in infrastructure are allowing regional transformation into self-sustainability. In addition, smart technologies and data solutions can be used to fine-tune resource efficiency and investments that promote recycling and waste minimization. It guarantees that materials are being recovered ensuring a low environmental footprint [8]. This pivot not only tackles urgent climate issues, but it sets the global standard for green innovation, leading green investment and dictate the course of long-term economic stability for the region [9].

SeaH2Land's distinctive feature is its ambitious plan to deploy a massive 1 GW electrolyzer capacity by 2030, which would make it one of the largest green hydrogen production facilities in the world [10]. This large-scale production capacity is a testament to Ørsted's commitment to driving the transition to a sustainable hydrogen economy. The SeaH2Land project cleverly integrates existing industrial infrastructure with new state-of-the-art hydrogen production facilities, much like adding bold new chapters to an old book [11]. Twenty leading industrial partners, including prominent chemical giants and steel manufacturers, have formed a sophisticated demand network that will collectively consume approximately 580,000 tons of green hydrogen annually [12].

### 3.2.3 The Spectrum of Hydrogen Production

Have you ever wondered why hydrogen, the simplest element in the universe, comes in so many "colors"? Although hydrogen itself is colorless, the industry has assigned color codes to represent the different production methods. These range from the least eco-intensive to the highest technological advances [13].

To start off explaining the spectrum as shown in Figure 3.6, there is brown and black hydrogen at the darker end, which are from coal gasification, a 1800s technology that spews substantial $CO_2$ (about 18–20 kg $CO_2$/kg $H_2$). The current market leader, which dominates 95% of global production, is typified by gray hydrogen, which comes from steam methane reforming (SMR) of natural gas. This produce about 9–10 kg $CO_2$/kg $H_2$ [14].

Emerging pathways include pink hydrogen from electrolysis with nuclear power and turquoise hydrogen because of methane pyrolysis; all have their own technological hurdles and pros. Solar-powered electrolysis generates yellow hydrogen, which is similar to green hydrogen, however with an emphasis on solar as a specific renewable source.

Green hydrogen, the focus of Gigastack, GFDK, and SeaH2Land projects, stands apart through its production method. It uses water electrolysis powered exclusively by renewable electricity. The process splits water molecules ($H_2O$) to produce hydrogen and oxygen with zero direct $CO_2$ emissions [15]. The Technical

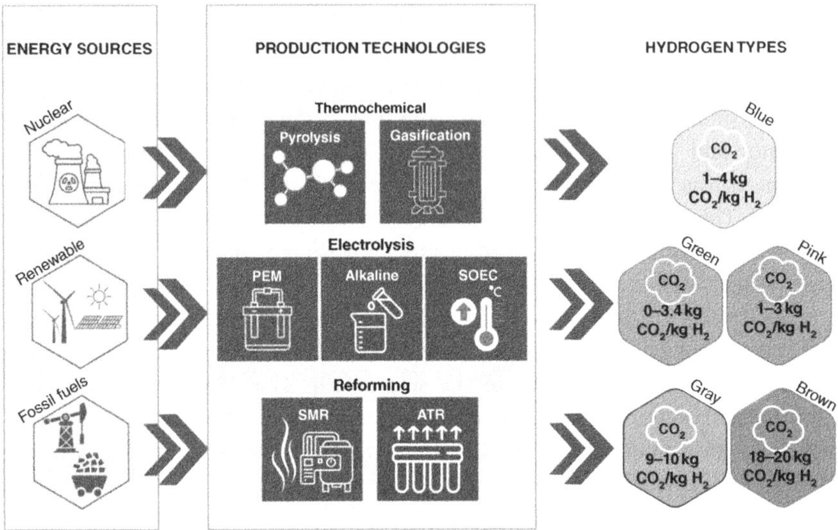


**Figure 3.6** The hydrogen spectrum and its equivalent production processes. *Source:* Author's own work.

Specifications for EU's RED II (Renewable Energy Directive II) and CertifHy's green hydrogen certification, as follows:

1. Maximum carbon intensity: 3.4 kg $CO_2eq/kg\ H_2$.
2. 100% renewable energy source verification.
3. Grid connection requirements with temporal correlation.
4. Water consumption monitoring and reporting.
5. Detailed production chain documentation.

The certification process, overseen by bodies like TÜV SÜD and CertifHy, involves rigorous monitoring various assessments [16]. These credible certification bodies carry out a four-tiered monitoring structure. This system begins with power source authentication that guarantees the source of power is 100% renewable. This is followed by a comprehensive verification of the production process to guarantee that all technical specifications and operational parameters are equal or superior to what is required. The third part is a detailed chain of custody documentation, from production to delivery that provides every bit of transparency and traceability for hydrogen. Lastly, an environmental assessment is made to determine the ecological footprint of the production process in general. Factors such as water consumption and carbon emissions are included.

The green hydrogen certification framework operates through this structured framework as illustrated in Figure 3.7. Production devices include the foundation

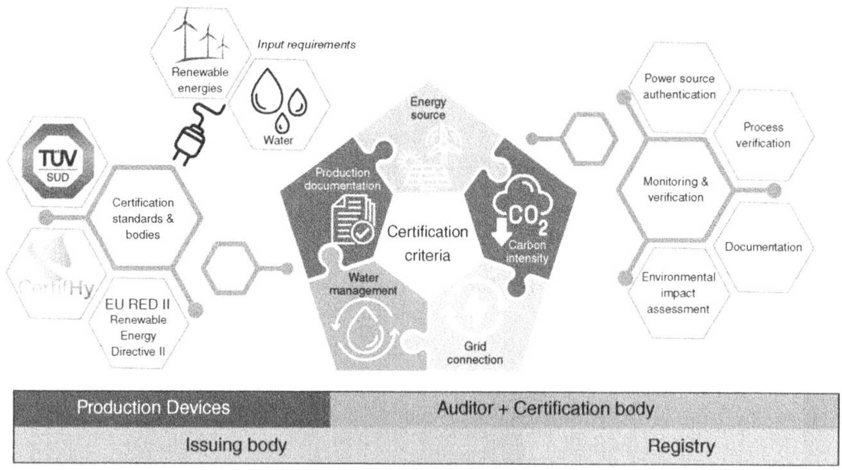


**Figure 3.7** Green hydrogen certification framework. *Source:* Adapted from [17].

level physical infrastructure where only hydrogen is produced such as electrolyzers and relevant renewable power feeders which need individual identification levels, and technical documentation. Next to this, auditors and certification bodies play a crucial role performing independent verification, site monitoring and proof that technical requirements are met on a regular basis. Accreditation is the key, as these bodies (such as TÜV SÜD) have to be assessed themselves to ensure their competence and fairness. Certification must conform with standards such as EU RED II or CertifHy. Finally, the registry acts as the registration system that logs and tracks all certified hydrogen production for transparent proof of production, supports valid certificate's period, and avoids double-counting of green hydrogen volumes. This certification framework provides a strong verification and traceability throughout all the certification processes.

These flagship projects chose green hydrogen for several compelling technical and strategic reasons. The rapidly declining cost of renewable electricity (now below €50/MWh in optimal locations) combines with improving electrolyzer efficiency (now reaching 70–80% LHV) to make green hydrogen increasingly competitive. Modern PEM and Alkaline electrolyzers achieve remarkable response times (PEM <1 second, Alkaline <10 seconds), enabling dynamic operation with variable renewable energy sources. Table 3.2 shows the technical comparison of the three projects.

The technical sophistication of these projects is evident in their integration strategies. For instance, Gigastack's ITM Power PEM electrolyzers operate at 80°C with an efficiency of 75% LHV, while GFDK's combined use of Nel's atmospheric alkaline and Haldor Topsøe's SOEC technologies represents a hybrid approach.

**Table 3.2** Technical comparison of north European flagship green hydrogen projects (2020–2030).

| | Gigastack | Green Fuels for Denmark | SeaH2Land |
|---|---|---|---|
| **Electrolyzer Technology** | PEM (ITM Power) | Alkaline (Nel) and SOEC (Haldor Topsøe) | Mix of technologies |
| **Power Source** | Hornsea Wind Farm (Offshore) | Offshore wind + onshore renewables | North Sea Wind + Onshore Wind |
| **Electrolyzer Capacity** | Initial: 100 MW → 1 GW | Phase 1: 10 MW → Phase 3: 1.3 GW | Initial: 1 GW → Potential: 3–4 GW |
| **$H_2$ Production** | ~900 tons/year (initial) | Phase 1: ~1,000 tonnes/year Phase 3: ~275,000 tons/year | ~20,000 tons/year (initial) |
| **End Products** | Pure hydrogen | E-methanol | Pure hydrogen |
| | Industrial heat | Sustainable aviation fuel (e-kerosene) | Industrial feedstock |
| | Power generation | e-diesel | Process heat |
| **End Users** | Phillips 66 Refinery | Maersk (shipping) | Dow Chemical |
| | Industrial cluster | Copenhagen Airport | Yara |
| | Local gas grid | SAS (aviation) | ArcelorMittal |
| | | Road transport | North Sea Port industries |
| **Storage Solution** | Pressurized tanks | Liquid fuel storage | Pipeline network |
| | Pipeline injection | Cryogenic storage | Underground storage potential |
| **Grid Connection** | Direct connection to Hornsea | Copenhagen grid + dedicated offshore connection | North Sea Port grid infrastructure |
| **Carbon Source** | N/A (pure $H_2$) | Captured $CO_2$ + biogenic sources | N/A (pure $H_2$) |
| **Water Source** | Municipal supply + Recycled | Desalinated seawater | Mix of industrial + municipal |
| **Process Integration** | Grid balancing | $CO_2$ capture | Industrial cluster integration |
| | Refinery operations | Power-to-X | Cross-border infrastructure |
| **Auxiliary Systems** | Water treatment | Methanol synthesis | Regional pipeline network |
| | Compression | Fuel upgrading | Industrial symbiosis |
| | Heat recovery | Air separation | Water management |

| | Gigastack | Green Fuels for Denmark | SeaH2Land |
|---|---|---|---|
| **Control Systems** | ITM Power's proprietary | Integrated multi-process control | Distributed control architecture |
| **Safety Systems** | Leak detection | Explosion protection | Advanced monitoring |
| | Emergency shutdown | Fire suppression | Emergency response |
| | Ventilation | Spill containment | Process safety |

*Source:* Author's own work.

This demonstrates the evolution toward higher temperature and efficiency systems (SOEC operating at 600–850°C with potential system efficiency >90% LHV) [18].

These three important projects represent significant milestones for green hydrogen scaling in Europe. The first flagship initiative shows the commercial competitiveness of renewable hydrogen production using the most advanced electrolyzer technologies which is running on offshore wind. While Gigastack focuses on decarbonizing industrial processes in the UK's Humber region, GFDK aims to establish a hydrogen hub with landmark size in Germany's northern coast. And the SeaH2Land project intends to build a large-scale renewable hydrogen network in Dutch–Flemish North Sea Port cluster. Together, these projects showcase different approaches to industrial-scale green hydrogen implementation [19].

## 3.3 Think and Solve

Analyzing large-scale green hydrogen projects such as GFDK, Gigastack or SeaH2Land is a unique challenge as these are early in their journey of implementation. Although not operational yet nor have final performance metrics, valuable learnings can be taken from their planning stage, the roadblocks faced during development processes, and preliminary results. Ørsted's projects represent a leap toward the integration of renewable energies, particularly offshore wind into the hydrogen economy. Nevertheless, the analysis in this case study is made with the implementation of ongoing developments and available information.

Ørsted's hydrogen projects seek to expedite the move toward a sustainable hydrogen economy but substantial technical, economic and infrastructural bottlenecks do exist. If we are to achieve the success of these projects, the key to this will be the optimization of electrolysis efficiency, reliable infrastructure development, and delivery of cost reduction plans. These needs will continue to be addressed through deeper research, technological progress and policy support.

To sum up, the reader's understanding of the complexities and critical aspects of these green hydrogen initiatives, here is a series of analytical questions. These questions are designed to test comprehension of both technical and strategic elements of large-scale hydrogen projects. They are followed by a practical engineering problem that addresses real-world scaling considerations. The questions simulate key decision points and challenges faced during project development and implementation phases.

**Q 3.1** What is the primary challenge in integrating offshore wind energy with hydrogen production?

**A** Low water availability for electrolysis.
**B** Variability and intermittency of wind power.
**C** High carbon emissions from wind turbines.
**D** Excessive heat generation in electrolyzers.

**Q 3.2** What is the most commonly used process for green hydrogen production in Ørsted's projects?

**A** Steam methane reforming (SMR).
**B** Coal gasification.
**C** Electrolysis.
**D** Partial oxidation.

**Q 3.3** Why is hydrogen storage critical for Ørsted's hydrogen projects?

**A** Hydrogen must be stored to balance production fluctuations.
**B** Hydrogen is naturally found in underground reservoirs.
**C** Hydrogen cannot be transported through pipelines.
**D** Hydrogen does not require storage due to continuous production.

**Q 3.4** Which aspect of GFDK's design most directly influenced SeaH2Land's development?

**A** Electrolyzer technology selection.
**B** Multi-stakeholder governance model.
**C** Geographic location.
**D** Production capacity targets.

**Q 3.5** Ørsted operates an electrolysis facility powered by offshore wind energy. The system uses a PEM electrolyzer. If the facility receives 400 MW of electricity from wind turbines, how much hydrogen energy can be produced? Refer to the given facts below and solve the problem.

**Given:**
Offshore wind farm capacity: 400 MW
Electrolyzer efficiency: 70%
Required hydrogen production: 50,000 kg/day
Wind capacity factor: 45%
System availability: 95%

**Calculate:**

**A** The actual power available for hydrogen production
Capacity × capacity factor × system availability = actual power (MW)
**B** The maximum daily hydrogen production capacity
Use standard conversion of 20 kg $H_2$/MWh at 100% efficiency
**C** Determine if the system meets the required hydrogen production target?

**Hint:** the answer is based on the calculated maximum daily hydrogen capacity from (b).

## 3.4 Future Pathways for Large-Scale Green Hydrogen Projects

Ørsted's hydrogen projects shows the capacity of what scaling in green hydrogen production can accomplish and some of the related challenges. This shows that the successful operation of large-scale hydrogen projects is not only down to the steady hand of balancing technology innovation. Most importantly, it is also about strategic partnerships and supportive policy frameworks.

The technical insights gained from these projects highlight several critical aspects. Energy efficiency remains paramount, with electrolysis optimization being crucial for economic viability in large-scale operations [20]. Grid integration challenges have necessitated sophisticated solutions for managing renewable energy variability and pushing the boundaries of current technology [21]. Infrastructure development, particularly in storage and transportation, requires significant investment and planning to ensure reliable distribution networks. Additionally, cost competitiveness with conventional hydrogen production methods continues to be a critical hurdle that must be overcome through technological advancement and economies of scale.

The success of these initiatives has been largely attributed to three key strategic factors. For one, multi-stakeholder partnerships have proven critical industrial competences combined with public sector backup, research institutions for incremental breakthroughs, common risk, and investment pools across project participants. Second, a phased approach has enabled sequential scaling from pilots to commercial operations, risk mitigation through serial development from one phase to the next, and transfer of learning between project phases.

Third, alignment with policy frameworks is fundamental so far as being in-line with national and EU hydrogen strategies. It helps to use existing subsidies and support mechanisms, as well as engagement in creating regulatory frameworks. Figure 3.8 synthesizes the main findings in a visual perspective that effectively summarizes the complex interplay of factors discussed in this case study. The structure shows core success factors in the inner ring, the current achievements and challenges in the middle ring while future developments as the outer ring which also symbolizes the growth of this circle. It is a key to the future of large-scale green hydrogen projects.

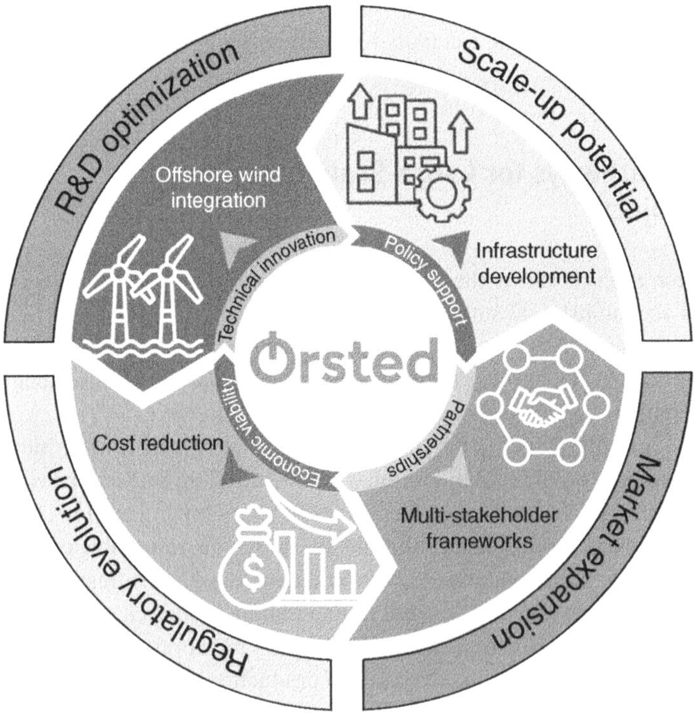


**Figure 3.8** Key success factors and future pathways for large-scale green hydrogen projects. *Source:* Author's own work.

Looking ahead, these projects will likely become the roadmap for future green hydrogen projects around the globe. Although there is no lack of challenges to attain cost-competitive green hydrogen production. The practical realization of industrial-scale offshore wind-powered green hydrogen generation is an important milestone in the evolution of decarbonization.

To ensure continued progress, several key recommendations emerge from this case study. Long-term success will depend on sustained research and development investment to improve electrolyzer efficiencies. Standardizing the process for infrastructure integration would lead to faster rollouts and lower costs. It also enables a sound regulatory setting for green hydrogen projects via policy advocacy. The improved sharing of knowledge among projects and regions would accelerate the learning and optimization. Last, but not least, a laser-like focus on supply chain optimization and cost reduction strategies will be required to achieve commercial viability.

These initiatives highlight that while technical and economic challenges exist, the pathway to commercial-scale green hydrogen production is becoming increasingly clear. The lessons learned from these projects will be of high value to kickstart the global shift toward sustainable hydrogen production especially for hard-to-abate sectors in decarbonization. The knowledge and solutions garnered from these pioneering projects will continue to guide the path of renewable energy integration for industrial decarbonization efforts globally.

## 3.5 Discussion Points

What if the future of energy wasn't just about replacing fossil fuels, but about redefining how we produce, store, and use power altogether? Imagine a city where every vehicle, factory, and home runs on clean hydrogen, produced using offshore wind energy. This vision is at the heart of Ørsted's hydrogen projects, but achieving it comes with complex challenges and opportunities that need further discussion.

1. **Scalability:** This is one of the primary questions. Is renewable hydrogen production expandable fast enough to cover the needs of industry? Wind and electrolysis projects from Ørsted show the possibilities of combining wind power, but the energy supply versus grid stability balance is a key challenge. What could help to scale-up production of hydrogen efficiently and at a more affordable price?
2. **Infrastructure Readiness:** Hydrogen needs specialized storage and transport with pipeline networks being considered but the cost and conversion of current natural gas infrastructure presents significant challenges. Will the pathway for future hydrogen investments be several smaller decentralized

production centers nearby consumption areas? or should large-scale production facilities dominate the industry?

3. **Policy and Regulatory Frameworks:** These play a defining role in shaping the success of hydrogen adoption. Ørsted is pioneering projects in regions where government support for green hydrogen is strong, but global policy alignment is still fragmented. Could international cooperation and standardization accelerate progress, or will national interests and energy independence create roadblocks?
4. **Public and Industry Adoption:** This is a very critical discussion point. Hydrogen is extremely promising in efforts to decarbonize heavy industries and transport but for now the costs are higher than conventional energy sources. How can we best motivate various industries to move to green hydrogen and what should government be doing to promote green hydrogen now?

As Ørsted continues to refine its projects, these discussions will shape the long-term trajectory of renewable hydrogen. Today's decision will decide if hydrogen stays a niche solution or gets to be the backbone of a very real renewables energy future.

## 3.6 Limitations of the Case Study

While this case study provides valuable insights into Ørsted's hydrogen projects, several limitations must be acknowledged:

1. **Ongoing Project Status:** Since projects like GFDK, Gigastack, and SeaH2Land are still at different stages, the analysis runs on early results from prototype, projection and existing documentation. These are not fully operational data tested. It therefore restricts the ability to evaluate long-term efficiency, performance, and economic viability.
2. **Evolving Technological Landscape:** The green hydrogen sector is maturing rapidly with electrolyzer efficiency improvements, energy storage advances and grid integration being pushed continuously ahead. Consequently, some of the technological considerations covered in this case study may become outdated with better solutions emerging.
3. **Regulatory and Policy Uncertainty:** Government policies, subsidies and regulatory frameworks which drive hydrogen development are prone to change over time. Changes in legislation or carbon pricing, and other energy policies would have significant implications for the viability and scale of Ørsted's projects.

4. **Market and Economic Factors:** Energy prices and investment in infrastructure is another driver for green hydrogen commercial success, as is the large demand swings. Uncertainty still remains on possible changes to market conditions as this study does not focus on economic analysis.
5. **Limited Access to Proprietary Data:** Some aspects of Ørsted's operational strategies, internal risk assessments and detailed engineering designs are not disclosed publicly. Consequently, the analysis depends on only public information and industry reports, which may not capture the full scope of the project's complexity. Nevertheless, this does not affect the intention of the case study because the aim is to provide a comprehensive overview and bring out further discussion and critical thinking.
6. **Geographical and Infrastructure Constraints:** The findings of this case study are specific to Ørsted's projects, which are largely based in Europe and influenced by regional energy infrastructure. The conclusions drawn may not be directly applicable to hydrogen projects in other geographical contexts with different grid systems, renewable energy availability, and industrial demand.

Despite these shortcomings, this case study provides a starting point into the complexities and opportunities of large-scale green hydrogen production that will serve as the backbone for future research and analysis as these projects evolve.

## References


1 Ørsted Annual Report. "CEO's letter to stakeholders," 2020: 8.

2 Thompson, S. *Green and Sustainable Finance: Principles and Practice in Banking, Investment and Insurance*, vol. 9, Kogan Page Publishers, 2025.

3 Department for Business, Energy & Industrial Strategy. "Gigastack phase 2: pioneering UK renewable hydrogen public report," ITM power PLC, August 2021, https://assets.publishing.service.gov.uk/media/61f957ca8fa8f5388b582d82/Phase_2_Report_-_ITM_-_Gigastack.pdf (accessed February 24, 2026).

4 Ørsted. "Leading Danish companies join forces on an ambitious sustainable fuel project," 26 May 2020, https://orsted.com/en/media/news/2020/05/485023045545315 (accessed February 24, 2026).

5 Northseaport, O.W. https://www.northseaport.com/en (accessed February 24, 2026).

6 Roque, A.J., Paleologos, E.K., O'Kelly, B.C., Tang, A.M., Reddy, K.R., Vitone, C., Mohamed, A.M.O., Koda, E., Goli, V.S.N.S., Vieira, C.S. and Fei, X. "Sustainable environmental geotechnics practices for a green economy," *Environmental Geotechnic*, 9(2), 68–84, 2021.

7 Ørsted. "Ørsted official website," www.orsted.com (accessed February 24, 2026).

8 Kumar, S. and Singh, D. "Smart cities waste minimization, remanufacturing, reuse, and recycling solutions using IoT and ML," *In IEEE International Conference on Computing, Power and Communication Technologies (IC2PCT)*, (vol. 5), 2024.

9 Fücks, R. *Green Growth, Smart Growth: A New Approach to Economics, Innovation and the Environment*, Anthem Press, 2015.

10 Terlouw, T., Rosa, L., Bauer, C. and McKenna, R. "Future hydrogen economies imply environmental trade-offs and a supply-demand mismatch," *Nature Communications*, 15(1), 7043, 2024.

11 Ørsted SeaH2Land, https://orsted.nl/duurzame-energie/groene-waterstof/seah2land (accessed March 18, 2026).

12 Bailey, M. "Ørsted developing major green-hydrogen hub serving Dutch and Belgian industrial players," *Chemical Engineering Online*, 2021. www.chemengonline.com (accessed February 24, 2026).

13 Incer-Valverde, J., Korayem, A., Tsatsaronis, G. and Morosuk, T. "Colors of hydrogen: Definitions and carbon intensity," *Energy Conversion and Management*, 291, 117294, 2023.

14 *Global Hydrogen Review*, https://www.iea.org/reports/global-hydrogen-review-2023 (accessed March 19, 2026).

15 Boretti, A. "Green imperialism a barrier to equitable progress in the hydrogen economy," *International Journal of Hydrogen Energy*, 105, 137–147, 2025.

16 Certifhy official website, www.certifhy.eu (accessed February 24, 2026).

17 RED II, "Recast renewable energy directive (EU 2018/2001)."

18 Cavaliere, P. "Electrolysis Economy," In *Water Electrolysis for Hydrogen Production*, 793–830, Springer International Publishing, 2023.

19 Ghosh, A. and Chhabra, S. "Case for a global green hydrogen alliance," 2021.

20 Franco, A. and Giovannini, C. "Recent and future advances in water electrolysis for green hydrogen generation: Critical analysis and perspectives," *Sustainability*, 15(24), 16917, 2023.

21 Kataray, T., Nitesh, B., Yarram, B., Sinha, S., Cuce, E., Shaik, S., Vigneshwaran, P. and Roy, A. "Integration of smart grid with renewable energy sources: opportunities and challenges – a comprehensive review," *Sustainable Energy Technologies and Assessments*, 58, 103363, 2023.

# 4

# First Solar's Cadmium Telluride Revolution: Disrupting Silicon Dominance

From Laboratory Breakthrough to Commercial Thin-Film Technology Deployment

## 4.1 Synopsis

Can a leap in solar panel technology redefine the future of renewable energy? Imagine a marathon where silicon-based solar panels have long been leading the race, and then, out of the blue, this competitor appears – quicker, lighter and more effective. This challenger became the industry's disruptor. First Solar, the company who have gone on to change the face of commercial thin-film photovoltaic technology through their alternative: cadmium telluride (CdTe) solar cells.

Silicon-based panels have been the market leaders for decades mostly because of their performance, and long-established industry processes [1]. As the appetite for solar that is not only cleaner and cheaper but also more scalable grew, First Solar chose to push the bet button on CdTe. CdTe panels are very different from a normal silicon panel, because they have the potential for lower-cost energy generation as well as higher energy yields, even in hot or cloudy conditions [2]. Additionally, these panels boast a dramatically shorter energy payback period. The increasing global demand for affordable solar energy has been a major driver of CdTe technology.

CdTe panels have shown as much as 30% greater energy yields in hot or low-light environments [3]. Manufacturing CdTe panels require up to 90% less energy use (compared with silicon) so they can pay back in less than a year [4]. These benefits make CdTe a very attractive candidate for mass solar applications that are constrained on price-performance in different climate regions [5].

First Solar has taken advantage of these competencies and has capitalized the efficiency of its CdTe solar modules, from about 10% in 1990s to over 22% today [6]. Founded in 1999, the company commercialized the alternative to classical crystalline silicon technology, CdTe panels. This innovation not only increased efficiency but made solar power more accessible to industrial-sized customers.

First Solar has turned CdTe into a challenging material through a combination of relentless R&D support, sophisticated manufacturing and strategic alliances in the industry.

Although the headquarters is in Tempe, Arizona (US), the company has evolved into a global manufacturing powerhouse. In fact, First Solar has a multi-national manufacturing presence outside the United States including Malaysia, Vietnam, and India. Such an international scale of operations is supported with strong financial results. In 2023, First Solar recorded net sales of USD $3.3 billion, with a total nameplate production capacity of 16.6 GW [7]. This metric refers to the maximum cumulative power rating of solar modules the company can manufacture in one year. To put this scale into perspective, this translates to producing approximately 35 million individual solar modules annually, which is sufficient to power roughly 3 million average American homes. These results show that CdTe technology has moved from laboratory potential to a commercially dominant force in global utility-scale solar power.

This case study dives into First Solar's journey with the development of CdTe solar technology. Think of it as the superhero story of photovoltaic solutions. It uncovers the epic battles against challenges, the eureka breakthroughs, and the market twists that turned it into a leading solution. Along the way, the case study also discusses the scientific advances, policy changes, and long-term ripple effects on the renewable energy world.

## 4.2 The Evolution of Solar Power

Can the sun, an energy source as old as time, truly power the future? Picture a world in which our energy requirements are not from fossil fuels but by the same force that sustains life on Earth. While the concept of harnessing sunlight to generate electricity is very much like a modern innovation, its origins are centuries old and intertwined with scientific breakthroughs, technology and drive to achieve higher efficiency.

One such game-changing moment in the history of solar power was the emergence of thin-film photovoltaic technologies, led by the cadmium telluride solar cell (CdTe). Silicon-based solar cells were the norm in the market traditionally, but their high pricing and energy-intense production meant they were never going to become the go-to for many.

### 4.2.1 From Early Discoveries to the Photovoltaic Effect

Long before solar panels were yet to line rooftops and power entire cities, the ancient civilizations already understood the power of the sun. Passive solar heating means that during winter the Greeks and Romans built houses to take advantage of sunlight. Already in the seventh century BCE, humans utilized magnifiers to focus sunlight for fires as an early example of solar energy use [1].

The true scientific breakthrough, however, came in 1839, when the French physicist Edmond Becquerel discovered the photovoltaic effect, which absorbs light in some materials and produces an electric current [8]. Becquerel's discovery was the fundamental building block of today's solar power technology, although it was nearly 100 years later that the first pioneering applications and commercial use of photovoltaics emerged. Following Becquerel's discovery, many notable scientific discoveries and innovations came. Such early discoveries and inventions would ultimately form the basis for the subsequent explosion in solar technology and paved the way for solar to be a realistic as well scalable energy source.

Figure 4.1 shows American inventor Charles Fritts, who built the first crude solar cell out of selenium in 1883, with an efficiency lower than 1% [10]. This

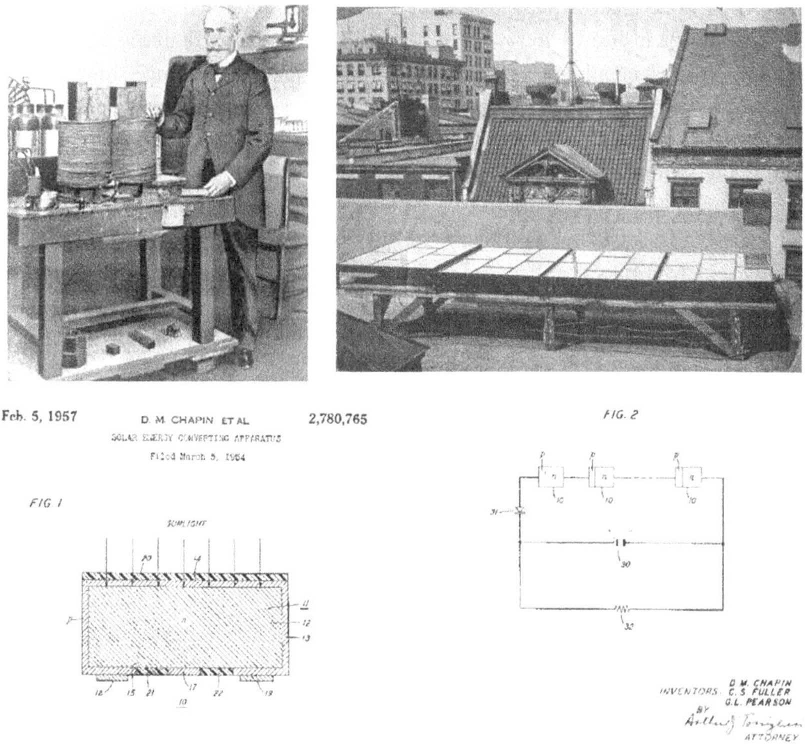


**Figure 4.1** (Top left) Edmon Becquerel. (Top right) The first solar installation by Charles Fritts on a rooftop in New York (1884) (Bottom) Patent of D. M. Chapin et al.'s "Solar energy converting apparatus," February 5, 1957. *Source:* [9] / Intellectual Property Magazine.

inefficiency meant it could not be implemented widely. But the concept was radical: that sunlight can be directly converted to electricity. The real breakthrough came in 1954. The first commercial silicon solar cell (6% efficiency) was developed by a team of researchers at Bell Labs: Daryl Chapin, Calvin Fuller and Gerald Pearson. This was the start of contemporary photovoltaic (PV) era [11].

By the early 1960s, solar cells found their first practical application – not on Earth, but in space [12]. Solar panels were first used to power satellites (including Vanguard 1, the first solar satellite launched by NASA in 1958). It showed photovoltaics could function properly where the condition change more drastic [13]. The space market drove progress in efficiency as engineers sought to enhance solar cells performance for use in space. But due to the high costs of these space-based solar technologies, solar energy would not be competitive in mainstream electricity production on Earth throughout this period.

The rise of silicon-based PV technology began to dominate during the 1970s and 1980s. Growing concerns with fossil fuel dependency and environmental concerns led many countries down the path of renewable energy, particularly solar power. Governments, especially in the US and Germany started to finance solar research along with incentives for development. The ability of silicon-based solar cells to improve efficiency and gain reliability with declining production costs made it the norm in the industry.

Silicon solar panels are generally divided into two types: monocrystalline, and polycrystalline. These traditional crystalline silicon solar panels faced several limitations [14]. Monocrystalline silicon (mono-Si) panels, are made of a single, continuous crystal structure. They are the most efficient and robust but have a higher cost due to manufacturing constraints [15]. Polycrystalline silicon panels (poly-Si), on the other hand are lower cost due to their simpler manufacturing process but are usually significantly less efficient [16]. All of these developments, however, still faced obstacles of material availability (such as high-purity silicon), and performed poorly under high temperatures and low-light conditions.

On top of that, the fragile nature of silicon wafers made large-scale deployment both expensive and difficult. These negative aspects eventually made way for competing photovoltaics technologies, primarily thin-film solar cells that seek to bridge the cost and scalability gap for solar power production. Figure 4.2 shows the milestones in the evolution of PV technologies with comparison of efficiency and various regulatory developments.

The timeline in Figure 4.2 illustrates the progression of photovoltaic technologies, from discovery of the PV effect to the beginning of solar energy usage via silicon-based solar cells, then the entrance of CdTe thin-film panel

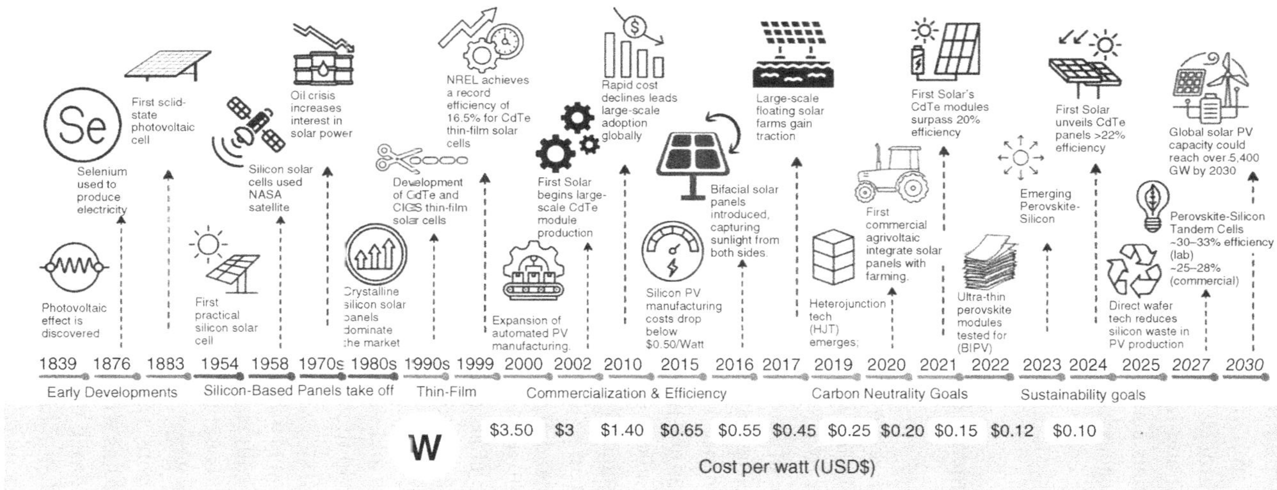


**Figure 4.2** Selected milestones in the evolution of photovoltaic (PV) technologies and historical cost of solar production (1839–2030). *Source:* Author's own work.

for commercialization. The evolution also illustrates the historical cost of solar production from the year 2000–2023 showing the significant reduction in cost per watt over the years as solar technology advances and becomes more efficient. Through the years substantial progress has been made, mostly in manufacturing and improvements in efficiency as well as cost reductions. Silicon-based panels continue to reign as the most popular solar panel technology even if decades have passed since its commercialization; however, CdTe film offers strong competition claiming to be less costly to manufacture and more effective.

### 4.2.2 The Emergence of a New Contender: The Thin-Film Technology

In the early 1990s, researchers started with efforts to develop thin-film solar cell technologies employing non-silicon semiconductors [17]. Some of these, such as cadmium telluride (CdTe) and copper indium gallium selenide (CIGS) solar cells, showed a glimpse of eventual success [18]. The materials were also cheaper to make, requiring easier production methods and less material. They also showed improved performance in actual applications, especially when the conditions were tough, such as low-light or medium efficiency and high-temperature regions where traditional silicon cells consistently performed poorly.

One big advantage of thin-film solar cells having over conventional solar cells is that they can be deposited onto rollable flexible surfaces such as plastic or metal foils instead of having to be scaled to rigid glass substrates [19]. This flexibility enabled the creation of lightweight solar power modules. These include portable chargers, integrated photovoltaics (IPV), and potentially wearable solar technology. They laid the groundwork for almost anything imaginable in terms of usage from powering remote sensors to use on surfaces of buildings and vehicles. Table 4.1 provides a comprehensive technical comparison to help illustrate the tradeoffs between thin-film and crystalline silicon solar panels.

This transition paved the way for First Solar, the leading company that transformed the solar industry with its breakthrough CdTe thin-film technology [20]. First Solar showed that CdTe could not only compete but, in some applications, outperform silicon-based technology.

Among the various thin-film technologies, First Solar's cadmium telluride solar cells emerged as a particularly promising and commercially viable option. The cadmium telluride thin-film solar cells had several factors that made them very appealing, such a direct bandgap that is close to the solar spectrum, high absorption coefficient, relatively simpler and less energy-intensive manufacturing process as compared to silicon cells [21]. First Solar rapidly expanded its CdTe

**Table 4.1** Technical comparison of thin-film solar cells vs. conventional silicon solar cells.

| Category | Parameter | Thin-Film Solar Cells (CdTe, CIGS, a-Si) | Conventional Silicon Solar Cells (Mono-Si, Poly-Si) |
|---|---|---|---|
| **Material and Composition** | Material | Cadmium telluride (CdTe), Copper indium gallium selenide (CIGS), Amorphous silicon (a-Si) | Monocrystalline silicon (Mono-Si), Polycrystalline silicon (Poly-Si) |
| | Material Abundance | Uses less semiconductor material | Requires high-purity silicon, which is resource-intensive |
| | Recycling Potential | CdTe and CIGS can be recycled, but Cd requires careful handling | Easily recyclable, but energy-intensive process |
| **Performance and Efficiency** | Typical Efficiency (%) | 10%–22% (CdTe: 17–22%, CIGS: 13–20%, a-Si: 6–12%) | 16%–24% (Mono-Si: 20–24%, Poly-Si: 16–20%) |
| | Temperature Coefficient | Better performance in high temperatures (−0.20 to −0.30%/°C) | Efficiency drops significantly in high heat (−0.35 to −0.45%/°C) |
| | Low-Light Performance | Better in low-light conditions (CdTe and CIGS) | Moderate to poor performance in low light |
| | Degradation Rate | Slightly higher (CdTe: ~0.3%–0.6% per year) | Lower (Mono-Si: ~0.2%–0.4% per year) |
| | Lifespan | 25–30 years | 25–40 years (Mono-Si has longer lifespan) |
| **Manufacturing and Cost** | Manufacturing Cost | Lower due to simpler, faster production | Higher due to complex wafer production and purification |
| | Manufacturing Process | Thin-film deposition techniques (vapor deposition, sputtering) | High-temperature crystal growth and wafer slicing |
| | Production Time | ~4 hours per module (CdTe) | Several days per module |
| | Energy Payback Time | ~1–2 years | ~2–3 years |

*(Continued)*

**Table 4.1** (Continued)

| Category | Parameter | Thin-Film Solar Cells (CdTe, CIGS, a-Si) | Conventional Silicon Solar Cells (Mono-Si, Poly-Si) |
|---|---|---|---|
| **Physical and Structural** | Weight | Lighter and flexible (e.g., A-Si on plastic substrates) | Heavier due to thick silicon wafers and glass layers |
| | Durability and Fragility | More resilient and shatterproof (CdTe and CIGS) | Brittle and prone to microcracks |
| **Environmental Impact** | Energy Use in Production | Lower energy-intensive production | Energy-intensive silicon refining |
| | Toxicity Concerns | CdTe has toxicity concerns due to cadmium | No toxicity concerns |
| **Applications** | Best Applications | Large-scale solar farms, BIPV (Building-Integrated PV), flexible panels | Rooftop installations, residential and commercial solar projects |

*Source:* Author's own work.

production facility in the early 2000s, growing to one of the top solar module producers in the world and bringing down the cost of solar.

A comparison of thin-film solar cells with conventional silicon solar cells for their respective applications, efficiency ranges, cost-effectiveness, and environmental impact is shown in Figure 4.3. Thin-film technologies are like the yoga instructors of the solar world: lightweight, flexible, and perfect for cost-sensitive situations like BIPV, space tech, and portable solar gadgets. Meanwhile, silicon solar panels are the gym bros: bulky, powerful, and dominating in rooftop installations, high-efficiency power plants, and commercial energy systems. The comparison also highlights the eternal tradeoffs, for example, efficiency vs. cost, and durability vs. climate compatibility. These factors will drive the future adoption of each technology.

As the solar sector moved from prototype energy solution to widespread source of global energy a key moment arose. The obvious question was: could a new technology displace silicon's long-standing dominance and redefine the future trajectory of solar power? The answer to this question lies through the pioneering journey of a company that has made a profound impact on the solar energy industry.

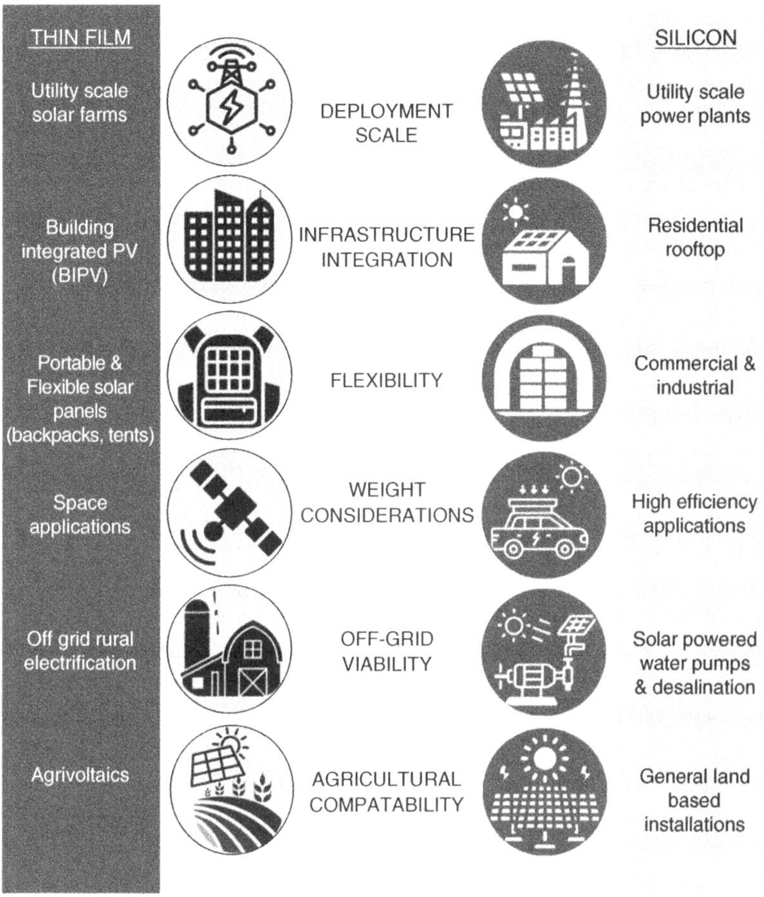


**Figure 4.3** Applications of thin-film CdTe solar panels vs. conventional silicon solar cells. *Source:* Author's own work.

### 4.2.3 First Solar's Entry into the Market

How did a bold vision and relentless innovation propel First Solar to the forefront of the solar industry? In the end of 1990s as people started realizing the negative effects of fossil fuels on planet Earth, solar energy emerges as a potential solution. First Solar founded in 1999, set out to commercialize the thin-film PV technology with cadmium telluride (CdTe). Aiming to provide cost effective and scale competitive with traditional silicon-based solar panels. CdTe solar cells were suggested to have both technical and economic feasibility by researchers at the turn of the millennium.

Recognizing the limitations of existing silicon-based photovoltaics in terms of cost and scalability, First Solar was driven to pursue cadmium telluride as a promising alternative. CdTe looked to offer more affordable production and more consistent performance in all environmental settings. The CdTe technology and First Solar's commitment to building it into an economically viable source for harnessing solar energy for the entire globe. From the outset, First Solar's approach is driven by relentless focus on innovation and cost reduction. A combination of advances in manufacturing technology and innovative processes allowed the company to drive down prices. This made it increasingly competitive with conventional silicon-based solutions.

As First Solar ramped up its production capabilities through the early 2000s, it was able to leverage economies of scale and refine its processes to significantly drive down the cost per watt for its CdTe modules. Transitioning from concept to large-scale production presented significant challenges. During its formative years, First Solar hit technical roadblocks in scaling CdTe module manufacturing at acceptable cost per watt output. This would require constant innovation to improve module efficiency and production costs, in order for the company to compete with the matured silicon photovoltaic (PV) market. Therefore between 2004 and 2009, First Solar was able to reduce manufacturing costs from over $3 per watt to $0.98 per watt. "This achievement marks a milestone in the solar industry's evolution toward providing truly sustainable energy solutions," said Mike Ahearn, First Solar's CEO [22]. With this, the company continues to drive the transition to clean, affordable solar power as a viable alternative to fossil fuels.

Additionally, the company heavily invested in research and development for improved efficiency of the module. Commercial monocrystalline wafer-based silicon modules have grown in efficiency over the last decade by 22% or more (up from 6–7%). And the CdTe module's efficiency has also leapt from around 9% to 20% [23]. This is the indication of technological maturity in CdTe, which shows consistent improvements in performance.

An important factor of First Solar's success is its vertical integration manufacturing. Instead of using multiple facilities and long production processes like the conventional silicon PV manufacturing, First Solar created a very different process. They can convert sheets of glass to working solar panels in about four hours using their in-house process and production line. Not only is this method faster but it also lowers the manufacturing carbon footprint by trimming time spent in production. By 2023, First Solar's global annual manufacturing capacity reached 16.6 gigawatts (GW), with plans to expand to over 20 GW by 2025 [6].

Aside from increasing manufacturing efficiencies, the company has advanced its module performance drastically. The company's CdTe modules have shown comparable conversion efficiencies with more than 22% of research cells. To put

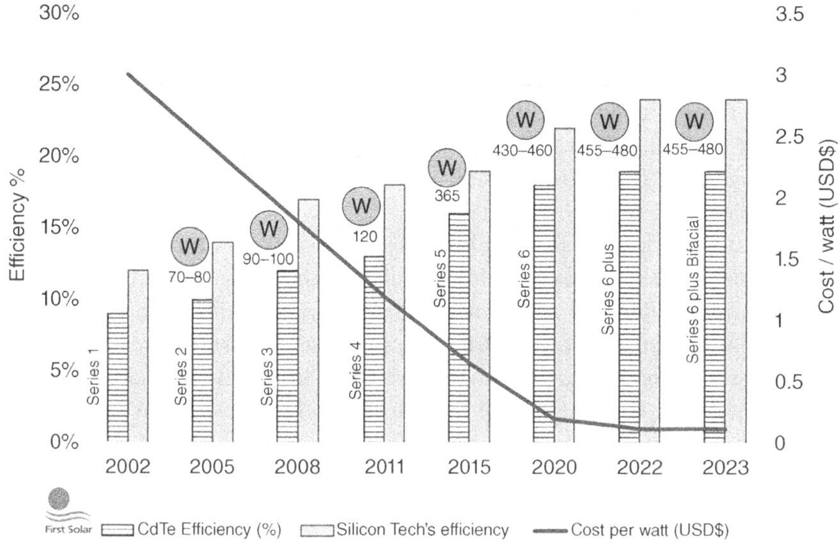


**Figure 4.4** First Solar's CdTe panel and efficiency comparison with silicon technology panels. *Note:* W = watts. *Source:* [23, 24].

this in perspective, these advancements have firmly placed First Solar as market leader in thin-film solar.

First Solar's CdTe technology has made significant progress in efficiency over the years. Figure 4.4 presents the series of CdTe panels that the company launched since 2002 and comparing efficiency to silicon tech. The figure also shows the downward push of cost per watt to produce solar energy as technology gets more efficient. While it started lower in efficiency than the conventional silicon-based solar panels, it has steadily progressed and was approaching nearly ~19.5% with the Series 6 Plus in 2022. Though silicon has hit a high efficiency at (~22–24% band gap), First Solar has delivered competitive costs through the much lower manufacturing of CdTe panels. CdTe is inherently thin film, which means you need less material in order to harness the light for solar, making it cheaper than traditional crystalline for the same size. Its affordability focus, with incremental efficiency gains, is well-positioned to continue as the go-to provider for utility-scale projects, where cost is everything.

The Series 6 Plus optimized for power density and actual conditions can probably lead to wider utility-scale solar plants deployments. CdTe will be in play over large-scale renewable energy projects due to its greater scalability and lower cost per watt once silicon technology hits its efficiency cap for big cell size panels. The evolution of solar technology will likely be marked by a continued evolution of CdTe and silicon technologies side by side, each one meeting different market requirements and various applications (i.e., residential-commercial vs. large-scale utility).

Through visionary leadership and innovative manufacturing processes, First Solar has reinvented itself with dedication toward advancements in CdTe technology. It is a journey that illustrates the power of competitive innovation in addressing industry bottlenecks and reducing barriers for sustainable energy solutions.

The future of photovoltaics stands on the improvement of the next generation materials – perovskite solar cells. It is anticipated that when used in tandem cell configurations, these will allow for >30% efficiency together with silicon. The improved development of manufacturing processes, market launch, and environmental benefits are also likely to increase with clean energy policies, and even regulatory tailwinds from a global regulatory standpoint. This will set the path for future global energy transition aspect that solar power will hold over coming decades.

## 4.3 Challenges in Scaling CdTe Solar Technology

Although First Solar was the pioneer in the technology of thin-film CdTe, the company faces a significant challenge. This was to prove that they are better than the established crystalline silicon (c-Si). This scenario pins down to making the decision for scaling up production, lowering cost, and increasing efficiency while retaining sustainability and CdTe competitive edge.

The biggest hurdle between CdTe versus c-Si was their efficiency gap. In the early 2000s, CdTe modules reached 10–12% efficiency while crystalline silicon panels were already at 14–16% efficiency, giving silicon a performance lead. Additionally, the supply chain of CdTe was another critical factor. While silicon is cheap in the market; cadmium and tellurium are by-products from zinc/copper mining, consequently there is a concern about their long-term availability and price stability. This dependency on limited raw materials created a potential bottleneck in scaling up production.

In addition, market competition and consumer perception further complicated the company's growth. Crystalline silicon has dominated the solar industry for decades, benefiting from stable supply chains and wide adoption. Investors and consumers alike, in general hesitate to change from their comfort zone. Especially to a new technology with no proven efficiency and questionable future longevity. Environmental regulations sustainability was also problematic. Although CdTe panels had less carbon footprint, required less energy for processing, the toxicity of cadmium necessitated the establishment of strict recycling measures and end of life management.

## 4.4 Think and Solve

The questions that follow summarize some of these concerns and challenges, focusing on First Solar's market positioning, technological advancements, and environmental sustainability.

**Q 4.1** What was the primary advantage of CdTe thin-film solar cells compared to crystalline silicon solar cells?

**A** Higher efficiency under all conditions.
**B** Lower production cost and better performance in high temperatures.
**C** Availability of raw materials without supply chain constraints.
**D** Longer lifespan than silicon-based solar cells.

**Q 4.2** What was one of the biggest challenges in scaling CdTe solar panel production?

**A** The lack of demand for solar energy.
**B** High cost of silicon materials.
**C** Limited availability of tellurium and supply chain dependency.
**D** Low-temperature sensitivity of CdTe modules.

**Q 4.3** How did First Solar address the efficiency gap between CdTe and silicon-based solar panels?

**A** By improving deposition techniques and reducing material defects.
**B** By shifting to monocrystalline silicon technology.
**C** By increasing the size of solar panels without improving efficiency.
**D** By replacing cadmium telluride with perovskite materials.

**Q 4.4** Why was cadmium telluride considered an environmentally sustainable alternative to traditional silicon solar panels?

**A** It used non-toxic materials that required no special handling.
**B** CdTe panels had a lower carbon footprint and required less energy to manufacture.
**C** CdTe was an unlimited natural resource with no environmental concerns.
**D** CdTe panels degraded faster, making them easier to recycle.

**Q 4.5** Assuming that a CdTe solar module has an efficiency ($\eta$) of 19.5% under standard test conditions (STC) and receives an average solar irradiance (G) of $1000\,W/m^2$. If the total surface area of the module (A) is $1.5\,m^2$, calculate the maximum power output ($P_{max}$) of the module. Formula: $P_{max} = \eta \times G \times A \times P$

## 4.5 Conclusion and Key Takeaways

First Solar's journey in advancing cadmium telluride (CdTe) solar technology underscores the intricate balance between technological innovation, market competition, and sustainability. From 1999 to the present, the company has been driving at full speed to shatter the monopoly of crystalline silicon (c-Si) solar panels. The company has proven thin-film solar technology is here to stay by continuously pushing boundaries.

First Solar to date has achieved some of its biggest wins around massive manufacturing cost reductions. Having fine-tuned its fully integrated production flow, the firm lowered the cost of modules from over $2/watt in 2009 to less than $0.20/watt, which is the lowest in the industry. The cost advantage of First Solar has allowed the company to win critical large-scale utility projects in particular regions where levelized cost of electricity (LCOE) plays an important role in driving solar power.

Despite CdTe panels not being able to compete with silicon when it comes to efficiency, the company set a record 22.3% efficiency for a single CdTe cell in 2023 closing the gap with state-of-the-art high-efficiency monocrystalline silicon panels. This was made possible by diligently advancing bandgap engineering, module innovation and manufacturing automation.

Sustainability is another pillar of First Solar's strategy. CdTe modules have much lower energy payback time than conventional silicon panels. CdTe modules approximately 1 year vs. c-Si panels = 1.5–3 years. Furthermore, First Solar is one of only a handful of large-scale PV manufacturers with a functioning recycling program in operation. It provides for >90% recovery of semiconductor materials and >95% glass from decommissioned panels. The closed loop will not only reduce its environmental footprint but also enhance its long-term sustainability.

But First Solar still struggles. The solar market globally is still very competitive with continuous innovation for silicon-based technologies driving down costs. Discreet material supply constraints were also raised concerns over the availability of tellurium, a lesser used byproduct of copper refining. The response of the company has been signing long-term material sourcing contracts and developing modernized methods that help to increase efficiency in tellurium extraction.

The biggest lesson from this case study is that technology is not always enough for industry leadership. First Solar's success is rooted in the convergence of R&D strategy investments, vertical integration of manufacturing, and sustainability innovations. Overall, this case demonstrates the need for being market responsive, having a strong supply chain, policy benefits and strong environmental stewardship in the renewable space. In context of the continued growth of solar, lessons for engineers, industry, and policy-makers from the experience of First Solar is highly relevant at robust innovation edge and in the cost-efficient sustainable global transition toward clean energy.

## 4.6 Discussion Points

The case of First Solar's cadmium telluride (CdTe) technology presents several critical discussion points relevant to the evolution of the solar industry, the challenges of scaling emerging photovoltaic (PV) technologies, and the broader implications of sustainability and energy policy. These discussion points help to contextualize the key learnings from First Solar's journey and highlight the fundamental engineering, economic, and environmental considerations in the renewable energy sector.

Crystalline silicon (c-Si) solar panels have been the dominant technology for decades, benefiting from a well-established supply chain, high efficiency, and continuous cost reductions. CdTe panels from First Solar had to overcome a lot of push back from the market. Though relying on lower cost production methods, early CdTe modules were generally less efficient than silicon-based PVs [25]. But, with continuous improvement, CdTe has been able to close the efficiency gap significantly and have achieved module efficiencies over 19¢/W while still leading in terms of cost per watt [26].

A key discussion point is whether thin-film solar technology can continue to gain market share against the ever-evolving silicon PV sector. As monocrystalline silicon panels push efficiency beyond 22–24%, First Solar must maintain a balance between efficiency gains and cost competitiveness to sustain its advantage.

1. How should the company continue to differentiate its products in an increasingly competitive market?
2. Should sustainability be a greater priority in PV manufacturing? How can First Solar leverage its sustainability initiatives as a competitive advantage, especially as more companies begin integrating circular economy principles?
3. Should First Solar invest in tellurium recycling and alternative semiconductor materials, or should it seek partnerships with mining companies to secure supply?
4. The global solar market is also shaped by trade policies, tariffs, and geopolitical factors. US restrictions on Chinese solar imports have helped First Solar gain a stronger foothold in the domestic market. But how sustainable is this advantage? Would First Solar be competitive in an open market without trade protections?
5. Looking ahead, advanced manufacturing and next generation solar deployment are in alignment with First Solar's investments in tandem solar cells? Another area the company is investigating is perovskite thin-film blending, which shows promises in research but is still commercially unproven.

A vital question remains: Will thin-film solar technologies remain competitive with silicon? Or will emerging technologies such as perovskite–silicon hybrids render CdTe irrelevant in the future? What should First Solar continue to invest in R&D for the next decade as an aggregate to remain number one in renewables?

First Solar's quest for thin-film solar dominance is like trying to win a dance-off while wearing roller skates. It's exciting, ambitious, and full of unexpected twists. From outshining silicon PV rivals to juggling material constraints like a solar-powered circus act, the company's journey is anything but boring. They also had to leverage sustainability advantages and navigate global trade policies. Whether they go on to conquer or just fall on the roller skates, they can decide whether they will be the Beyoncé of renewables or just another one-hit wonder.

## 4.7 Limitations of the Case Study

Although this case study provides some useful lessons for the growth of First Solar CdTe technology, a few limitations should be noted. These are technology limits, geopolitics, economics as well material availability and long-term sustainability concerns.

1. **Technological Focus Limitation:** First Solar's CdTe technology is the focus of this case study. Other upcoming photovoltaic technologies like perovskites or tandem cells from contenders such as Solar Frontier were not compared. Those innovations could shake up the market and change First Solar's future down the line.
2. **Geopolitical and Economic Factors:** Trade policies and global economic factors, such as tariffs on Chinese imports had large impacts on First Solar. But the case study does not include these factors.
3. **Supply Chain and Material Risks:** First Solar's reliance on tellurium, a rare material, presents a potential bottleneck in scaling production. Despite long-term agreements, the limited global supply of tellurium could pose risks to First Solar's growth.
4. **Rapid Technological Change:** The quick developments of solar technologies mean that the findings here are hardly more than a specific point in time when this case study was written. Future innovations may dramatically change the competitive landscape.

## References

1 Sundaram, S., Benson, D. and Mallick, T.K. *Solar Photovoltaic Technology Production: Potential Environmental Impacts and Implications for Governance*, Academic Press, 2016.

2 Dallaev, R., Pisarenko, T., Papež, N. and Holcman, V. "Overview of the current state of flexible solar panels and photovoltaic materials," *Materials*, 16(17), 5839, 2023.

3 Mathews, I., Kantareddy, S.N.R., Liu, Z., Munshi, A., Barth, K., Sampath, W., Buonassisi, T. and Peters, I.M. "Analysis of CdTe photovoltaic cells for ambient light energy harvesting," *Journal of Physics D: Applied Physics*, 53(40), 405501, 2020.

4 Solar Energy Technologies Office (SETO), "Cadmium telluride," U.S Department of Energy, 2023. https://www.energy.gov/eere/solar/cadmium-telluride (accessed February 24, 2026).

5 Wang, A., Chang, N.L., Sun, K., Xue, C., Egan, R.J., Li, J., Yan, C., Huang, J., Rong, H., Ramsden, C. and Hao, X. "Analysis of manufacturing cost and market niches for $Cu_2ZnSnS_4$ (CZTS) solar cells," *Sustainable Energy and Fuels*, 5(4), 1044–1058, 2021.

6 First Solar, "First Solar: a leading global provider of comprehensive PV solar solutions," 2024. www.firstsolar.com (accessed February 24, 2026).

7 O'Neill, W. "Seeing the light: key inventions that illuminate our world," *Journal of Physics: Conference Series*, 2877(1), 012068. IOP Publishing, 2024.

8 Chu, E. and Tarazano L.D. "A brief history of solar panels," *Smithsonian Magazine*, April 2019. https://www.smithsonianmag.com/sponsored/brief-history-solar-panels-180972006 (accessed February 24, 2026).

9 Kumar, K., Finnegan, H., Farabow, G. and Dunner, L. "A history of the solar cell, in patents," *Intellectual Property Magazine*, 2020, https://ipo.org/wp-content/uploads/2020/06/Solar-Cells_Article_IPO_KarthikKumar.pdf (accessed February 24, 2026).

10 Shahabuddin, M., Asim, M. and Sarwar, A. "Evolution and modeling of solar photovoltaic cells: from early to modern concepts," in *Photovoltaic Systems Technology*, ed. M.A. Husain, M.W. Ahmad, F.I. Bakhsh, P. Sanjeevikumar and H. Malik), 27–41. Wiley, 2024.

11 Iles, P.A. "Evolution of space solar cells," *Solar Energy Materials and Solar Cells*, 68(1), 1–13, 2001.

12 Raffaelle, R.P. "An introduction to space photovoltaics: technologies, issues, and missions," in *Photovoltaics for Space*, ed. S.G. Bailey, A.F. Hepp, D.C. Ferguson, R.P. Raffaelle, and S.M. Durbin, 3–27. Elsevier, 2023.

13 Battaglia, C., Cuevas, A. and De Wolf, S. "High-efficiency crystalline silicon solar cells: status and perspectives," *Energy & Environmental Science*, 9(5), 1552–1576, 2016.

14 Edalati, S., Ameri, M. and Iranmanesh, M. "Comparative performance investigation of mono-and poly-crystalline silicon photovoltaic modules for use in grid-connected photovoltaic systems in dry climates," *Applied Energy*, 160, 255–265, 2015.

15 Ballif, C., Haug, F.J., Boccard, M., Verlinden, P.J. and Hahn, G. "Status and perspectives of crystalline silicon photovoltaics in research and industry," *Nature Reviews Materials*, 7(8), 597–616, 2022.

**16** Maalouf, A., Okoroafor, T., Jehl, Z., Babu, V. and Resalati, S. "A comprehensive review on life cycle assessment of commercial and emerging thin-film solar cell systems," *Renewable and Sustainable Energy Reviews*, 186, 113652, 2023.

**17** Rahman, M.A., Hasan, S. and Ferdous, M.J. "An investigation into efficiency improvement of thin film solar cell," Doctoral dissertation, Department of Electrical, Electronics and Communication Engineering (EECE), MIST, 2011.

**18** El-Atab, N. and Hussain, M.M. "Flexible and stretchable inorganic solar cells: progress, challenges, and opportunities," *MRS Energy and Sustainability*, 7, E19, 2020.

**19** Bulavko, G. "Organic photovoltaics: a journey through time, advancements, and future opportunities," *History of Science and Technology*, 14(1), 10–32, 2024.

**20** Bosio, A., Rosa, G. and Romeo, N.J.S.E. "Past, present and future of the thin film CdTe/CdS solar cells," *Solar Energy*, 175, 31–43, 2018.

**21** "First Solar reaches US $1 per watt milestone," *Renewable Energy World*, 2009. https://www.renewableenergyworld.com/solar/first-solar-passes-us1watt (accessed February 24, 2026).

**22** Fraunhofer ISE. "Photovoltaics report," 2024: 7. https://www.ise.fraunhofer.de/en/publications/studies/photovoltaics-report.html (accessed February 24, 2026).

**23** First Solar, "First Solar products," 2024. www.firstsolar.com (accessed February 26, 2026).

**24** "Solar PV prices," Our World in Data, 2024. http://www.ourworldindata.org/grapher/solar-pv-prices (accessed February 26, 2026).

**25** National Renewable Energy Laboratory (NREL). "PV efficiency records 2023," 2023.

**26** "First Solar Q4 Financial Report," 2023. https://investor.firstsolar.com/financials/annual-reports/default.aspx (accessed February 24, 2026).

# 5

# Blacksmith to Smart Wind: Vestas' AI Transformation

Integrating IoT, Big Data, and Predictive Analytics in Modern Wind Turbine Systems

## 5.1 Synopsis

Picture a fleet of wind turbines that can predict mechanical failures, change their blade angles in real-time for optimal efficiency and work together with solar and battery storage. Just as self-driving cars are powered by AI to be better and safer, wind energy is undergoing a digital transformation, one led by Vestas, a Danish company and one of the world's leading wind turbine manufacturers.

Founded in Denmark in 1945, Vestas has grown into a global leader in renewable energy solutions, with operations in over 80 countries. The company turned to AI-driven technology to tackle the challenges in harnessing wind energy, fluctuating wind conditions, rising maintenance costs, and the ever-elusive quest to keep renewable power from throwing tantrums on the grid. Armed with big data analytics, IoT sensors, and aerodynamics so advanced that Vestas is on a mission to squeeze every watt out of the wind, slash costs, and keep the turbines spinning.

The path to smart wind energy, on the other hand, has not been without obstacles. Unpredictable weather patterns, grid stability issues, and the complexity of integrating AI into aging infrastructure is no easy equation. Moreover, the large development and deployment costs of using AI for predictive maintenance have put Vestas in a tightrope act between technological aspirations and commercial viability. In fact, in 2022, Vestas reported an Earnings Before Interest and Taxes (EBIT) margin of minus 8%, heavily impacted by increasing costs, product impairments, and elevated warranty provisions [1]. In the same year, the company's overall financial performance saw a significant loss of approximately €1.15 billion, a steep decline compared to an operating EBIT of around €428 million in 2021 [2]. Even the area of cybersecurity threats in connection with wind farms has emerged as a growing concern.

Despite all these hurdles, Vestas is still striving to be on the leading edge of innovation. Leveraging machine learning algorithms to predict potential failures, optimizing blade design for variable wind speeds, and enhancing hybrid energy systems for improved stability. Are the technological and economic barriers of smart wind turbines holding back their potential to redefine sustainable energy for the future, as the world looks for cleaner and more reliable power? This case study delves into the breakthroughs and challenges of AI-powered wind energy and offering a glimpse into the future of intelligent renewable solutions.

## 5.2 Background and Case Scenario

Can a company's DNA fundamentally transform while retaining its core identity? The evolution of Vestas offers a compelling answer to this question. Like a seed that grows into a tree bearing unexpected fruit – perhaps a pineapple on an oak – this Danish enterprise transformed from a humble metalworking shop into a global renewable energy leader. It proves that with enough adaptability, you can not only change your company's trajectory but also leave an entire industry wondering, "Wait, how did they pull that off?"

### 5.2.1 From Agricultural Equipment to Wind Energy Pioneer

In the quiet Danish countryside of 1945, as World War II drew to a close, a small blacksmith shop opened its doors in the village of Lem. H.S. Hansen founded it as a humble enterprise called "Vestjysk Stålteknik A/S in West Denmark (West Jutland Steel Technology).

Initially, the shop focused on household appliances and agricultural tools. The business then changed as time passed to respond to the immediate needs of Denmark's predominantly agricultural economy. Few could have predicted that this humble beginning would evolve into the Vestas we know today. This pragmatic beginning established two traits that would become synonymous with Vestas throughout its history: technical craftsmanship and the ability to respond to market demands.

The 1970s marked a pivotal split in its history. As the global oil crisis of 1973 sent shockwaves through energy markets, Denmark found itself particularly vulnerable because they are heavily dependent on imported oil. This external situation catalyzed re-evaluation of their national energy policy and created opportunities to explore alternative energy options. At that time, Vestas demonstrated strategic foresight as they started testing wind turbines in 1979 [3]. That action was not merely opportunistic but represented a fundamental structuring of the company's purpose and technical focus.

The transition wasn't immediate or without challenges. Early wind turbine designs were primitive by modern standards. They consists of simple horizontal-axis machines that were low in efficiency and highly unreliable. Vestas would have some mix of successes and setbacks through the 1980s. Then came the California wind boom. It initially provided big opportunities but when these subsidies collapsed in 1986, Vestas would nearly go bankrupt [4]. This crisis threatened the company's existence. Ultimately, they strengthened their resolve and focused their purpose exclusively on wind energy technology.

In the 1990s, Vestas started to develop into a clear organization in wind energy with better engineering capabilities. They pioneered innovations in blade design, materials science, and control systems. These revolutionized the efficiency and reliability of turbines. Vestas released its OptiSlip® solution in 1995, which had turbines that were able to deliver a similar electrical output capability despite wind conditions [5]. This was a significant improvement in grid compatibility.

As global awareness of climate change intensified in the early 2000s, Vestas positioned itself at the vanguard of the emerging sustainable energy transition. The company's 2004 merger with NEG Micon solidified its status as the world's largest wind turbine manufacturer, expanding both its technological portfolio and global market reach [6]. This period marked Vestas transformation from a manufacturer into an integrated energy solutions provider.

Table 5.1 highlights Vestas' evolution with selected milestones. The timeline also summarizes the turbine model introduced and the progress of technology that Vestas incorporated. The example of Vestas's growth from a small Danish post-war electrical appliance manufacturer to the biggest wind turbine producer in the world illustrates how wind energy itself went through an incredible transformation. In the beginning of its technological process, it was a rather modest 30 kW unit standing just 18 meters high in 1979. Then, by 1981 its first commercial wind farm in California grew to using 55 kW turbines. But the biggest developments took place in the late 1980s when the variable slip and pitch regulation of OptiSlip® were introduced, establishing the basis of future success for the company.

By the mid-1990s, Vestas had introduced its revolutionary OptiTip® blade angle optimization system on 500 kW turbines reaching 40 meters high, before crossing the significant megawatt threshold in 1999 with the V66-1.65 MW that towered 67 meters above the landscape. The new millennium brought accelerated scaling with the breakthrough V80 platform in 2002 and the company's strategic offshore market entry in 2004 with 3.0 MW turbines.

Subsequent technological leaps included the V112 platform (2009), the formation of MHI Vestas offshore joint venture (2014), and the introduction of the highly versatile 4 MW platform optimized for low-wind sites with towers reaching up to 166 meters. The company's innovation trajectory culminated in the modular

**Table 5.1** Vestas' journey of innovation (1945–2023).

| Year | Milestone | Turbine Model | Technological Innovation | Capacity | Rotor Diameter | Hub Height | Market Position |
|---|---|---|---|---|---|---|---|
| 1945 | Company founded as "Vestjysk Stålteknik A/S" | – | Steel fabrication | – | – | – | Local Danish manufacturer |
| 1979 | First wind turbine | Vestas turbine (30 kW) | Fixed-pitch blades, simple control systems | 30 kW | 10 m | 18 m | Early market entrant |
| 1981 | First wind farm<br>First export orders to US | Vestas 55 kW | Improved blade design | 55 kW | 15 m | 20 m | Emerging international player |
| 1986 | Bankruptcy and restructuring<br>Operations paused | Vestas V17-75 kW | – | 75 kW | 17 m | 23 m | Market setback after California subsidy collapse |
| 1987 | Re-emergence as wind turbine -only company | Vestas V17-75 kW | – | Pitch-regulated blades | – | – | Focused wind energy specialist |
| 1989 | Pitch regulation | V20-100 kW | | 100 kW | 20 m | 24 m | |
| 1995 | OptiTip® technology | V39-500 kW, V44-600 kW | Variable slip technology for grid stability | 500 kW | 39 m | 40 m | Technology leader |
| 1997 | MW-class turbine | V44-600 kW | – | 600 kW | 44 m | 45 m | – |
| 1998 | IPO on Copenhagen Stock Exchange | V66-1.65 MW | | Larger scale commercial turbines | Growing market presence | | |

| | | | | | | | |
|---|---|---|---|---|---|---|---|
| 1999 | First MW turbine | V66-1.65 MW | – | 1.65 MW | 66 m | 67 m | – |
| 2000 | Launch of MW platform | V80-2.0 MW | – | Advanced pitch control systems | Industry pioneer in MW-class turbines | – | – |
| 2002 | V80 platform | V80-2.0 MW | – | 2.0 MW | 80 m | 78 m | – |
| 2004 | Offshore entry<br>Merger with NEG Micon | V90-3.0 MW | Combined technological portfolios | 3.0 MW | 90 m | 80 m | World's largest wind turbine manufacturer |
| 2007 | Introduction of V90-3.0 MW | V90-3.0 MW | – | Advanced blade materials, improved efficiency | Market leader | – | – |
| 2009 | V112 platform<br>Offshore wind focus intensifies | V112-3.0 MW | Designed for harsh offshore conditions | 3.0 MW | 112 m | 84 m | Competing with Siemens in offshore market |
| 2014 | MHI Vestas JV | V117-4.2 MW, V136-4.2 MW | Enhanced power electronics, larger rotors | 8.0 MW | 164 m | 105 m | Technology differentiation against GE and Goldwind |
| 2017 | Launch of 4 MW platform | V150-4.2 MW | Vestas Intellimax®<br>AI-based condition monitoring | 4.2 MW | 150 m | 105–166 m | Pioneer in smart turbine technology |
| 2019 | EnVentus platform<br>Carbon neutrality commitment | V162-5.6 MW | Modular turbine architecture | 5.6 MW | 162 m | 119–166 m | Sustainability leadership |

(Continued)

**Table 5.1** (Continued)

| Year | Milestone | Turbine Model | Technological Innovation | Capacity | Rotor Diameter | Hub Height | Market Position |
|---|---|---|---|---|---|---|---|
| 2021 | Introduction of V236-15.0 MW | V236-15.0 MW | Record-breaking offshore turbine size | 15.0 MW | 236 m | 149 m | Competing with Siemens Gamesa and GE Haliade-X |
| 2022 | Acquisition of MHI stake<br>Digital twin technology deployment | – | Vestas Cirrus™<br>Full turbine simulation capabilities | – | – | – | Digital transformation leader |
| 2023 | Advanced materials research<br>Recyclable blade technology | V172-7.2 MW | Zero-waste blade manufacturing | 7.2 MW | 172 m | 148–199 m | Circular economy pioneer |

*Source:* Adapted from [7].

EnVentus architecture (2019), the announcement of the massive V236-15.0MW offshore giant with a total height of 267 meters (2021). Its full reinstatement of the offshore business, as well as single onshore flagship V172-7.2MW that tops nearly 200 meters when fully configured. Vestas has been consistent throughout all of this as a pioneer, driving technological breakthroughs which have shaped the industry's journey from niche technology to mainstream energy alternative (Figure 5.1).

The historical trajectory of wind turbine evolution itself parallels Vestas' development. Over the centuries, wind power technology has been improved from the ancient Persian vertical-axis mills to electricity-producing models of nineteenth century, thanks to inventors like James Blyth and Charles Brush who performed early experiments. Vestas has added to this evolutionary story by rationally applying scientific ideas of aerodynamics, materials engineering and today computational intelligence to collect the energy whilst also minimizing loading on structure and maintenance.

Today's Vestas bears little resemblance to the small blacksmith shop of 1945, yet certain organizational characteristics persist such as their adaptability, technical innovation, and market responsiveness. The farm supplier that once sold agricultural implements now makes artificial intelligence systems used for wind pattern prediction and blade pitch in real-time optimization. Its transformation reflects both corporate paradigms shift, as well as more general societal move to sustainable technological development.

As we trace Vestas' evolution from agricultural equipment maker to a pioneer in wind energy, we do more than follow the journey of a company. It is the evolution of this innovative tech vs. market forces with environmental pressure that are writing the story in renewable energy.

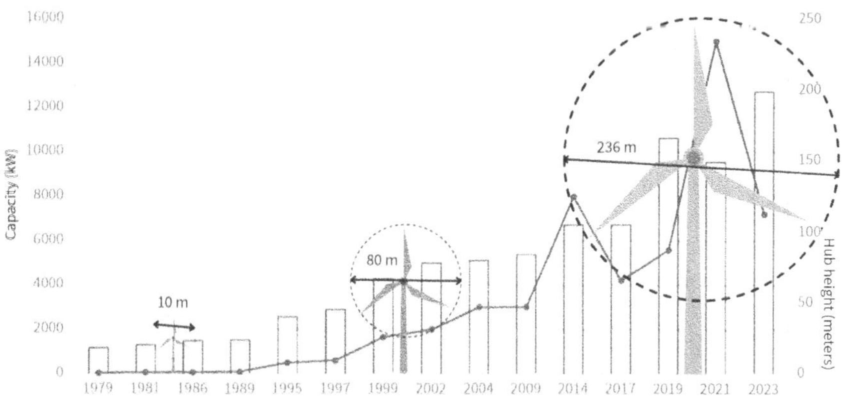


**Figure 5.1** Development of Vestas' wind turbine height, radius and capacity. *Source:* Author's own work.

### 5.2.2 The Smart Wind Revolution – Technology and Implementation

Wind power's history runs thousands of years back from the earliest Persian windmills in 500 BCE grinding grain to the iconic Dutch structures used for draining polders in the fourteenth century. It took until 1887 for the first wind turbine to be built by Scottish Professor James Blyth to power his holiday home. The early twentieth century was about Danish pioneers like Poul la Cour with rougher early designs. The 1970s oil crisis helped to kick start the interest in alternatives to fossil fuels. The first generation of commercial turbines on the market in the 1980s were small, with a capacity of 50–100 kW and often made of wooden or fiberglass-blades. Their diameter spanning around 15–20 meters compared to the modern turbines today that can span up to 170 meters [8]. Over the coming decades these technological seedlings will expand at an exponential rate.

The wind energy sector changed quite a lot in just over two decades. As detailed in Table 5.2, which tracks the advance of turbine models by top global manufacturers (2000–2023), the industry shifted from 1.5 MW rotors and ~65 m diameter in 2000, to gigantic 15 MW giants exceeding 230 m diameter by 2023 [9]. Heralding an increase in power capacity by ten times and rotor-sweep by nearly four times. This technological progress represents increasingly larger offshore platforms.

Vestas continued to innovate with modular design and blade technology, and being at the frontier as a pioneer. While Siemens Gamesa differentiated itself with direct drive optimized for the worst offshore conditions. GE scaled up motors later and aggressively into the ultra-large offshore sector with its revolutionary Haliade-X. Meanwhile, Goldwind's evolution from licensing foreign technology to developing indigenous innovations that exemplifies the globalization of wind technology, and Enercon's unwavering commitment to direct drive systems and distinctive design aesthetics established its unique market position.

By 2023, all major manufacturers had aligned to similar specifications for their top of stack offshore models (14–16 MW with 220–242 m rotors), but in different ways regarding drivetrain design, blade technology and grid integration capabilities. This reflects the industry's maturity while continuing to push the boundaries of scale and efficiency.

Vestas' current portfolio illustrates this evolution, with each platform tailored to distinct deployment scenarios. The V236-15.0 MW offshore beast generates enough electricity in a single rotation to power an average home for more than three days. The 115.5-meter-long blade (longer than a football field) sweeps an area of 43,743 $m^2$, capturing wind energy across a surface greater than six soccer fields [14]. The V150-4.2 MW, on the other hand, has been the hammer for medium wind regimes with its 150-meter rotor that delivers great performance at wind speeds as low as 6.5 m/s. For sites with complex terrain or very land-limited,

**Table 5.2** Advance turbine models by top global manufacturers (2000–2023).

| Year | Vestas | Siemens Gamesa | GE Renewable Energy | Goldwind | Enercon |
|---|---|---|---|---|---|
| 2000 | **V66-1.65 MW** | **Bonus 1.3 MW** | **GE 1.5s** | **No significant presence** | **E-66/1.8 MW** |
| | Rotor: 66 m (width) | Rotor: 62 m (width) | Rotor: 70.5 m (width) | | Rotor: 66 m (width) |
| | Hub height: 78 m | Hub height: 68 m | Hub height: 65 m | Licensed technology only | Hub height: 98 m |
| | Total height: 111 m | Total height: 99 m | Total height: 100.3 m | | Total height: 131 m |
| | ~350 units installed | ~200 units installed | ~150 units installed | | ~300 units installed |
| | First pitch-regulated MW-class turbine | Active stall regulation | Variable speed technology | | Direct drive, gearless design |
| 2005 | **V90-3.0 MW** | **Siemens SWT-2.3-93** | **GE 1.5sle** | **Goldwind 750 kW** | **E-82/2.0 MW** |
| | Rotor: 90 m (width) | Rotor: 93 m (width) | Rotor: 77 m (width) | Rotor: 48 m (width) | Rotor: 82 m (width) |
| | Hub height: 105 m | Hub height: 80 m | Hub height: 80 m | Hub height: 50 m | Hub height: 108 m |
| | Total height: 150 m | Total height: 126.5 m | Total height: 118.5 m | Total height: 74 m | Total height: 149 m |
| | ~600 units installed | ~450 units installed | ~2,500 units installed | ~300 units installed | ~800 units installed |
| | OptiTip® pitch regulation | IntegralBlade® technology | WindRIDE-THRU® grid support | Licensed Repower technology | Storm control system |
| 2010 | **V112-3.0 MW** | **SWT-3.6-120** | **GE 2.5XL** | **GW 2.5 MW** | **E-126/7.5 MW** |
| | Rotor: 112 m (width) | Rotor: 120 m (width) | Rotor: 100 m (width) | Rotor: 100 m (width) | Rotor: 127 m (width) |
| | Hub height: 119 m | Hub height: 90 m | Hub height: 100 m | Hub height: 85 m | Hub height: 135 m |
| | Total height: 175 m | Total height: 150 m | Total height: 150 m | Total height: 135 m | Total height: 198.5 m |
| | ~1,200 units installed | ~350 units installed | ~1,000 units installed | ~1,500 units installed | ~25 units installed |
| | Full converter technology | First major offshore turbine | NetConverter® system | PMDD technology | World's largest onshore turbine |

*(Continued)*

Table 5.2 (Continued)

| Year | Vestas | Siemens Gamesa | GE Renewable Energy | Goldwind | Enercon |
|---|---|---|---|---|---|
| 2015 | **V164-8.0 MW** | **SWT-7.0-154** | **GE Haliade 150-6 MW** | **GW 6.0 MW** | **E-126/4.2 MW** |
| | Rotor: 164 m (width) | Rotor: 154 m (width) | Rotor: 150 m (width) | Rotor: 150 m (width) | Rotor: 127 m (width) |
| | Hub height: 105 m | Hub height: 106 m | Hub height: 100 m | Hub height: 100 m | Hub height: 135 m |
| | Total height: 187 m | Total height: 183 m | Total height: 175 m | Total height: 175 m | Total height: 198.5 m |
| | ~50 units installed | ~30 units installed | ~20 units installed | ~30 units installed | ~200 units installed |
| | Offshore-specific design | Direct drive offshore technology | Direct drive offshore design | Offshore prototype | Enhanced EP1 platform |
| 2017 | **V164-9.5 MW** | **SG 8.0-167 DD** | **GE Haliade-X (prototype)** | **GW 6.7 MW** | **E-126 EP4** |
| | Rotor: 164 m (width) | Rotor: 167 m (width) | Rotor: 150 m (width) | Rotor: 155 m (width) | Rotor: 127 m (width) |
| | Hub height: 105 m | Hub height: 109 m | Hub height: 100 m | Hub height: 110 m | Hub height: 135 m |
| | Total height: 187 m | Total height: 192.5 m | Total height: 175 m | Total height: 187.5 m | Total height: 198.5 m |
| | ~100 units installed | ~75 units installed | Development announced | ~50 units installed | ~150 units installed |
| | World's most powerful offshore turbine | Direct Drive technology | 12 MW target capacity | Enhanced offshore design | Segmented blade design |
| 2019 | **EnVentus V162-5.6 MW** | **SG 10.0-193 DD** | **GE Haliade-X 12 MW** | **GW 8.0 MW** | **E-160 EP5** |
| | Rotor: 162 m (width) | Rotor: 193 m (width) | Rotor: 220 m (width) | Rotor: 184 m (width) | Rotor: 160 m (width) |
| | Hub height: 149 m | Hub height: 118 m | Hub height: 138 m | Hub height: 110 m | Hub height: 166 m |
| | Total height: 230 m | Total height: 214.5 m | Total height: 248 m | Total height: 202 m | Total height: 246 m |
| | ~200 units installed | ~50 units installed | First prototype installed | ~25 units installed | ~30 units installed |
| | Modular platform architecture | Flex technology platform | World's largest turbine | Offshore-specific design | Low wind specialist |

| | | | | | |
|---|---|---|---|---|---|
| 2021 | **V236-15.0 MW Prototype** | **SG 14-222 DD** | **GE Haliade-X 14 MW** | **GW 10 MW** | **E-160 EP5 E3** |
| | Rotor: 236 m (width) | Rotor: 222 m (width) | Rotor: 220 m (width) | Rotor: 210 m (width) | Rotor: 160 m (width) |
| | Hub height: 149 m | Hub height: 128 m | Hub height: 138 m | Hub height: 130 m | Hub height: 166 m |
| | Total height: 267 m | Total height: 239 m | Total height: 248 m | Total height: 235 m | Total height: 246 m |
| | Prototype announced | Prototype testing | Serial production begins | Prototype testing | ~75 units installed |
| | World's largest swept area | 14 MW offshore platform | Upgraded capacity | Largest Chinese turbine | Enhanced EP5 platform |
| 2023 | **V236-15.0 MW commercialization** | **SG 14-236 DD** | **GE Haliade-X 14.7 MW** | **GW 16.0-242** | **E-175 EP5** |
| | Rotor: 236 m (width) | Rotor: 236 m (width) | Rotor: 220 m (width) | Rotor: 242 m (width) | Rotor: 175 m (width) |
| | Hub height: 149 m | Hub height: 130 m | Hub height: 138 m | Hub height: 146 m | Hub height: 170 m |
| | Total height: 267 m | Total height: 248 m | Total height: 248 m | Total height: 267 m | Total height: 257.5 m |
| | Commercial deployment | Commercial orders | Dogger Bank installation | Announced | Prototype testing |
| | Industry-leading capacity factor | Upgraded rotor size | Uprated power rating | Largest Chinese offshore design | Largest onshore rotor |

*Source:* [7], [10–13].

V162-6.2 MW has outstanding component flexibility (pBay™) of EnVentus™ architecture in which, the components are optimized to specific site conditions [15].

When compared to competitors, Vestas' turbines show some overlap, but also some differentiated advantages. Siemens Gamesa's SG 14-222 DD matches Vestas' V236 in power class (14 MW, upgradable to 15) but is built on direct drive technology as opposed to Vestas' medium speed geared drivetrain. There is a key difference in design philosophy that may influence maintenance cycles and power density. GE's Haliade-X platform (12–14 MW), as a first mover in the ultra large segment, it offers slightly smaller rotor diameters (220–222 m) than Vestas' 236 meters. Hence, it has ~10% less swept area, thus, annual energy production is somewhat lower though wind conditions plays a vital role [16].

In the onshore segment, Vestas' V162-6.2 MW competes directly with Goldwind's GW 6 S.X and Enercon's E-160 EP5, with all three targeting the growing "super-sized" onshore market. While similar in overall dimensions, Vestas differentiates its offering through superior power density (watts generated per square meter of swept area) and its EnVentus™ platform's modular architecture. It allows more precise customization to site-specific conditions than competitors' more standardized approaches [17].

But what truly distinguishes these modern turbines from their predecessors isn't visible to the naked eye. It's the invisible neural network of sensors, processors, and communication systems that transform these mechanical giants into "intelligent" responsive energy systems.

### 5.2.3 AI and IoT Sensor Integration Architecture

Modern wind turbines have evolved into a very advanced sensor platform. Think of this: one V162-6.2 MW turbine has more than 1,000 sensors monitoring everything from nano-vibrations in bearings to hundreds-of-millisecond variations in grid frequency. The "sensors" are the nervous system of the turbine, providing concurrent data streams to local and cloud processing systems [18].

Sensor architecture is based on hierarchical design that emulates some of the most successful solutions in nature such as the human nervous system [19]. As shown in Figure 5.2, at the lowest level, simple threshold sensors watch for the basic parameters, such as temperature or fluid levels [20]. Higher level sensors track vibration signatures, stress patterns, and electrical characteristics [21]. Top level integrated systems such as LIDAR (Light Detection and Ranging) measures incoming wind patterns of 300 meters, and more ahead so that the turbine can adjust for changes that may come [22]. This is much like the process reminiscent of how your brain anticipates catching a ball by tracking a trajectory. Figure 5.2 shows a wind turbine and the placement of different sensors in different parts of

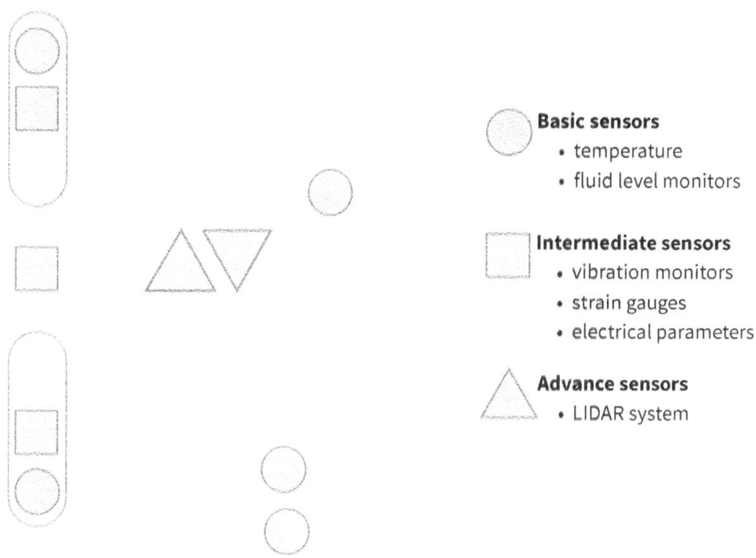


**Figure 5.2** A turbine sensor architecture. *Note:* Placement of the sensors is only for illustration purposes and does not indicate exact location. *Source:* Author's own work.

the complete system. Over 1,000 sensors form a comprehensive nervous system that enables predictive response to environmental conditions, much like the human body anticipates and reacts to external stimuli [23].

This sensory information feeds into a multi-tiered AI architecture. Edge computing handles immediate control responses within milliseconds, while more complex pattern recognition occurs in on-site servers. The most sophisticated analytics, those comparing performance across entire fleets of thousands of turbines, take place in cloud data centers [24].

How much data does a modern wind farm generate? A modern wind farm is one of the highest data-intensive industrial operations in the renewable energy sector. An average of 100-turbine operations create 400 terabytes of operational data per year – the equivalent of 400 million books. This sprawling data ecosystem begins with the real-time edge computing (RTEC) nodes placed in each turbine, designed for harsh conditions on site that focus on first-pass data for processing. After which data streams through hardened redundancy communication channels, which generally comprise of fiber-optic backbone network with various cellular and satellite layers in case of bad weather. The intricate data finally ends up in the clouds, as always on platforms such as Microsoft Azure and other services where the data-extracting power is finally realized [25].

The analytics layer uses a multi-layered AI methodology approach. First, the supervised learning algorithms use this point of vibration signatures and the performance anomalies as reference of known failure patterns. Then the unsupervised learning method uncovers previously unknown correlations that might indicate potential issues. Next is reinforcement learning that helps to continuously tune the response of turbine control systems in order to capture more energy and at the same time, reduce the mechanical stress. Lastly, digital twin simulations generate digital representations of physical assets, and operators can observe real-time status and even try out operational changes on these virtual replicas before real-time implementation. The comprehensive data infrastructure turns wind farms into intelligent IOT enabled self-optimizing energy systems. These smart systems have the capabilities of predictive maintenance and adaptive performance beyond simply power generators.

Figure 5.3 shows a comprehensive diagram of a wind farm data ecosystem. It represents the data flow in modern wind farm operations. The blue hardware infrastructure components collect data from turbine sensors, which flow through green communication systems to orange analytical processes. Data moves hierarchically from turbine level through farm level to cloud level, ultimately feeding the below purple application components that enable predictive maintenance, performance optimization, and executive decision-making. The bidirectional mesh network at the applications level facilitates integrated management of the entire wind farm fleet.

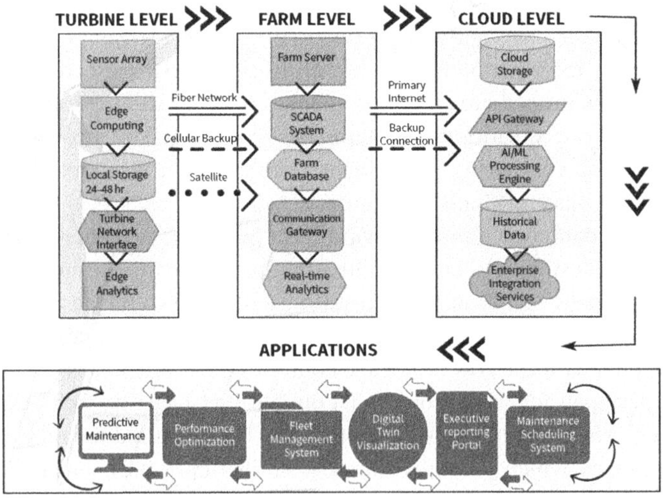


**Figure 5.3** Wind farm data ecosystem architecture. *Source:* Author's own work.

The diagram uses the standard flow chart shapes for component types and specific color for functional categories. Computing hardware or software modules are represented by rectangles, storage systems are cylinders, hexagons denote analytics engines, and rounded shapes showcase user interfaces. The directional arrows show on what order the data goes around the ecosystem. This architecture keeps the flow of raw sensor data progressively as it works in a way through the system.

The four integrated levels process data from wind farms under different scopes and level of analysis. The turbine level has on-site sensors, edge computing and local storage for real-time capturing data from individual turbines, focusing on primary data collection and processing. Then, the farm level collects this data through central servers, SCADA systems, and communication gateways for monitoring and controlled operation. The cloud level uses the remote infrastructure of cloud storage, AI/ML engines and data lakes to conduct further analytics and maintain historical records. Finally, the applications (cloud) layer translates this processed data to be used as insights in specific interfaces like maintenance dashboards, optimization tools, and digital twins. This complete systematic approach catering to a spectrum of end users from operators, technicians to other stakeholders and management.

Although renewable energy globally is on an upward curve, the wind turbine manufacturing sector is battling with some major headwinds. In the last 2021–2022, the industry experienced a great deal of financial strain. Major players such as Vestas, Siemens Gamesa and GE Renewable Energy reporting operational losses stemming from supply chain problems, cost inflation, and project deferrals. The stresses have driven a re-shuffle of strategy in the industry with manufacturers repositioning themselves to adapt to evolving market dynamics.

Supply chain vulnerabilities represent a critical concern, particularly regarding rare earth metals for permanent magnet generators and specialized composite materials. Price volatility in raw materials, notably steel, copper, and carbon fiber has compressed profit margins and complicated long-term contract pricing. Leading manufacturers have responded by diversifying supplier networks, increasing vertical integration, and developing alternative material solutions to reduce dependency on volatile commodities.

With wind farms getting increasingly digitalized, security threats are a considerable operational risk. Industrial control systems are under threat of advanced attacks which could cause power generation disruption or data compromise. The sector has leveraged greater security frameworks, as well as network segmentation and periodic vulnerability assessments to harden their critical infrastructure.

The integration of wind power with complementary technologies represents both challenges and opportunities. Several manufacturers aimed at mitigating the intermittency issue by developing hybrid energy systems for use on land that combines wind and solar, battery storage, and green hydrogen production. The integrated solutions are dealing with these complex grid management challenges, and most markets are at an early stage of developing the requisite regulatory frameworks.

For market leaders like Vestas, maintaining technological leadership while ensuring commercial viability presents a delicate balance. As illustrated in Figure 5.4, market leadership positions have shifted significantly over two decades. Vestas has maintained its prominent position through strategic acquisitions, substantial R&D investments, and a comprehensive service portfolio that generates recurring revenue streams. This approach has helped the company navigating market fluctuations while continuing to advance turbine technology toward higher capacity factors and lower levelized cost of energy.

The industry's future success depends on addressing these interconnected challenges through technological innovation, operational excellence, and strategic partnerships across the renewable energy ecosystem. Manufacturers who

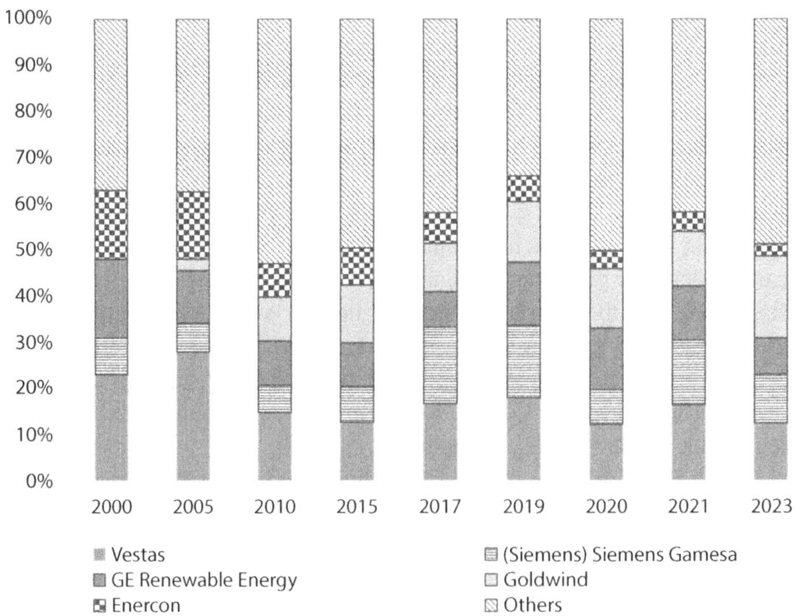


**Figure 5.4** Global wind turbine market share comparison (2000–2023). *Source:* [26–29].

can effectively balance the immediate market pressure with long-term sustainability goals be likely to emerge as leaders in the next generation of wind energy development.

## 5.3 Analysis and Problem Statement

The renewable energy sector is rapidly evolving, and wind power is playing an important role. Yet, optimizing the efficiency and reliability of wind turbines while keeping maintenance costs on lower end is essential now. One of the largest challenges faced by Vestas is how to optimize smart wind turbines using special aerodynamics and AI predictive maintenance. It is mainly the challenge of technology versus price and cybersecurity risks.

Several factors contribute to this challenge. First of all, one needs intelligent turbine blade designs to dynamically harness the power of fluctuating wind. Secondly, AI and IoT Sensors based predictive maintenance system, is costly but offers a solution to reduce operational downtimes. Lastly, digitalizing wind farms gives rise to new cybersecurity threats. Resolving these issues may entail increasing the operational inefficiencies and add security vulnerabilities to Vestas as well as other stakeholders relying on wind power as an affordable resource.

## 5.4 Think and Solve

The questions below summarize some of the key concerns and technical considerations in smart wind energy solutions.

**Q 5.1** Which of the following factors primarily influences the aerodynamic efficiency of a wind turbine blade?

**A** The material used for the turbine tower.
**B** The color of the turbine blades.
**C** Blade length and shape optimization.
**D** The type of foundation used for offshore wind turbines.

**Q 5.2** What is a major benefit of AI-driven predictive maintenance in wind farms?

**A** Increased mechanical failures due to reliance on automation.
**B** Reduced downtime and maintenance costs.
**C** Elimination of the need for human oversight.
**D** Increased energy storage capacity.

**Q 5.3** How does grid integration impact the efficiency of wind energy systems?

**A** It helps to balance supply and demand, preventing power fluctuations.
**B** It reduces the need for wind energy storage.
**C** It increases turbine wear and tear.
**D** It eliminates the need for predictive maintenance.

**Q 5.4** Which cybersecurity risk is most relevant to connected wind farms?

**A** Physical damage from extreme weather.
**B** Hacking of control systems leading to turbine malfunctions.
**C** Inefficiencies in blade design.
**D** Overproduction of electricity.

**Q 5.5** A Vestas offshore wind turbine has a rotor diameter of 150 meters and operates at a wind speed of 12 m/s. The air density at sea level is approximately 1.225 kg/m$^3$. Assuming an ideal power coefficient (Cp) of 0.45, calculate the theoretical power output of the turbine.

**Given:**
Rotor diameter = 150 m
Wind speed = 12 m/s
Air density (p) = 1.225 kg/m$^3$
Power coefficient (Cp) = 0.45
Formula for wind power output: $P = \frac{1}{2} \times p \times A \times V^3 \times C_p$
where A is the swept area of the blades, given by $A = \pi \times (D/2)^2$

## 5.5 Conclusion and Key Takeaways

From a tiny Danish metal shop to global wind energy solutions leader, Vestas is an example of what innovation and adaptability can do. Vestas continually tested the limits of wind energy by growing with technological progress, market hurdles, and strategic realignments through several decades. Still, advancing toward smart wind energy solutions has never been easy. AI-driven predictive maintenance, aerodynamic optimization, and hybrid energy systems integration are opportunities as well as complexities.

The economic challenges faced by Vestas, including a significant financial loss in 2022, speaks to the fine line of advancing technology and financial recovery. The high development and deployment costs, as well as some concerns around cybersecurity of connected wind farms show that even experienced players need to find their way in the constantly evolving landscape on digital transformation of renewable energy sector.

Despite these obstacles, Vestas remains at the forefront of renewable energy, striving to redefine efficiency, sustainability, and reliability in wind power. With data analytics, IoT sensors, and state-of-the-art material scientists, the company keeps pushing boundaries to redefine the clean energy solutions outlook. The case of Vestas illustrates the common difficulties in renewable energy – while keep improving the performance and scaling, it is also important to stay within economic and grid stability bounds.

**Key Takeaways:**

- **Technological Evolution:** Vestas have evolved from manufacturing agricultural tools to becoming the world's leading wind turbine producer through strategic innovation and adaptation.
- **Smart Wind Solutions:** AI-driven predictive maintenance, blade optimization, and hybrid energy systems have revolutionized wind energy efficiency but come with high costs and integration challenges.
- **Economic Realities:** The company faced financial setbacks in 2022, showing the difficulty of balancing cutting-edge technological investment with profitability.
- **Cybersecurity and Grid Integration:** The increased connectivity of wind farms raises security concerns, emphasizing the need for robust cybersecurity measures and advanced grid compatibility solutions.
- **Future Outlook:** Despite challenges, Vestas remains committed to driving innovation in renewable energy, setting the stage for a more sustainable and intelligent wind energy future.

As Vestas continually enhances its digital ecosystem, the application of AI in wind energy is still developing. For a deeper understanding of the technical standards behind these developments, the industry reports of the Global Wind Energy Council (GWEC) and the International Energy Agency (IEA) contain vital information on grid integration and digitalization [30, 31]. Additionally, academic studies on digital twin technology provides the theoretical background for the predictive maintenance systems used by Vestas. Exploring these resources helps to realize how data analytics is changing the industry. As the world transitions toward cleaner energy solutions, the lessons learned from Vestas' successes and setbacks can serve as valuable insights for future advancements in smart wind energy systems.

## 5.6 Discussion Points

To fully grasp the complexities of Vestas' journey in smart wind energy, it is essential to examine the strategic, technical, and economic challenges the company has faced. In the perspective of the integration of AI in general, relevant costs as well as cybersecurity impacts on smart wind farms, Vestas experience is a great case study between innovation and commercial viability. This prompts for a more reflective discussion on the AI aspect, data-driven decision-making and direction of wind energy. How does Vestas' use of machine learning and IoT sensors impact operational efficiency, and what potential risks or limitations exist in relying on AI for predictive maintenance?

1. **Economic Viability vs. Technological Advancement:** Given the high costs of developing smart wind turbines, how Vestas can maintain profitability while pushing the boundaries of innovation?
2. **Cybersecurity in Smart Wind Farms:** What measures should Vestas implement to safeguard its AI-powered wind farms from cyber threats, and how can the industry address growing security concerns?
3. **Hybrid Energy Systems:** How do hybrid energy solutions (combining wind, solar, hydrogen and energy storage) enhance grid stability, and what are the key challenges in integrating these technologies?
4. **The Future of Wind Energy:** With continuous advancements in turbine technology and AI-driven energy solutions, what will the next decade look like for companies like Vestas, and how can they stay ahead of the competition?
5. **The Complexity of Digital Integration:** The transition from mechanical engineering to AI-driven management is technically demanding. Due to the complexity of this topic, further reading is suggested regarding digital twin simulations and Industrial Internet of Things (IIoT) architectures. How might these advanced frameworks evolve to address the cybersecurity concerns mentioned in the case?

These discussion points help to analyze the critical aspects of Vestas' transformation and provide insights into the broader implications of smart wind energy solutions.

## 5.7 Limitations of the Case Study

While this case study provides an in-depth analysis of Vestas' transition to AI-driven wind energy solutions, several limitations must be acknowledged. These limitations highlight areas where further research or additional data may be required for a more comprehensive understanding of the topic.

1. **Restricted Access to Proprietary Data:** Much of Vestas' AI-driven predictive maintenance strategies and turbine optimization processes are proprietary. As a result, this study relies on publicly available data, which may not fully capture the intricacies of their technological advancements.
2. **Evolving Industry Landscape:** The renewable energy sector is rapidly evolving, with continuous advancements in turbine design, AI integration, and grid management. Some of the findings in this study may become outdated as new technologies and regulatory frameworks emerge.
3. **Financial and Economic Constraints:** While this study discusses Vestas' financial challenges, it does not conduct a detailed financial modeling or in-depth economic feasibility analysis. Future research could provide a more quantitative approach to assessing the cost–benefit balance of AI-powered wind energy solutions.
4. **Cybersecurity Considerations:** The study touches on cybersecurity threats in smart wind farms but does not provide an exhaustive analysis of specific vulnerabilities, attack vectors, or mitigation strategies. A dedicated cybersecurity assessment could further explore these risks.
5. **Geographical and Policy Variations:** The impact of AI-driven wind energy solutions varies across different regions due to differences in government policies, grid infrastructure, and environmental conditions. This study primarily focuses on Vestas' global operations but does not provide an in-depth regional comparison.
6. **Lack of Long-Term Performance Data:** Since AI-driven wind energy optimization is a relatively recent development, long-term data on turbine efficiency, maintenance cost reductions, and AI reliability over extended operational periods remain limited. Future studies could examine real-world performance metrics over a decade or more.

Despite these limitations, this study provides a valuable exploration of Vestas' journey in integrating AI into wind energy, highlighting both opportunities and challenges in the field.

## References

**1** Vestas, "Vestas annual report 2022". https://www.vestas.com/en/media/company-news/2023/vestas-annual-report-2022---a-challenging-year-with-neg-c3710817 (accessed February 24, 2026).

**2** RTT News staff writer, "Vestas posts negative adj. EBIT in 2022," *RTT News*, 2 August 2023. https://www.rttnews.com/3342616/vestas-posts-negative-adj-ebit-in-2022.aspx (accessed February 24, 2026).

**3** Garud, R. and Karnøe, P. "Bricolage versus breakthrough: distributed and embedded agency in technology entrepreneurship," *Research Policy*, 32(2), 277–300, 2003.

**4** Karnøe, P. "When low-tech becomes high-tech: the social construction of technological learning processes in the Danish and the American wind turbine industry," in *Mobilizing Resources and Generating Competencies: The Remarkable Success of Small and Medium Sized Enterprises in the Danish Business System*, 139–184. Handelshøgskolens Forlag, 1999.

**5** Vestas Wind Systems A/S, "Annual report," Aarhus, Denmark, 2000. https://mb.cision.com/Main/18886/2965702/1144968.pdf (accessed March 23, 2026).

**6** Lewis, J.I. and Wiser, R.H. "Fostering a renewable energy technology industry: an international comparison of wind industry policy support mechanisms," *Energy Policy*, 35(3), 1844–1857, 2007.

**7** Vestas, "Vestas website," 2025. https://www.vestas.com/en (accessed February 24, 2026).

**8** Ackermann, T. and Söder, L. "An overview of wind energy-status," *Renewable and Sustainable Energy Reviews*, 6(1–2), 67–127, 2002.

**9** Yang, J. "Finite Element Modeling and Vibration Simulation of a 5-MW 61.5-Meter Composite Wind Turbine Blade for Debonding Damage Inference," 2023, https://ualberta.scholaris.ca/items/a0f45e51-a119-43b7-9702-b56280f7e4a0 (accessed February 24, 2026).

**10** Siemens Gamesa. "Offshore wind turbine," 2025. https://www.siemensgamesa.com/global/en/home.html (Accessed March 23, 2026).

**11** GE Vernova, "GE Vernova's offshore turbine," 2025, https://www.gevernova.com/wind-power (accessed February 24, 2026).

**12** Goldwind, "Goldwind equipment," 2025, https://www.goldwind.com/en/equipment (accessed February 24, 2026).

**13** Enercon, "Global manufacturer of wind turbines," 2025. https://www.enercon.de/en (accessed February 24, 2026).

**14** Vestas Wind Systems, "V236-15.0 MW technical specifications and performance data," 2022. https://www.vestas.com/en/energy-solutions/offshore-wind-turbines/V236-15MW (accessed February 24, 2026).

**15** DNV GL, "Technology Outlook for the Global Wind Industry," *DNV GL Clean Energy Research*, 2021.

**16** NREL., "Cost of Wind Energy Review," *National Renewable Energy Laboratory*, 2022.

**17** Vestas, "EnVentus™ platform," https://www.vestas.com/en/energy-solutions/onshore-wind-turbines/enventus-platform (accessed February 24, 2026).

**18** Pandit, R. and Wang, J. "A comprehensive review on enhancing wind turbine applications with advanced SCADA data analytics and practical insights," *IET Renewable Power Generation*, 18(4), 722–742, 2024.

**19** Song, D., Shen, G., Huang, C., Huang, Q., Yang, J., Dong, M., Joo, Y.H. and Duić, N. "Review on the application of artificial intelligence methods in the control and design of offshore wind power systems," *Journal of Marine Science and Engineering*, 12(3), 424, 2024.

**20** Teimourzadeh Baboli, P., Babazadeh, D., Raeiszadeh, A., Horodyvskyy, S. and Koprek, I. "Optimal temperature-based condition monitoring system for wind turbines," *Infrastructures*, 6(4), 50, 2021.

**21** Barszcz, T. *Vibration-based condition monitoring of wind turbines*, Applied Condition Monitoring, vol. 14. Springer International Publishing, 2019.

**22** Cooperman, A. and Martinez, M. "Load monitoring for active control of wind turbines," *Renewable and Sustainable Energy Reviews*, 41, 189–201, 2015.

**23** Cutmore, T.R. and James, D.A. "Sensors and sensor systems for psychophysiological monitoring: A review of current trends," *Journal of Psychophysiology*, 21(1), 51–71, 2007.

**24** IBM Research, "Multi-tiered AI architectures for renewable energy applications," 2023.

**25** Randhi, K. and Bandarapu, S.R. "Efficient resource allocation for generative AI workloads in cloud-native infrastructures: a multi-tiered approach," *International Journal of Science and Research Archive*, 13(2), 826–839, 2024.

**26** GWEC, "The global voice of the wind industry," *Global Wind Energy Council*, https://www.gwec.net (accessed February 24, 2026).

**27** BloombergNEF, "Energy transition investment trends," https://about.bnef.com/insights/finance/energy-transition-investment-trends (accessed February 24, 2026).

**28** IEA, "International Energy Agency – data and statistics," 2024. https://www.iea.org/data-and-statistics/data-tools/energy-statistics-data-browser?country=WORLD&fuel=Energy%20supply&indicator=TESbySource (accessed March 23, 2026).

**29** Wood Mackenzie, "Market insights,". https://www.woodmac.com/market-insights (accessed March 23, 2026).

**30** International Energy Agency, "Digitalization and energy," IEA, 2017, https://www.iea.org/reports/digitalisation-and-energy (accessed February 24, 2026).

**31** International Energy Agency, "Integrating solar and wind: global experience and emerging challenges," *IEA*, 2024, https://www.iea.org/reports/integrating-solar-and-wind (accessed February 24, 2026).

# 6

# London Array: NEXANS' Subsea Cable Breakthrough

From Subsea Installation to Power Optimization – Building Europe's Renewable Future

## 6.1 The Story of the London Array Offshore Wind Farm

As part of the strategic framework for 2030, Climate and Energy initiatives, Europe is gearing up to install 129 GW of new wind farms over the period 2023–2027 [1]. The planned installation capacity represents an enormous mission on green energy production facilities. A gigawatt (1 GW) is a massive amount of power, which is equal to one billion watts. To make this relatable, just imagine the power needed to light up more than 10 million of 60–100 W lightbulbs, all at once – that is 1 GW. It is also enough electricity to supply approximately 2.65 million households in the UK, powering everything from lights to appliances.[1]

It's like a giant game of "who can build the most windmills," and the UK, Sweden, Germany, Turkey, and the Netherlands are all in on the action. Sweden is going big with 2.1 GW of new onshore wind power. Meanwhile, the UK is leading the charge with 2.3 GW of total new wind installations, and most of them are offshore. Who knew the Brits are such fans of breezy ocean views? Figure 6.1 shows the statistics from 2021 about wind farm installations in Europe, their onshore and offshore capacity, and particularly, the UK's share of this market. As of February 2023, Europe has 255 GW of wind capacity and is expected to build 20 GW of new wind farms every year as an average over 2023–2027. Driving this growth are massive offshore wind projects, and one of the notable projects is the London Array, located on the outer Thames estuary off the coast of Kent, England.

1 According to the UK electric insights quarterly report (Q3, 2024), the UK as a whole uses around 300 TWh of electricity a year. Power flow is measured in gigawatts (GW). The volume of energy is measured in gigawatt-hours (GWh) or terrawatt-hours (TWh). Producing 1 GW constantly for an hour would give 1 GWh = 0.001 TWh. Producing 1 GW constantly for a year would give 8760 GWh = 8.76 TWh, which is enough to power 2.65 million homes.

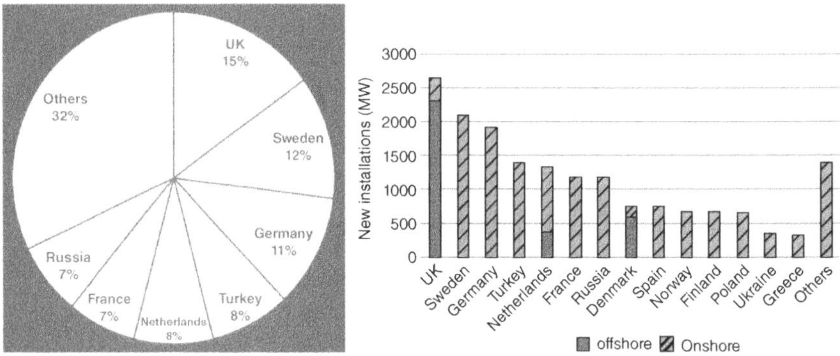


(a) (b)

**Figure 6.1** (a) Wind farm installations in Europe (2021), (b) onshore and offshore output in MW. *Source:* Adapted from [1].

The London Array Offshore Wind Farm stands out as a remarkable engineering achievement that made waves in the renewable energy world. This case study shows how creative problem-solving can tackle big environmental issues head-on. Located in the UK off the coast of Kent, this wind farm is not just another run-of-the-mill energy project. It was a game-changer that set a new standard for energy production and showcased some serious engineering skills.

The concept for the London Array emerged in the early 2000s. The plan was to make use of the wind in the Thames estuary area, where the River Thames flows into the North Sea in southeast England. Once finished, it was expected to generate an enormous 1000 MW of power – enough to keep the lights on in hundreds of thousands of homes. For several years, it held the title of the world's biggest wind farm, which is impressive.

The London Array project was not just about generating energy; it was a complex undertaking that required the collaboration of leading industry players. Companies like DONG Energy (now Ørsted), EON, and Masdar joined forces to bring this ambitious vision to life. The project's success also relied on NEXANS' advanced cable technology expertise to establish a robust network for this offshore energy giant.

NEXANS, a global leader in cable technology, was tasked with a critical role in this historic project. They were responsible for supplying and installing the high-voltage subsea cables that would connect the wind turbines and enabling the transmission of clean electricity to the mainland. The cables had to withstand the harsh marine environment, including strong currents, extreme weather, and the corrosive effects of seawater, while maintaining optimal performance.

As the project progressed, NEXANS's engineers and technicians worked tirelessly to design, manufacture, and install these specialized cables. The scale and technical challenges were unprecedented, pushing the boundaries of offshore wind farm development. The success of the London Array does not only contribute to the UK's renewable energy targets but also serve as a blueprint for future offshore wind projects around the globe.

In the following years, the London Array Offshore Wind Farm became a remarkable renewable energy project by demonstrating the feasibility and benefits of large-scale offshore wind power generation. It has inspired the development of numerous other offshore wind farms, contributing to the global effort to combat climate change and create a sustainable future.

This case study explores into the historical context, challenges, and achievements of the London Array project. It highlights NEXANS's crucial role in powering this renewable energy venture and shaping the future of sustainable energy infrastructure.

Figure 6.2 is an illustration of how the wind farm works. Over 200 km of array cables were used in this project to link between wind turbines and the offshore substations. These cables, buried 1.5 m under the seabed, carry 33 kV electricity and include fiber optics for remote monitoring and control. Each array cable was individually tailored to their specific location, ranging from 720 m to 3110 m in length. These cables weigh approximately 50 kg per meter.

The London Array consists of 175 Siemens 3.6 MW wind turbines with a total capacity of 630 MW, arranged across 100 sq. km. Each turbine stands 87 m above sea level, with blades extending to 147 m at their highest point, and features a 120 m diameter swept area. The turbines operate at wind speeds of 3 m/s to 25 m/s, using a motorized yaw mechanism to face the wind and independent pitching mechanisms to optimize performance and reduce stress. The rotational energy generated is converted to electricity via a three-stage gearbox and generator, then stepped up to 33 kV for export. Designed to run 24/7 for over 20 years, the turbines are remotely monitored and controlled.

Four high-voltage export cables, each weighing over 4,500 tons and manufactured by Nexans, are like the heavyweight champions of the electrical world.

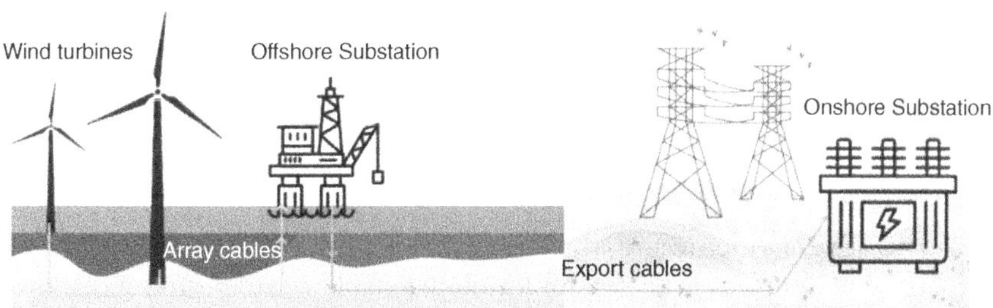


**Figure 6.2** Wind farm's operational overview. *Source:* Author's own work.

They carry 150 kV of electricity from the offshore substations to the onshore substation at Cleve Hill in north Kent, where they plug into the national grid. These cables, stretching over 50 km, have copper core conductors and fiber optic cores for monitoring – because even cables need to keep an eye on things. The offshore construction began in March 2011, with the first turbine installed in January 2012, and the project was wrapped up by December 2012, just in time for a New Year's celebration [2].

The onshore substation at Cleve Hill connects the London Array wind farm to the National Grid, transforming 150 kV electricity brought ashore via four high-voltage undersea cables to 400 kV for distribution. This is where electricity gets a makeover before hitting the National Grid runway. Key components in this makeover include transformers in noise-enclosed housings, and the static VAR compensator (SVC) pods, which regulate voltage by controlling reactive power. SVC provides fast-acting reactive power on high-voltage electricity transmission networks, like efficient traffic controllers of the electrical world, managing power flow and maintaining system stability. SVC can quickly and reliably control line voltages which is very important in this case. The 150 kV filters are on hand to remove waveform distortions. Plus, there's a control building for monitoring and remote communication. The National Grid substation receives electricity via underground cables and circuit breakers. Owned by Blue Transmission London Array Limited since 2013, the transmission system is maintained by London Array under contract [3].

## 6.2 Can Nexans Step Up to This Formidable Challenge?

In the heart of the North Sea, approximately 20 kilometers off the coast of Kent, a monumental battle against nature was about to unfold. The London Array Offshore Wind Farm project aimed to harness the fierce winds of the Thames estuary, but through the very environment it sought to conquer posed immense challenges. The numbers were staggering: 175 wind turbines, each standing tall at up to 147 meters, capable of generating 3.6 MW of electricity, totaling a massive 630 MW of power – enough to supply nearly half a million homes. But the real test lay beneath the waves. See Figure 6.3 for the height comparison between a wind turbine at London Array versus the London Eye Ferris wheel.

The subsea condition was both the project's lifeblood and its greatest adversary. This environment is known for its unpredictable weather, strong currents,

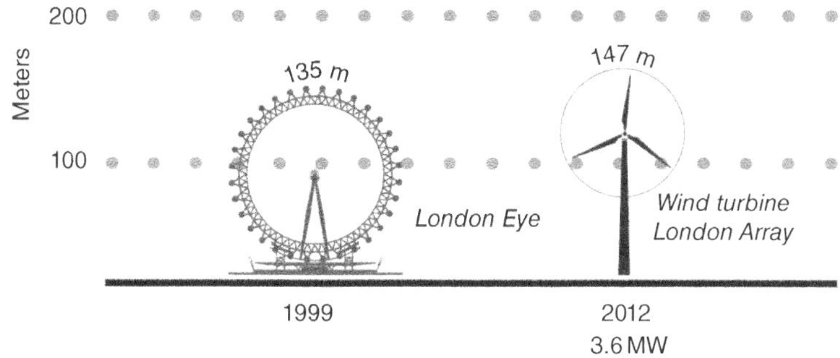


**Figure 6.3** Height comparison of a wind turbine in the London Array wind farm vs. London Eye Ferris wheel. *Source:* Author's own work.

and harsh conditions, demanded a robust and resilient power transmission system. Over 300 kilometers of high-voltage subsea cables were required to interconnect the wind turbines and transmit the generated electricity to an onshore substation. This intricate network had to withstand the relentless forces of nature while maintaining optimal performance and minimizing power loss.

NEXANS, a renowned cable technology specialist, stepped up to this formidable challenge. They were tasked with supplying and installing these critical subsea cables, ensuring the wind farm's electricity reached the mainland safely and efficiently. The cables had to endure not only the physical stresses of the sea but also the corrosive effects of seawater and the test of time. The stakes were high, as any failure in the cable system could jeopardize the entire project's success and reliability.

As the project moved forward, the sheer scale and complexity of the London Array became evident. It was not just about erecting wind turbines; it was about engineering a sustainable energy solution amidst one of nature's most unforgiving environments. The project team grappled with technical challenges, logistical nightmares, and the relentless march of time and tides. The success of the London Array is not only a triumph over the elements but also a beacon of hope for the future of renewable energy. The stage was set for an epic battle between human ingenuity and the mighty sea, where every cable lay and every turbine erected, brought the world one step closer to a sustainable energy future. See Figure 6.4, and its map showing where the London Array wind farm is located, its proximity to London and its location on the overall map of UK [2].

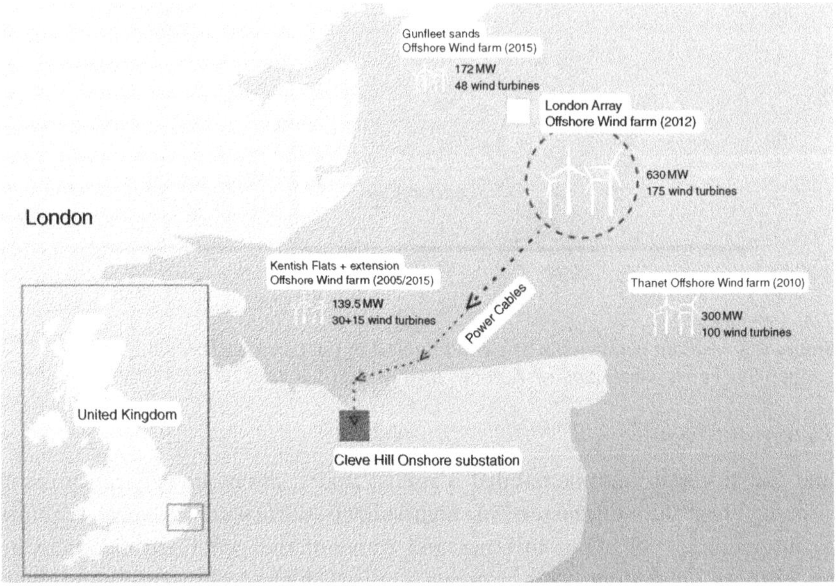


**Figure 6.4** Location of the London Array wind farm and its location on the map of UK. *Source:* Author's own work.

## 6.3 Background on Power Systems and Cable Technology

This project resembles a buffet of multidisciplinary courses. From the perspective of electrical engineering, we are stretching into the challenges of Power Systems Analysis and High Voltage Engineering. You can think of Power Systems Analysis as the head chef overseeing power flow studies, making sure our energy distribution remains stable. It helps to handle the unpredictable nature of wind power, which can be as tricky as trying to herd cats. The system also faces the complex task of managing power flow from numerous offshore turbines through undersea cables to the onshore grid. It's comparable to attempting to juggle lit torches while balancing on a unicycle – difficult but thrilling.

Next important subject is about High Voltage Engineering that creates high-voltage undersea cables. These cables are the lesser-known champions, bravely carrying electricity from offshore wind farms to the mainland. They are like dedicated pizza delivery drivers navigating through traffic on a rainy day to bring you your cheesy treat. To really get the London Array project, it's important to understand the bigger picture of power systems and cable tech. This section provides an overview of the key concepts relevant to case study.

Electricity, unlike other commodities, cannot be easily stored. This demands a constant balance between production and consumption. Renewable energy like wind power makes this tricky since it comes and goes with the wind [4]. Unlike hydro-power, which can be regulated to match demand, wind power is unpredictable because of wind patterns. So, grids need smart management to keep power flowing steady. Voltage levels in power systems are standardized according to IEC 60038, these are key for making sure everything is compatible and safe across the grid [5].

Power systems often face challenges related to the distance between generation and consumption points, managing peak demand, and accommodating power export/import needs [6]. These challenges are particularly relevant in the context of integrating large-scale renewable energy projects like offshore wind farms. Industrial IoT with LoRaWAN, wireless sensors, cloud data storage, and advanced algorithms work together with edge computing to create comprehensive monitoring systems. Smart power management and advanced power electronics further enhance these capabilities. This integrated approach helps monitor, convert, control, and optimize power flow in the electrical supply and demand ecosystem. Additionally, these technologies enable predictive maintenance, all while maintaining the lowest marginal cost [7]. Renewable energy sources like wind power typically have low marginal costs due to the absence of fuel and labor costs associated with incremental power generation [8]. Advancement in technology encourages the use of renewable energy sources.

### 6.3.1 Cable Technology and Optimization

The technical aspects of cable technology and optimization strategies relevant to offshore wind farms include cable construction and ensuring correct material properties to withstand thermal and mechanical challenges. Subsea cables used in offshore wind farms typically consist of copper or aluminum conductors insulated with XLPE (cross-linked polyethylene) [9]. Figure 6.5 illustrates the construction of an XLPE cable, which includes multiple conductors, XLPE insulation, a fiber optic cable for communication and SCADA systems, fillers for structural integrity, and an armored screen for protection against mechanical stress [10]. DC cables share a similar construction but typically use a single conductor configuration.

The choice between copper and aluminum conductors is often driven by the trade-off between conductivity and cost considerations [11]. Comparing the resistance properties of these materials is essential for understanding transmission efficiency. Copper has better electrical conductivity with lower electrical resistivity. The smaller cable size for a given current rating is offset by higher cost and weight, especially when compared to aluminum. While aluminum has 61% of the electrical conductivity of copper. This lower conductivity means aluminum has greater electrical resistivity than copper. Resistivity and conductivity are inversely

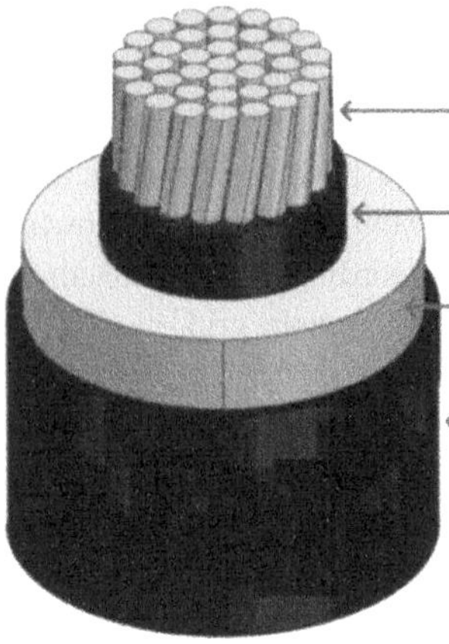


**Figure 6.5** Simplified representation of XLPE cable – cross sectional view. *Source:* [10]/SINTEF Foundation.

related, and therefore aluminum is less conductive than copper. To compensate, aluminum cables need a larger cross-sectional area for the same current but with substantial reduction in weight and savings in cost [11].

XLPE and EPR (ethylene propylene rubber) are the standard insulation materials, each with its own characteristics. A significant challenge in cable design is managing heat generated due to current overload, which can degrade the insulation and shorten the cable's lifespan [12]. XLPE cables soften and expand at temperatures above 90 °C, leading to deformation and reduced performance, whereas EPR exhibits greater thermal stability and less expansion under similar conditions, resulting in lower physical stress on the cable structure [13].

Optimizing cable routes and placement is just as critical as the cable itself. It is a complex mathematical challenge essential for minimizing capital expenditure and operational losses. Offshore wind farm cable layout optimization involves advanced mathematical modeling to solve for the shortest feasible path while avoiding seabed obstacles. This process significantly reduces material costs and energy losses.

Table 6.1 presents a comparison between a standard cable route and an optimized counterpart, highlighting potential savings in acquisition costs, annual energy loss reductions, and total cable length [14].

Another way to cut down on cable lengths and costs is by finding the best spot for the substation [15]. A well-placed substation can help reduce cable length, which means less energy lost, a more efficient system, and lower installation costs. To figure out the best location, it is important to look at things like where the power is needed most, what the land looks like, and how easy it is to get to [16]. By doing this, it ensures electrical network works greatly, and at the same time the cost of the electrical network is minimized [17]. All in all, knowing the ins and outs of cable tech and how to set things up in the best way possible is super important for offshore wind farms to do well.

**Table 6.1** Overview of costs and energy losses for inter-array cables compared to optimized cable routing.

| | Inter-array cables | | |
|---|---|---|---|
| | Standard cable route | Optimized cable route | Savings |
| Acquisition cost (M€) | 91.92 | 86.34 | 6.07% (cost) |
| Installation cost (M€) | 19.71 | 18.06 | 8.37% (cost) |
| Cost of energy loss (M€) | 27.38 | 25.15 | 8.14% (cost) |
| Annual energy loss (MWh) | 17112 | 15716 | 8.16% (energy loss) |
| Cable length (km) | 155.73 | 142.73 | 8.35% (cable length) |

*Source:* Adapted from [14].

## 6.4 The Technical Conundrum of Power Delivery Through Subsea

The primary engineering challenge of the London Array Offshore Wind Farm project was ensuring efficient and reliable power transmission from the offshore turbines to the onshore grid. This task presented a complex set of technical requirements that the project team, including NEXANS, had to address. For educational discussion, this section focuses on sample calculation regarding power loss and optimization of cable voltage.

The spacing between subsea cables is critical for minimizing electromagnetic interference. Excessive interference can lead to increased thermal losses and potential disruptions to communication systems [18]. Engineers use electromagnetic field theory (EMFT) to calculate optimal spacing, considering the cable's current-carrying capacity, voltage levels, and the thermal resistivity of the surrounding seabed [19]. In the London Array project, cable spacing was determined using specialized software like Computational Modeling Solutions (COMSOL) Multiphysics, which employs finite element analysis (FEA) to model electromagnetic fields and calculate interference levels [20]. This approach ensured compliance with the international standards like IEC 60287 and minimized the environmental impact by optimizing cable route and reducing seabed area affected by the installation.

Power loss calculations are very important for the overall design of the wind farm to be optimized. Reducing power losses enables one to deliver more energy onto the grid and thus make an economically favorable project with a reduced environmental impact as a result [21]. In the London Array project, detailed power loss analyses were conducted using specialized software like Power System Simulator for Engineering (PSSE). These tools process numbers based on cable parameters, location topology, and load flow data to create a high-fidelity grid model. This data allowed engineers to determine the optimal conductor sizing, balancing the cross-sectional area against materials costs and to plot cable routes that minimize voltage drops. Power losses are generally lower at low currents for both AC and DC transmission. But the mega power projects like London array use high currents over long distances.

In the context of the London Array Offshore Wind Farm and its subsea infrastructure, power transmission efficiency is governed largely by cable resistance (R) and current (I). These values depend on factors like what the cable is made of, how long it is, how warm it gets, and how it is built.

**Resistance (R):**

- Cable resistance basically refers to how much the cable fights against electricity flowing through it. We measure this in ohms (Ω).
- For subsea cables, resistance changes based on how long the cable is, how thick it is, and what it's made of (usually copper or aluminum).
- If a cable is longer or thinner, it will have more resistance.
- Depending on how it's made and how long it is, a subsea power cable might have anywhere from a low resistance to high resistance.

**Current (I):**

- Current, is the rate of flow of electric charge, and it is measured in amperes (A).
- In a wind farm scenario, the current is derived from the power output and the system voltage.
- For a theoretical illustration, considering a wind farm with a total capacity of 630 MW operating at a voltage of 33 kV, the aggregate current demand can be estimated. While in practice this power is distributed across multiple circuits to prevent thermal overload, the total theoretical current (I) can be calculated as:

$$I = P/V = 640\,\mathrm{MW}/33\,\mathrm{kV} \approx 19\,\mathrm{kA}$$

So, the aggregate current for the system would be approximately 19 kiloamperes (kA).

These values play a key role in figuring out how much power is lost and how effectively the whole power transmission system works. When designing the system, engineers need to think carefully about these factors and make sure the cables can handle the necessary current while keeping energy losses from resistance as low as possible. It is important to remember that in operational scenarios, the actual values for (R) and (I) might be different, depending on how the cable is made, what the environment is like, and dynamic load factors.

## 6.5 Think and Solve

The key issue was to minimize power loss over the long distances between the wind turbines and the onshore substation. The subsea cables had to transmit high-voltage electricity efficiently, typically at 33kV or higher, and reduce energy losses during transmission. The power loss ($P_{loss}$) in a cable can be approximated using the formula:

$$P_{loss} = I_2 \cdot R$$

Where:
$I$ is the current flowing through cable in amperes,
$R$ is the cable resistance in ohms.
The goal was to keep $P_{loss}$ as low as possible to maximize energy efficiency.

**Q 6.1**

- Estimate the power loss for 66 Kv transmission at $R = 0.887\,\Omega/\text{km}$.
- Can you estimate the most accurate power loss $P_{loss}$ using the formula:

$$P_{loss} = I^2R \text{ (or) } V^2/R?$$

- Discuss about other parameters of cable (other than $R$) that influence the power loss?
- Estimate the power loss for 66 Kv transmission line at $R = 0.887\,\Omega/\text{km}$, $A = 400\,\text{mm}^2$, $L = 100\,\text{Km}$, $n = 3$ and $\rho = 1$?

Since power cable plays an important role in the loss calculation, the following formula gives more precise power loss calculation. While $n$, $I$, $\rho$ and $L$ are directly proportional to the loss, cross sectional area, $A$ has opposite effect. It means if the cross-sectional area of conductor increases, then power loss decreases but it affects the cost of the conductor as well.

$$P_{loss} = n \cdot I^2 \cdot \rho \cdot \frac{L}{A}$$

Where,
$n$ = number of cores in cable
$I$ = current flowing in conductor
$\rho$ = resistivity of conductor
$L$ = length of conductor
$A$ = cross-sectional area of conductor

### 6.5.1 AC vs. DC Transmission Considerations

Figure 6.6 presents a simplified block diagram of a wind farm's complete technical structure. When it comes to offshore wind farms, choosing between AC and DC transmission involves weighing factors such as cost, efficiency, and distance. The cables used underwater to transmit power are classified as either high-voltage alternate current (HVAC) transmission cables or high-voltage direct current (HVDC) transmission cables. AC cables typically have three conductors, while DC cables can be either bipolar or single conductors. The decision between AC and DC largely depends on how far the power needs to travel, which decides which option is most cost-effective. Generally, DC transmission is better for longer distances because it loses less power, even though it requires more complex and expensive converter stations. The best choice depends on balancing power loss against the investment needed for conversion equipment.

For shorter distances, AC transmission is usually preferred because it costs less to set up initially. However, AC systems lose power due to the cable's capacitance and the alternating current's frequency. These losses get worse at higher frequencies, often requiring the use of coils and capacitors to make up for it. One way to reduce these losses is to use lower frequencies. For example, some train networks use a 16.7 Hz with Low Frequency Alternating Current (LFAC) system, which can cut down on reactive power by about 66% compared to the standard 50 Hz frequency.

DC power transmission has some advantages over AC, such as no reactive power losses due to its zero-frequency operation. But it needs equipment for converting AC to DC and back at both ends, which increases the initial cost. Even so, DC becomes more economical for longer routes and higher power levels. Recent improvements in power converter technologies have brought power conversion losses down to under 1%.

Many subsea cable projects, like those linking Norway with other nations, often go for DC transmission. While AC cables are sometimes used for longer stretches, such as the 163 km line powering Martin Linge, DC is typically the go-to choice for substantial distances and high-power needs.

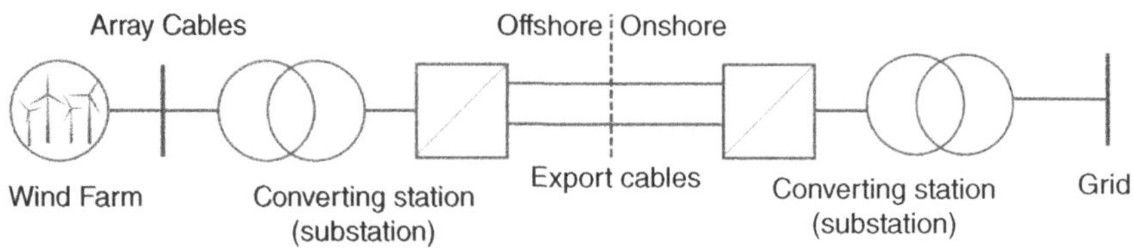


**Figure 6.6** Block diagram of the technical structure of a wind farm. *Source:* Adapted from [22].

**Table 6.2** General determination of cable type depending on distance.

| Distance | Cable type |
|---|---|
| <50 km | AC cable |
| >50 km/<100 km | AC or DC cable |
| >100 km | DC cable |

Table 6.2 shows a general determination of cable type depending on distance. There has been considerable debate regarding which type of cable is best suited for various transmission distances and optimal interface locations. It is also important to consider other factors like technical efficiency and economic factors while choosing the cable type for a particular project. More concretely, when the power loss becomes significant over longer distances, and as a long-term solution, it is profitable to invest in conversion stations and use DC. Most commonly, companies like Semco Maritime carry out a detailed study on the mapping of the economic intersection between HVAC and HVDC transmission for a large-scale offshore wind farm [23]. It helps the stakeholders to decide the solution.

**Q 6.2**

- How does the power loss impact the overall efficiency of the wind farm, especially over long cable distances?
- Estimate the power losses for 33 Kv transmission line at different lengths. Here, $R = 0.887\,\Omega/\text{km}$, $A = 400\,\text{mm}^2$, $n = 3$ and $\rho = 1$. Assume $L = 30\,\text{km}$, 60 km, 120 km and 180 km.
- Plot a graph showing the relationship between cable length and power loss for a given voltage and cable length (On the x-axis, plot the cable length (in kilometers).
- On the y-axis, plot the power loss (in megawatts).

A graph can be created to visualize the relationship between cable length and power loss for a given voltage and cable type. This graph can help to identify the point at which power loss becomes significant. For a typical subsea cable, the voltage of 33 kV can be used.

**Q 6.3** Higher voltage levels can reduce power loss in transmission. For the London Array, a higher voltage, such as 66 kV or 132 kV, could be considered.

- Discuss the advantages and disadvantages of using higher voltage levels (e.g., 66 kV or 132 kV) for power transmission. Determine the operational current levels at 66 kV and 132 kV for the given system parameters.

- Calculate equivalent power losses for both HVAC and HVDC systems for these given voltages. Assume, DC resistance of conductor $R = 0.1\,\Omega/\text{km}$, and AC resistance of conductor at 50 Hz, $R = 0.129\,\Omega/\text{km}$. Also, remember the discussed project's power capacity (London Array capacity: approximately 630 MW), assume distance from offshore to onshore as 20 km and power factor as 0.95 for AC calculations.
- Tips for simple calculations and understanding of the systems:
  For current calculations:

  $I = P/(\sqrt{3} \times \text{V} \times \cos\varphi)$ for AC three-phase, and
  $I = P/V$ for DC

- For power loss calculations:

  Power loss $= \text{I}^2 \times R \times$ distance (both HVAC and HVDC systems)
  What did you observe?

NEXANS employed its advanced cable technology and engineering expertise to address the challenges. They designed and manufactured specialized high-voltage subsea cables with enhanced insulation and protection against the marine environment. These cables could transmit high-voltage electrical power over long distances with minimal power loss.

The installation process required careful planning and execution due to the complex underwater terrain and the need to avoid damage to the cables during laying. NEXANS utilized specialized cable-laying vessels and equipment to ensure precise installation.

## 6.6 Conclusion and Key Takeaways

Has this green project unlocked the power of renewable energy through engineering?

The London Array Offshore Wind Farm was successfully commissioned. A big part of why it works so well is because of the cables made by NEXANS. They are very important for getting power from the wind turbines out at sea to the electrical grid on land. Their contribution is notable in making the entire project operational, and running smoothly and efficiently.

The exciting fact is that this project has shown everyone that large-scale offshore wind farms can totally work. It's not just a dream anymore. Because it has been so successful, more people are getting interested in putting money into renewable energy projects like this. It's practical to see how engineering can make a real difference in creating cleaner energy sources.

London Array's success has blown in the winds of change and sparked the development of other mega offshore wind projects. While we can't exactly draw a dotted line between them, it's safe to say that larger wind farms like Walney Extension, Hornsea One, and Hornsea Two did not just pop up because someone had a sudden craving for wind turbines. These subsequent projects were built upon London Array's pioneering scale, demonstrating the industry's commitment to large-scale development. It shows that they took a few notes from London Array's diary on offshore logistics and power grid connections – because who doesn't love a good cheat sheet?

There are several key take aways from this chapter. The case study motivates engineers and shows that large-scale offshore wind power projects are technically feasible and economically viable, particularly, when properly engineered. The specialized transmission solutions (like HVAC or HVDC systems) from industries like Nexans play a crucial role in delivering power from distant offshore installations. The project also proved that "bigger is better," and large installations achieve better economies of scale and lowering per-MW costs in the long term. The study also shows through London Array project that offshore wind could become a mainstream energy source and highlights the transition to a sustainable future. The following provides more comprehensive information.

**Engineering Expertise is Critical:** NEXANS' work on the London Array project shows how crucial engineering skills are for big renewable energy projects, just like how important coffee is needed on a Monday morning. Their expertise in cable tech and setting up offshore systems helped tackle the challenges of the sea environment – because, let's face it, the ocean isn't exactly the easiest coworker.

NEXANS didn't just solve technical problems; they set a high bar for future green energy projects. By using advanced cable technology and carefully installing offshore equipment, they managed to be both eco-friendly and efficient to save the planet with style. This teamwork shows that engineering innovation isn't just about fixing issues; it's about bridging the imagination and reality. As renewable energy projects continue to pop up like mushrooms after rain, expertise like NEXANS' is the secret sauce in the recipe for sustainable energy goals.

**High-Voltage Subsea Cables – A Key to Reliability:** To ensure consistent and reliable delivery of electricity from the offshore wind turbines to the onshore grid, high-voltage underwater cables play a main role. These cables are purpose-built to handle the ocean's mood swings and salty attitude. As this case study proves, selecting the appropriate materials and technologies is critical based on the project requirements.

The improved underwater power cables for high voltage were not just durable, they were also very efficient. They cut down on power loss over long distances and also made sure that the electricity from offshore wind farms made it to the land without wasting energy. These cables also used very good insulation to handle the intense pressure underwater. They also must stay safe from sea creatures and other interference from the ocean. This careful planning in how the cables were made and put in place shows how important smart engineering is for clean energy projects. When you think about it, having a solid foundation – both in the physical sense and in how we plan things – is crucial for a future powered by sustainable energy.

**Offshore Wind Farms – A Viable Renewable Energy Source:** The London Array project is a standout performer in the energy sector. Proving that offshore wind farms are a robust weapon against clean and sustainable energy production. Offshore wind could produce energy while providing environmental benefits and the worldwide effort of moving toward clean sources. The London Array did not just impress its critics on this occasion: investors are being inspired to put more bucks toward offshore wind technology as well. They see the positive future of renewable energy.

But as this case study also shows in its way, it comes with a certain amount of turbulence. Offshore wind farms also have their own set of issues such as dealing with inconsistent ocean characteristics and making infrastructure age well in tough environments. As offshore wind technology matures, engineers and scientists are doing their best to solve these problems. Manufacturers have been refining turbine design with tougher materials. The London Array project put more effort into reducing marine conditions impacts. Moreover, government and private players are also contributing to reducing bureaucracy. It aims at increasing the grid infrastructure so that the power coming from offshore windfarms can reach onshore seamlessly.

The London Array's achievements are motivating other countries to boost their offshore wind goals. Nations such as the US, China, India and Germany are now setting ambitious targets for offshore wind power, viewing it as a key part of their clean energy plans. This worldwide push highlights how important harnessing offshore wind is in fighting climate change and cutting down on fossil fuel use.

Despite some obstacles, the direction is obvious: offshore wind is not just a passing fad but a crucial part of tomorrow's energy mix. The London Array shows what's possible when creativity, money, and care for the environment come together to use wind power effectively.

**Inspiring a Greener Future:** The success of the London Array has sent a positive ripple effect across the globe. It is encouraging the development of similar wind farms and renewable energy projects. It has proven that renewable energy isn't just a fictional story, it's the real deal: economically smart, environmentally friendly, and way better than fossil fuels. The project has inspired both governments and industries. Together, everyone works toward a more sustainable future – it's like Mother Nature finally got her own fan club to act against climate change.

The UK government's enthusiasm for offshore wind is as clear as a sunny day at sea. Programs like contracts for difference and setting ambitious goals show that they are serious about riding the wind to a greener future. These policies give companies the green signal to dive in and invest in, making offshore wind capacity grow faster than a kite in a hurricane [24]. The London Array project has not only generated a whirlwind of investment into the offshore wind industry but has also boosted growth and economic benefits to coastal regions and related sectors. According to the Offshore Wind Industry Council, over 26,000 people were employed in the UK's offshore wind sector in 2020.

This high-profile project has also likely increased public awareness and acceptance of offshore wind energy. The public support encourages more future offshore wind projects, politically positive, and making it easier to move away from fossil fuels without causing a national meltdown. The London Array Offshore Wind Farm is a shining example of another engineering marvels to harness renewable energy. By studying projects like this, we can continue brainstorming ways to transition the world toward sustainable energy.

## 6.7 Discussion Points

While focused on the London Array, the underlying principles of cable design, installation, and maintenance are broadly applicable. This broader perspective makes case study a valuable resource for both students and professionals throughout their careers. It serves as a refresher on fundamental principles, offers cross-disciplinary insights, provides practical examples of real-world problem-solving, and can be used for mentoring and training junior engineers. The challenges faced and solutions implemented by NEXANS offer valuable insights for experienced engineers from four various perspectives:

1. **Advanced Project Management:** The project's execution provides insights into managing complex logistics, coordinating diverse teams, and mitigating risks in challenging environments.
2. **Cross-Disciplinary Collaboration:** The collaboration between electrical, mechanical, and environmental engineers highlights effective communication and integrated problem-solving.
3. **Optimizing Cable Performance:** The specialized cable technology demonstrates its technical strength to work with diverse environments, from subsea to underground transmission.
4. **Staying Current with Innovations:** The case study exposes engineers to cutting-edge wind-farm technology and relevant techniques.

Here are some suggested discussion points and further analysis based on the scenario from this case study:

1. Discuss the importance of renewable energy and the role of offshore wind farms in the energy transition.
2. Explore the technical challenges of transmitting electricity in subsea environments and the solutions provided by specialized cable technology.
3. Analyze the impact of successful renewable energy projects on the environment, energy security, and the economy.
4. Encourage students to research and present other innovative cable technologies for renewable energy applications.
5. Add an exercise where students calculate the appropriate cable size based on factors like current carrying capacity, voltage drop limits, thermal considerations, and installation constraints. Provide realistic parameters and encourage students to consider the trade-offs involved in cable selection.
6. Include a fault scenario analysis exercise. Present students with different fault scenarios, such as short circuits or insulation failures, and ask them to discuss the appropriate protection settings and analyze the system response.
7. Introduce a simplified power flow exercise where students analyze the power flow from the offshore wind farm to the onshore grid, considering cable impedance and load characteristics.
8. Facilitate a class debate on the economic and environmental trade-offs of offshore wind energy compared to other renewable energy sources (solar, hydro, etc.). Consider factors such as land use, intermittency, and cost.
9. Have students research and present on recent advancements in subsea cable technology, such as dynamic line rating and HVDC transmission, and their potential applications in other renewable energy projects (wave energy, tidal energy, etc.).

## 6.8 Limitations of the Case Study

This case study on the London Array project has inherent limitations that should be acknowledged. While it provides valuable insights into the technical aspects of offshore wind farm development and cable technology, it does not comprehensively address all related factors. Specifically:

1. **Environmental Impact:** While the case study mentions environmental considerations, it does not go into a complete environmental impact and pollution analysis of the wind farm. Looking deeper into these areas would give a better understanding of how sustainable the project really is.
2. **Economic Factors:** The case study does not provide a detailed economic analysis. Factors such as investment costs, repayment schedules, levelized cost of energy, and government subsidies are not explored. A comprehensive

economic analysis would be beneficial for a more holistic understanding of the project's viability.

3. **Socioeconomic Impacts:** The case study does not address the broader socioeconomic impacts of the project, such as job creation, and workforce development. More research on these topics would give a clearer picture of how the project affects everything around it.
4. **Potential Conflicts of Interest:** The study does not directly address possible conflicts with other industries or environmental concerns related to the project. It is important to think about how offshore wind might affect fishing, maritime transport, shipping routes, military operations, and sea life. Taking these issues into account is key for ensuring the sustainable and responsible development of offshore wind energy.
5. **Energy Loss Optimization:** While cable technology is discussed, it does not go into detail about ways to make the most of energy use and cut down on energy loss in the wind farm's actual operations. Further exploration of these optimization strategies would enhance the technical analysis.

Acknowledging these constraints allows readers to have a more balanced view of the London Array project and the challenges of offshore wind farm development. Future research and evaluation are suggested to resolve these limitations so that a more complete assessment of the full impact on the project is possible.

## References


1 Constanzo G., Brindley G. and Cole P. "Wind energy in Europe: 2022 statistics and the outlook for 2023–2027," *Wind Europe*, Belgium, 2022: 8.

2 Sofia. "London array," 2022, www.lindyenergy.com/london-array (accessed February 25, 2026).

3 "London Array Offshore Wind Farm," https://www.londonarray.com (accessed February 25, 2026).

4 Ren, G., Liu, J., Wan, J., Guo, Y. and Yu, D. "Overview of wind power intermittency: Impacts, measurements, and mitigation solutions," *Applied Energy*, 204, 47–65, 2017.

5 Mustajärvi, R. "Standard design criteria for electrical substations," Master's Thesis, *Lappeenranta-Lahti University of Technology LUT*, 2024.

6 Ahmed, T. "Future Grid Expansion Planning Towards Cost Effective and Sustainable Asean Power Grid Development," Doctoral Dissertation, University of Malaya (Malaysia), 2018.

7 Appunn, K. "German utilities and the Energiewende," 2015. *Clean Energy Wire*, https://www.cleanenergywire.org/factsheets/german-utilities-and-energiewende (accessed February 25, 2026).

**8** Timilsina, G.R. "Are renewable energy technologies cost competitive for electricity generation?" *Renewable Energy*, 180, 658–672, 2021.

**9** Gulski, E., Anders, G.J., Jongen, R.A., Parciak, J., Siemiński, J., Piesowicz, E., Paszkiewicz, S., and Irska, I. "Discussion of electrical and thermal aspects of offshore wind farms' power cables reliability," *Renewable and Sustainable Energy Reviews*, 151, 11580, 2021.

**10** Lesaint, C. "Høyspente sjøkabler: reduserte kostnader med et enklere design," Sintef, 2021, www.blogg.sintef.no/energi/hoyspente-sjokabler-reduserte-kostnader-med-et-enklere-design (accessed February 25, 2026).

**11** Sriraman, A. "Thermal aging studies and lifetime prediction of commercial filled ethylene propylene rubber (EPR) cable insulation," Doctoral Dissertation, Iowa State University, 2020.

**12** Rafati, A., Mirshekali, H., Shaker, H.R. and Bayati, N. "Power grid renovation: a comprehensive review of technical challenges and innovations for medium voltage cable replacement," *Smart Cities*, 7(6), 3727–3763, 2024.

**13** Prysmian Group. "66 kV submarine cable systems for offshore wind," *Prysmian Group, Milano,* 2024.

**14** Lerch, M., De-Prada-Gil, M. and Molins, C. "Collection grid optimization of a floating offshore wind farm using particle swarm theory," *Journal of Physics: Conference Series*, 1356(1), 012012, 2019.

**15** El Mokhi, C. and Addaim, A. "Optimization of wind turbine interconnections in an offshore wind farm using metaheuristic algorithms," *Sustainability*, 12(14), 5761, 2020.

**16** Pérez-Rúa, J.A. and Cutululis, N.A. "Electrical cable optimization in offshore wind farms—a review," *IEEE Access*, 7, 85796–85811, 2019.

**17** Cazzaro, D., Koza, D.F. and Pisinger, D. "Combined layout and cable optimization of offshore wind farms," *European Journal of Operational Research*, 311(1), 301–315, 2023.

**18** Song-Manguelle, J., Todorovic, M.H., Chi, S., Gunturi, S.K. and Datta, R. "Power transfer capability of HVAC cables for subsea transmission and distribution systems," *IEEE Transactions on Industry Applications*, 50(4), 2382–2391, 2013.

**19** Ping, H., Yunchao, W., Ruiyong, Y. and Jing, Z. "The horizontal positioning and interference analysis of submarine buried cable. In *2016 First IEEE International Conference on Computer Communication and the Internet (ICCCI)*, 418–421, IEEE, 2016.

**20** Madsen, S.F. and Mieritz, C.F. "Current distribution and magnetic fields in complex structures using Comsol Multiphysics," *Comsol.com*, 2011, https://www.comsol.com/paper/current-distribution-and-magnetic-fields-in-complex-structures-using-comsol-multiphysics-10873 (accessed February 25, 2026).

**21** Msigwa, G., Ighalo, J.O. and Yap, P.S. "Considerations on environmental, economic, and energy impacts of wind energy generation: projections towards sustainability initiatives," *Science of The Total Environment*, 849, 157755, 2022.
**22** DNV GL. "Qualification procedure for offshore high-voltage direct current (HVDC) technologies," 2014, https://www.dnv.com/energy/standards-guidelines/dnv-rp-0046-qualification-procedure-for-offshore-high-voltage-direct-current-HVDC-technologies (accessed February 25, 2026).
**23** SemcoMaritime, "HVDC and HVACtransmission," 2020. https://www.semcomaritime.com/news/hvdc-and-hvac-transmission (accessed March 23, 2026).
**24** Shields, C.A., Payne, A.E., Shearer, E.J., Wehner, M.F., O'Brien, T.A., Rutz, J.J., Leung, L.R., Ralph, F.M., Marquardt Collow, A.B., Ullrich, P.A. and Dong, Q., "Future atmospheric rivers and impacts on precipitation: overview of the ARTMIP Tier 2 high-resolution global warming experiment," *Geophysical Research Letters*, 50(6), p.e2022GL102091, 2023.

# 7

# Tesla's Approach to Grid-Scale Battery Storage

Engineering Solutions for Grid-Scale Renewable Integration

## 7.1 Synopsis

What happens when the sun sets on solar farms or when wind turbines stand still on a calm day? Much like how a backup generator keeps a hospital running during power outages, grid-scale energy storage serves as our power grid's insurance policy. This case study delves into the realm of electrical power systems engineering, specifically focusing on the integration challenges of renewable energy sources and the revolutionary role of grid-scale battery storage solutions.

To put that into a real-life dilemma, in 2003 the entire northeastern US and Canada were left in the dark by a cascading power failure for around 50 million folks at its peak, underscoring the weakness in the grid [1]. Now, 22 years hence as we move from those conventional power plants to renewables, the electrical grid faces an even complex problem. The International Energy Agency projects that renewable energy capacity must triple by 2030 to meet global climate targets. However, power system engineers face significant technical challenges in achieving this goal. The primary obstacle is the intermittent nature of renewable energy sources, which creates complex grid integration issues. Energy storage systems are now a vital piece of the puzzle for the next generation of smart grids, and they smooth out the variable output from renewable energy sources. It leads to a more reliable and efficient integration of these renewable sources into the existing grid infrastructure.

This engineering challenge intersects multiple stakeholders: utility companies striving for grid stability, renewable energy developers seeking reliable integration solutions, regulatory bodies establishing new grid codes, and consumers demanding both sustainable and reliable power. The power electronics industry, battery manufacturers, and system integrators play crucial roles in developing and implementing these solutions.

This case examines how grid-scale battery energy storage systems (BESS) are reshaping the power system stability. In 2023, the deployment of BESS in the power sector surged by 130% compared to 2022, contributing an additional 42 GW to global electricity systems [2]. This is one of the fastest-evolving sectors in power engineering. The technological advancement in this field continues to challenge traditional power system planning and operation paradigms, while opening new possibilities for grid modernization.

## 7.2 Background and Case Scenario

The technological development of modern power grids is the backbone for reliable electricity delivery to billions of people around the world. But as electricity demand increased grid complexities were hard to manage. Conventional power grids were traditionally designed for centralized power plants. For example, generators were large in fossil fuel or nuclear power plants, and they transmit through the grid of infrastructure (called transmission lines to distribution lines). While the traditional grid system functioned effectively for decades, it is now outdated due to aging infrastructure and limited automation capabilities. During 2000s, the system vulnerabilities were starting to come into focus. The North American blackout of 2003, one of the largest grid failures in history and it was an important wake-up call. This prompted urgent grid modernization and modern technologies like battery storage to make the system stronger and more reliable.

### 7.2.1 Overview of Battery Energy Storage Systems

How can we ensure a stable, reliable, and efficient energy system in a world that increasingly relies on intermittent renewable energy sources like wind and solar energy? BESS provides a critical answer to this question, emerging as a cornerstone of the modern energy landscape. These storage systems use batteries to store electrical energy. They are engineered to deliver on a wide array of critical services such as energy arbitrage, frequency regulation, peak load management, and backup. BESS are essential for the global energy transition. These systems are being deployed rapidly worldwide to address multiple challenges. It helps to balance supply and demand, maintains grid stability, and supports the UN's Sustainable Development Goals (SDG) for low-carbon energy transition. So, there is a need for BESS technology, and exploring its different forms of energy storage, and the concept of Long-Duration Energy Storage (LDES).

In 2015, the world took a monumental step toward global climate action with the adoption of the Paris Agreement at COP21. Deep decarbonization in all sectors was required to achieve the Paris agreement goal such as limiting global warming to no more than 2 °C above pre-industrial levels [3]. Recently, at COP28, the 28th annual Conference to the United Nations Framework Convention on Climate

Change (UNFCCC), governments pledged to speed up the shift from fossil fuels to renewable energy by committing to a target of tripling the world's renewable energy capacity by 2030. COP28, held in Dubai, United Arab Emirates, in 2023. It brought together world leaders, policymakers, scientists, and activists to assess global progress on climate goals and accelerate collective efforts to combat climate change. To achieve their decarbonization goal, the global renewable energy capacity must reach at least 11 terawatts [TW] over the decade [4].

Nations around the world are working to increase renewable energy's share in their power systems while meeting national emissions reduction targets. However, integrating variable renewable energy (VRE) sources such as wind and solar presents significant challenges. The intermittent nature of these energy sources leads to grid instability, making it increasingly complex to ensure a reliable and efficient energy supply. Advanced technologies, such as BESS, are emerging as critical solutions to address these challenges. By storing excess energy generated during periods of high renewable output, and discharging it during periods of low generation, BESS can transform VRE into reliable, and dispatchable baseload power.

The need for BESS is getting even more obvious as the demand for electricity increases along with base load demands from heating, transport electrification and the like. The International Energy Agency (IEA) states that its Net Zero by 2050 Scenario requires 1,500 gigawatts (GW) of energy storage capacity by 2030, including 1,200 GW battery storage [5]. This illustrates that BESS needs mass adoption for the decarbonization of power systems and to help the global energy transition.

Battery storage has become an indispensable technology for integrating VRE into power systems. Its ability to provide multiple services simultaneously, a concept known as value stacking, enables BESS to maximize revenue potential while supporting grid operations. For example, a single BESS installation can provide frequency regulation, peak shaving, and energy arbitrage, creating economic and operational synergies.

According to IEA, global energy storage needs to increase to six times its current level to deliver on COP28 optimism of tripling renewable capacity by 2030. Which would equate to around a 25% average annual growth rate for BESS deployments through to 2030 [6]. This growth requires numerous factors, like the sustainable implementation and substitution of public policies, and additional revolution of battery technologies forward with lower costs. In the last decade, the cost of lithium-ion batteries that power most BESSs has fallen by more than 85% due to technology evolution, manufacturing improvement, manufacturing output, and battery chemistries [7].

Undeniably the deployment of BESS for the power systems transformation is essential. The flexibility provided by BESS can significantly reduce greenhouse gas emissions and improve energy security. It forms the backbone of a global transition to sustainable energy future in general and helps the integration of high penetrations of renewables. As more sectors become electrified, including transportation, heating, and industrial processes, the demand for energy storage continuously grows. This trend firmly proves BESS as a fundamental component of the current energy system.

The versatility and scalability of BESS make them a critical enabler of the energy transition. These systems range from utility-scale installations that support grid operations to smaller-scale systems that empower individual consumers. As a result, BESS are poised to play a central role in achieving global climate and renewable energy targets. However, realizing the full potential of BESS require coordinated efforts across governments, industries, and research institutions to address challenges such as cost, scalability, and regulatory frameworks.

Exploring BESS technology's key features reveals why these systems offer such remarkable versatility and adaptability. All its components are vital to making the system safe and effective. Breaking down these core components can further help us to understand how BESS works, and its energy transition as a whole. Figure 7.1 summarizes the key components of a BESS which comprises of four main components: the battery cells, which store energy; the Battery Management System (BMS), which ensures safety and efficiency; the Power Conversion System (PCS), which handles AC/DC conversion for grid interfacing; and the Energy Management System (EMS), which optimizes overall operation by coordinating charging, discharging, and other functions.

As BESS are modular and scalable, they are suitable for a variety of applications, from residential-size substations to large utility-scale installations. Utility-scale BESS, with capacities ranging from 300 kWh to gigawatt-hours, complement

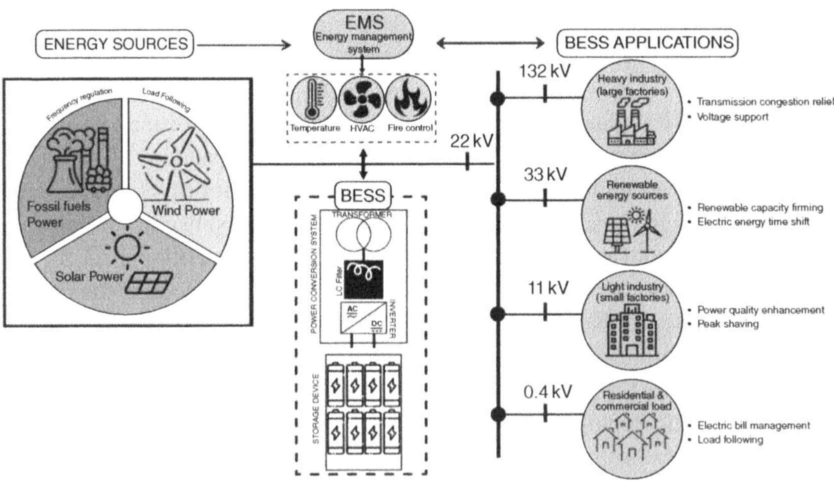


**Figure 7.1** Key components of a BESS. *Source:* Author's own work with data from [8, 9].

**Table 7.1** Applications and benefits of utility-scale battery energy storage systems (BESS).

| Application | Description | Benefit |
|---|---|---|
| Peaking Capacity | Discharges power during periods of peak demand, reducing reliance on fossil fuel-based peaking plants. | Reduces greenhouse gas emissions and lowers operational costs for energy providers. |
| Ancillary Services | Provides essential grid stability services such as frequency regulation, voltage control, and spinning reserves. | Ensures a stable and reliable power grid, preventing outages and maintaining power quality. |
| Black-Start Capability | Supplies initial power to restart the grid in the event of a grid outage. | Enhances grid resilience and ensures faster recovery after power disruptions. |
| Energy Arbitrage | Stores energy during low-demand, low-cost periods and discharges it during high-demand, high-cost periods. | Optimize economic returns for energy providers and consumers while balancing energy supply. |

*Source:* [10–13].

contemporary power grids by maintaining grid stability and resilience while supporting renewable energy integration. Table 7.1 highlights key applications and benefits of these systems, which play a critical role in enhancing economic efficiency and ensuring a steady state of the grid.

While BESS serves utility-scale applications, a growing number of smaller-scale systems are also being installed behind the meter to provide localized energy solutions. In addition to providing benefits for individual consumers through reduced electricity costs and backup power during outages, these systems furthermore reduce energy waste from local renewable energy sources [14]. Whether it be utility-scale or distributed energy storage, these are rapidly changing the face of energy and tackling challenges at all levels of the grid. BESS is one of the many technologies available for energy storage. To understand its role, it is essential to compare it with other energy storage types. Table 7.2 gives a comparison of key features across major energy storage technologies, categorized by energy density, efficiency, lifespan, applications, and maturity/adoption. Each technology has unique strengths and is suited to specific use cases, from high-density lithium-ion batteries to long-duration hydrogen storage.

To further explore the diverse landscapes of energy storage, it is important to examine the broader ecosystem of technologies and services that complement or compete with BESS. Energy storage technologies are not limited to batteries; they span a wide range of systems; each tailored to specific grid challenges and

**Table 7.2** Energy storage types and their characteristics.

| Storage Type | Energy Density | Efficiency | Lifespan | Applications | Maturity/ Adoption |
|---|---|---|---|---|---|
| Lithium-Ion Batteries | High | 90–95% | Moderate (5–15 yrs) | EVs, grid-scale storage, electronics | Dominates market; widely adopted |
| Flow Batteries | Low | 60–85% | Long (20+ yrs) | Large-scale, long-duration storage | Emerging; less common than Li-ion |
| Thermal Energy Storage | Variable | Moderate–High | Long | CSP, industrial processes, cooling/heating | Established in specific niches |
| Mechanical Storage | Low–Moderate | 70–85% | Very long (decades) | Grid-scale, peak shaving, frequency regulation | Most mature; >90% of global capacity |
| Hydrogen Storage | High (by weight) | 30–50% | Long | Seasonal storage, fuel, industrial uses | Emerging; gaining attention |

*Source:* [15–18].

applications. These technologies are at various stages of development and deployment, reflecting the dynamic and rapidly evolving nature of the energy storage market. The following figure highlights the ecosystem of energy storage technologies and services, showcasing their classifications and examples across electrochemical, thermal, mechanical, and chemical storage solutions (Figure 7.2).

A high degree of diversity within energy storage technologies exists across: electrochemical storage (e.g., BESS and flow batteries), thermal storage systems (molten salt), mechanical storage (e.g., pumped hydro), chemical storage (e.g., hydrogen, power-to-gas). Pumped hydropower storage is the most deployed energy storage technology, in which grid-connected to the existing infrastructure [20]. Lithium-ion batteries have reached maturity and are taking off in shorter-duration applications. Flow batteries and power-to-gas are considered emerging technologies, which are trying to address longer-duration needs but are in much earlier stages of development and commercialization. Energy storage is catching on at a fast pace due to government policies, improvements in technology and increased manufacturing abilities of the advanced components.

While BESS provides proven short to medium-term energy storage, other technologies are emerging to fulfill long-term storage requirements. LDES systems

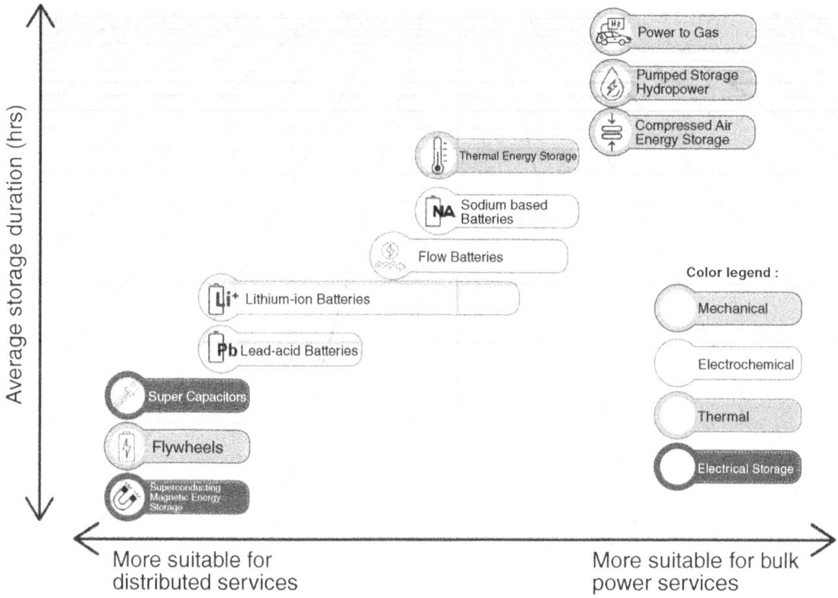


**Figure 7.2** Energy storage ecosystem. *Source:* Adapted from [19]. *Note:* Average storage duration is based on entries in the DOE global energy storage database.

are designed to complement BESS. It addresses seasonal energy variability and extended periods of low renewable generation. By exploring LDES technologies, it is possible to understand how they fit into the broader energy storage landscape and how they contribute to achieving a more reliable and sustainable energy system together with BESS.

LDES is for durable energy storage systems from eight hours to several days or even months. LDES is indispensable for dealing with the seasonal variability of renewables and ensuring grid reliability during prolonged periods of low wind or solar output. These are the systems needed to connect energy generation and consumption on long timescales. It forms the bedrock of decarbonization efforts. For example, LDES can store excess solar energy generated during summer months and release it in winter when solar output is significantly lower. While lithium-ion batteries dominate short-duration energy storage markets, they are less suited for long-duration applications due to cost and degradation issues. As a result, alternative LDES technologies have emerged to meet these demands.

Flow batteries, hydrogen storage and thermal storage are all varieties of LDES technologies. These are all technologies intended for long-term energy storage, making them well suited for applications that bridge hours or even days between when energy is produced and consumed. For example, seasonal energy storage or

**Table 7.3** Overview of long-duration energy storage (LDES) technologies.

| Technology | Description | Key Features | Examples | Applications |
|---|---|---|---|---|
| **Flow Batteries** | Rechargeable batteries where energy is stored in liquid electrolytes that flow through a cell stack. | • Independent scaling of power (cell stack) and energy (electrolyte volume).<br>• Long lifespan with minimal degradation.<br>• Suitable for large-scale, long-duration storage. | Vanadium Redox Flow Batteries (VRFBs),<br>Zinc-Bromine Flow Batteries | Grid-scale energy storage, renewable energy integration, microgrids. |
| **Hydrogen Storage** | Converts excess electricity into hydrogen via electrolysis, stores it for later use, and reconverts it into electricity using fuel cells or turbines. | • Can store energy for months (ideal for seasonal storage).<br>• Highly versatile: hydrogen can also be used as a fuel for transport, heating, and industry.<br>• High energy density. | Electrolyzers, Fuel Cells | Seasonal energy storage, de-carbonizing industrial processes, clean fuel for transport and heating. |
| **Thermal Storage** | Stores energy in the form of heat or cold, which can later be used for electricity generation or heating/ cooling. | • Cost-effective compared to other LDES.<br>• Can store large amounts of energy for extended periods.<br>• Commonly used with concentrated solar power (CSP) plants. | Molten Salt Systems, Ice-Based Cooling, Phase-Change Materials | Concentrated solar power plants, district heating/ cooling, industrial heat applications. |

*Source:* [17, 21–24].

long-term renewable energy variability. Table 7.3 summarizes their key features and main applications.

Despite their potential, LDES technologies face several challenges, for example high capital costs, efficiency penalties and reliance on supportive policy, as well as market structures. It is expecting to see the ongoing research and development of driving costs down and performance improvements as the scale-up runs its course. With these developments continuing, LDES is set to be a key enabler of the low-carbon renewables energy future.

With the changing energy landscape, complementing the relationship between BESS and LDES can be more apparent over time. While BESS is suitable for short to medium-duration, LDES is intended to greatly improve the long-term energy storage. It needs to meet the seasonal variation and prolonged renewable energy gaps. Thus, BESS and LDES make up a complete energy storage solution to solve different issues on timescales and use cases. Understanding their full potential requires analyzing how these technologies perform across various applications and contribute to energy system reliability, flexibility, and resilience. Figure 7.3 illustrates the key applications of BESS and LDES in the energy sector. This includes renewable energy integration, grid stabilization, energy arbitrage, backup power, and enabling microgrids for enhanced energy reliability and resilience.

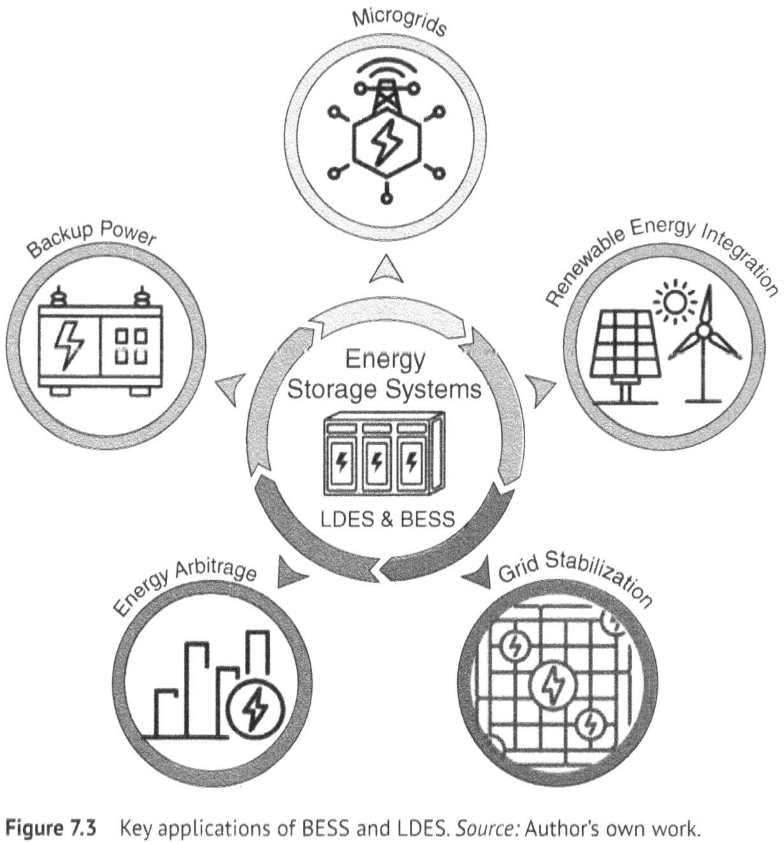


**Figure 7.3** Key applications of BESS and LDES. *Source:* Author's own work.

### 7.2.2 Evolution of Grid-Scale Energy Storage (2010–2017)

The traditional power grids had many constraints on integrating renewable resources, mainly because of bidirectional power flow and the inflexibility of dealing with variable generation in existing design. Integrating more wind and solar power puts an unprecedented level of stress on the global grid operators. In the US, the growth in renewable energy capacity rose from 121 GW in 2010 to over 200 GW by 2016 and revealed the shortcomings of legacy grid infrastructure [25]. As the share of intermittent resources continued to rise, grid operators confronted frequency regulation and voltage stability problems and required power quality solutions. This was especially true in heavily renewable regions such as California where the "duck curve" problem – the widening chasm between peak demand and available renewables became a major operational problem by 2013 [26].

The early efforts to tackle these problems started off with pilot projects throughout the world. The AES Laurel Mountain facility, established in 2011 was an example of one of the first big battery installations with wind power [27]. The project illustrated the power of battery storage both in renewable integration and frequency control applications. Followed by this success, several other commercial utility-scale trials formed, including Notrees Wind Storage Demonstration Project in Texas. It installed a 36 MW battery system in 2012 [28]. Japan put the 40 MW/40 MWh grid stabilization by lithium-ion battery system into operation at a Nishi-Sendai substation project in 2015 [29].

From 2010 to 2015, there was major progress in battery technology and a decline in costs. Lithium-ion batteries hit $1,200/kWh in 2010, but dropped to about $350/kWh, or slightly less over capacity in 2015. This was making utility-scale storage much more attractive [30]. The cost dropped, along with higher energy density and longer cycle life which started to drive significant larger storage projects around the world. "Cycle life" – refers to the number of charge/discharge cycles that battery can complete before its capacity degrades to a certain level. This threshold is typically set at 70–80% of original capacity, and it is a key measure of battery durability.

One such demonstration project in China was the Zhangbei National Wind and Solar Energy Storage & Transmission project, launched in 2011. It is China's 10 GW wind power base which includes a 100 MW wind power plant, a 40 MW PV farm and 14 MW/63 MWh lithium-ion BESS [31]. It is important to note that the Zhangbei National Wind and Solar Energy Storage & Transmission Demonstration Project is a pilot/demonstration project, located in the Zhangbei region. It served as part of a larger 10 GW wind power development plan for the area, rather than representing the entire 10 GW base itself. The total projects presented a broad range of battery storage applications like frequency regulation, voltage support and peak shaving to renewable smoothing.

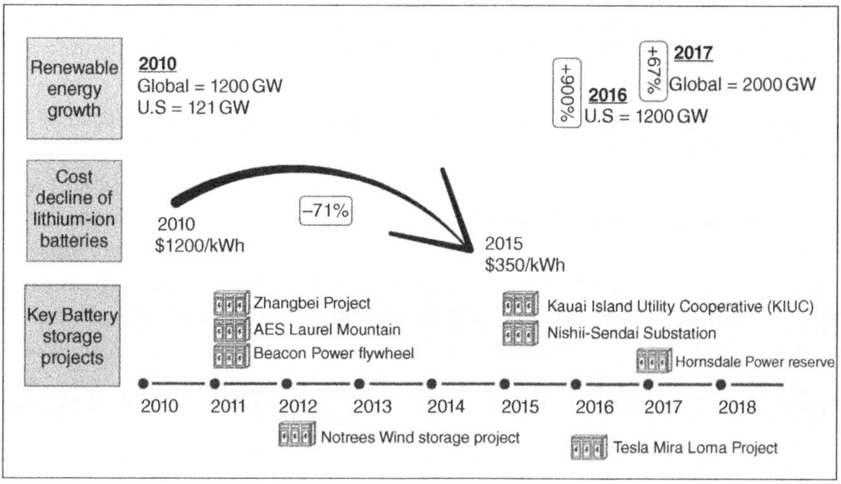


**Figure 7.4** Evolution of grid-scale energy storage (2010–2017). *Source:* Author's own work.

This evolution of grid-scale energy storage is shown in Figure 7.4, which focuses on the correlation between economic pressures and project execution. Global renewable energy capacity increased by 67% between 2010 and 2017, while US renewable energy capacity surged by over 900%. At the same time, energy storage became much more affordable as lithium-ion battery prices fell 71% from $1,200/kWh to $350/kWh by 2015. The timeline illustrates this progression, starting with a diverse cluster of projects in 2011 including Zhangbei, AES Laurel Mountain and the Beacon Power flywheel. This combination of declining costs and increasing demand for renewables triggered a second wave of even larger commercial projects in 2016/2017.

Among the firms that pounced on these developments, Tesla distinguished itself for its understanding of what battery technology could be for: not just in vehicles but anywhere. The world was slowly moving away from carbon; energy storage became imperative and flexible load able to respond quickly to prices. There is a sweet spot coming to grid-scale storage, driven by the decreasing costs of lithium-ion batteries and increasing Concentrated Solar Power (CSP) reGen+ (renewable energy integration). Heating up the utility-scale market, Tesla is looking forward to technology advantage in battery manufacturing and Energy Management systems.

Tesla's entry into the utility-scale storage market in 2015 marked a significant shift in the industry. The company leveraged its automotive battery expertise to develop the Powerpack system, designed specifically for commercial and utility applications. Their first major project, a 20 MW/80 MWh installation at Southern California Edison's Mira Loma substation. It demonstrated the rapid deployment capability of battery storage system [32]. However, it was the South Australia's power crisis, that provided Tesla with most significant opportunity to demonstrate the capability of battery storage technology.

On 28 September 2016, severe storms destroyed critical infrastructure in South Australia. It caused a blackout that affected 1.7 million people [33] in the region. The state, which had recently achieved its 40% renewable energy share goal, faced criticism over grid unreliability and stability issues. The economic impact was severe from storms, with businesses reporting damages totaling approximately AUD 367 million [34]. The blackout caused an intense debate on energy security, and the value of renewables in a grid. Independent analysis determined that the domino effect leading to blackout was a direct consequence of extremely severe weather and protection system settings and not the integration of renewable energy [35].

In March 2017, the South Australian government announced its AUD 550 million energy plan [36]. This would include the construction of grid-scale energy storage for grid system security and reliability. Tesla founder Elon Musk made a bold promise to build a 100MW battery system in South Australia or deliver it free of charge in case of failure within the promised deadline. This bold commitment led to the creation of the Hornsdale Power Reserve. This significant investment aimed to prove that utility-scale battery storage could effectively address grid stability challenges [37]. This project was completed on time, within 100 days as promised by Musk. It has been profitable for the operators and helped during several grid incidents. The investment was largely successful and proved to be another outstanding contribution to the world by Tesla.

### 7.2.3 Rising Renewables Driving Grid Storage Demand

In the last two decades, the deployment of renewable energies has really taken take off, mainly for solar and wind power. The International Energy Agency reports that renewable energy capacity has rapidly expanded due to improved technology and policy-motivated deployment with a focused reduction in carbon emissions [38]. But the problem is that renewable sources are intermittent as they produce energy only at specific instances (the sun for solar panels or wind for wind turbines). This intermittency sets a significant challenge for grid operators in balancing supply and demand to keep the grid in balance.

Grid-scale battery storage has emerged as a critical solution to address these challenges. By storing excess energy generated during periods of high production and releasing it during peak demand, battery systems can smooth out fluctuations and ensure a reliable supply of electricity. This capability is essential for integrating higher shares of renewable energy into the grid without compromising stability [38]. Energy storage also enhances the resilience and robustness of the grid infrastructure [39]. It can supply additional services such as frequency regulation and voltage support, that further stabilize grid operations [40].

The complete global shift to renewable energy has redefined views on power system stability and reliability. Global renewable energy sources generated around 1,223 GW in 2010, with wind and solar experiencing some of the fastest growth rates of any technology [41]. In just the way that power grids were created for one-way flows of centralized generation to customers, traditional power grids needed to adapt rapidly to the fluctuating nature of renewable sources. BESSs are increasingly vital in facilitating the integration of renewable energy by absorbing and releasing power across varying timeframes [42]. Tesla, well known for its electric vehicles, has become a major player in grid-scale battery storage with its Megapack system.

As grid-scale energy storage became an increasingly critical concept to tackle this challenge, global energy storage deployment expanded from 0.5 GW in 2011 to around 17 GW in 2020 [43]. The main storage technologies at large grid-scale were lithium-ion batteries. It is still evolving based on their falling cost (~89% between 2010 and end of 2020) as well as high degree of grid-compatible applications [44]. Tesla's evolution into utility-scale storage technology represents a critical inflection point for modernizing electrical grid infrastructure. Tesla's strategy to build grid-scale battery storage facilities was tested in one of the most challenging grid environments on Earth – South Australia.

### 7.2.4 Implementation and Performance of Hornsdale Power Reserve

Tesla's approach to grid-scale battery storage involves several key control strategies designed to optimize performance, extend battery life, and maximize revenue generation [45]. The Hornsdale Power Reserve (HPR), also known as the Tesla Big Battery, represents a groundbreaking implementation of large-scale energy storage technology. The system initially featured a 100 MW/129 MWh capacity utilizing Tesla's lithium-ion battery technology, specifically the Powerpack 1.0 system built around nickel-manganese-cobalt (NMC) chemistry [46]. NMC cathode materials are favored in stationary storage for their favorable balance of energy density, thermal stability, and long cycle life, exceeding 3,000 cycles at moderate depth of discharge [47].

Each Tesla Powerpack contains 16 individual battery pods with liquid cooling and onboard DC/DC converters. The converters regulate charge-discharge cycles and are integrated with Tesla's proprietary Powerpack inverter system for grid-scale applications [48]. The modularity of Tesla's Powerpack design allows for rapid deployment and scalability, crucial for meeting the tight deadline set by the South Australian government.

The HPR design is configured to provide millisecond response times (standby to full discharge within 150 ms), which is necessary to keep the grid at 50 Hz and frequency synchronized [49]. In contrast to traditional spinning reserves or thermal plants that require 6–10 minutes to respond, HPR systems are capable of fast, accurate and instantaneous regulation. For weak grids, the inverter works in four-quadrant mode and provides power support – active power, and reactive power for voltage and frequency support.

The Hornsdale project, which was first built out in late-2017 with an upgrade since then, has a total energy storage of at 150 MW/194 MWh, and was completed in 2020 [50]. This expansion featured Tesla Powerpack 2.0s with better energy density and greater round-trip efficiency (~88–90% on average). The upgrade added virtual inertia capability to the energy storage system. This feature allows the battery to emulate the inertial response typically provided by rotating mass in conventional generators. This capability is essential for maintaining stability in modern power systems dominated by renewable energy sources.

Economically, HPR has been a game-changer. Within the first year, it helped reduce South Australia's contingency Frequency Control Ancillary Services (FCAS) costs by over 90%, equating to tens of millions of dollars in savings [51]. It also participates in energy arbitrage by charging during off-peak times and discharging during peak demand periods, supporting grid economics and stability.

Although newer installations, such as the Moss Landing Energy Storage Facility in California (400 MW/1,600 MWh as of 2021), surpass HPR in scale, Hornsdale remains a landmark project due to its pioneering role in fast frequency response, its successful business model, and its robust real-world performance. It continues to serve as a reference for integrating large-scale BESS into power systems, dominated by variable renewable energy sources. HPR demonstrates to act as a dispatchable load by storing energy during periods of high wind generation and releasing it during periods of low wind generation.

This landmark status is not only due to its scale at the time of commissioning but also to its seamless integration into a real-world grid environment with high renewable energy penetration. Beyond its capacities and fast response times, the operational effectiveness of Hornsdale lies in its engineering design and system-level coordination. Understanding how the facility interacts with existing grid infrastructure offers valuable insight into the practical deployment of utility-scale battery energy storage systems.

The system's integration with South Australia's existing power infrastructure was achieved through sophisticated power electronics and control systems. The battery installation connects to the Hornsdale Wind Farm via the 275 kV transmission network, employing advanced inverter technology to ensure seamless power conversion and grid synchronization [52]. The facility's battery management system continuously monitors individual cell performance, temperature, and state of charge across thousands of battery modules to optimize system longevity and efficiency.

Operational performance data has demonstrated the HPR's exceptional capabilities in providing grid services. During its first two years of operation, the facility prevented numerous major grid events and saved consumers more than $150 million AUD [53]. The system has proven particularly effective in FCAS and responding to frequency deviations within seconds to maintain grid stability. In one notable instance, the HPR responded to a major coal plant trip in Victoria within 140 milliseconds, helping to prevent a potential widespread outage [54].

The facility has also performed well with peak shaving, able to discharge full capacity during peak and recharge during off-peak loads. Utility Control Panel (UCP) exceeds 98% availability in performance metrics, with the system able to handle its rated capacity after extensive cycling [55]. Its highly tuned control algorithms allow the facility to offer multiple market services in a year, for the highest economic value with necessary grid support services. The success of the HPR has spurred further investment in grid-scale battery storage, including projects in Victoria and other regions of Australia [56].

### 7.2.5 Technical Challenges and Future Considerations

The Hornsdale Power Reserve has demonstrated the impressive capabilities of utility-scale battery storage. It has shown how these systems can enhance grid reliability while reducing costs. However, several technical and operational challenges remain for the long-term success and scalability of such systems.

Thermal Management, one of the biggest concerns in lithium-ion battery installations. Lithium-ion cells work optimally at a temperature range of 15 °C to 35 °C [57]. Going above these limits can speed up degradation, and cause safety problems like thermal meltdown. In HPR and other places where active liquid cooling is used for thermal management, which keeps the machines cool. The systems, which eat up roughly 3–5% of the total energy throughput required to power the systems they are guarding. It constitutes a sizable expense in terms efficiency and yield [58]. Additionally, thermal management gets more difficult as the system scales up further, and in regions experiencing high ambient temperatures.

Another persistent technical challenge is battery degradation over time. Lithium-ion cells, particularly those based on nickel-manganese-cobalt chemistry used in Tesla's Powerpack systems, exhibit an average capacity fade of 2–3% per year under standard grid cycling conditions [59]. This degradation is influenced by multiple factors, including depth of discharge, charge/discharge rate (C-rate), ambient temperature, and operational frequency. Over a 10-year lifecycle, cumulative degradation can reduce usable capacity by 20–30%, necessitating partial module replacements or capacity augmentation to maintain performance guarantees.

One of the major complexities in grid interconnection is managing the dynamics introduced by large-scale BESS. These systems can respond almost instantaneously to grid signals. This rapid response capability is highly beneficial for grid stability and frequency regulation. However, it can also create challenges for legacy grid infrastructure that was not designed to handle such rapid changes in power flow. For example, studies have shown that rapid power dispatch from batteries can cause issues such as voltage instability and frequency overshoot, if not properly managed [46]. As battery deployments have increased, system operators have recognized the need for more sophisticated inverter firmware and AI based predictive control algorithms. These advanced controls are essential for dynamic power output management. They ensure that the plant's power output is continuously adjusted based on real-time grid conditions. This capability enables seamless integration of BESS into the grid while minimizing potential disruptions [60].

In economic terms, lithium-ion battery costs have declined by approximately 14% from 2022 to 2023. Figures from BloombergNEF placed 2023 costs at a record low of approximately $139/kWh [61]. A 100/400 MWh installation could require around $55–60 million USD in upfront costs. However, these investments are balanced by multi-stream revenue models, including participation in Frequency Control Ancillary Services, energy arbitrage, and capacity markets. At HPR, for instance, FCAS revenues alone contributed to payback within three years of commissioning. Going forward, emerging trends in battery technology, such as solid-state electrolytes, silicon anodes, and lithium-sulfur chemistries, offer potential improvements in energy density, cycle life, and safety [62].

While typically overlooked in the early phases of energy storage system deployment, maintenance requirements represent a significant operational burden. Operation and maintenance (OPEX) typically represent approximately 2–3% of annual costs for large-scale battery energy storage systems. Key maintenance activities include scheduled inspections, cell balancing, HVAC tuning, and firmware/software updates. These maintenance costs generally range from $2–4/MWh, depending largely on utilization levels and system size [63]. These costs are important to guarantee the system performance, lifetime, and safety under normal operation. Moreover, the compliance of the system with safety standards, cyber security frameworks, and changing grid code elements require constant technical support and upgrades. Regulatory standards demonstrate the growing complexity of BESS operations. For instance, the National Fire Protection Association (NFPA) Standard 855 for energy storage systems and increasing cybersecurity requirements for grid-connected assets, highlights evolving safety demands. These evolving requirements and standards emphasize the critical importance of maintaining strict operational procedures for BESS [64]. Moving on, there are some interesting emerging technologies and system architecture that may pave a way to solve the challenges:

- Several companies are investigating grid-scale deployments of solid-state battery technology. QuantumScape and Solid Power are among those predicting energy densities 50–70% higher than current lithium-ion systems. Additional advantages include much higher thermal stability and significantly longer cycle life [28]. To add to this, sophisticated power electronics and silicon carbide (SiC) inverters are getting greater attention as they possess over 98% efficiency rates on switching frequency when compared to current silicon IGBTs. Also, SiC's thermal loss of power is much lower [65].
- Likewise, advanced power electronics such as SiC based inverters are decreasing system cooling requirements and improving energy throughput [66].
- Integration of distributed storage units into Virtual Power Plant (VPP) networks is accelerating. Large-scale implementations like Tesla's VPP in South Australia harness hundreds of residential Power-walls to operate as the equivalent of a 250 MW peaking power plant units. These distributed networks offer dual operational modes that enhance grid resiliency. They also enable peer-to-peer energy trading infrastructure and more decentralized grid services [67].

To remain economically and technically viable, future grid-scale storage projects must continue to address these challenges through innovation in battery chemistries, intelligent control systems, modular maintenance practices, and grid harmonization technologies. The evolution of standards, supportive regulatory frameworks, and robust testing protocols are equally crucial in scaling battery storage across diverse grid contexts.

## 7.3 Think and Solve

The case study on the evolution, implementation, and challenges of grid-scale energy storage highlights the transformative role of BESS in modernizing power grids, integrating renewable energy, and improving grid stability. This study covers the spectrum from landmark projects like the Hornsdale Power Reserve to key technical challenges including battery degradation, thermal management, and grid harmonization. It demonstrates how advancements in battery technology, control systems, and economic models are reshaping the future of energy infrastructure. Below are questions for readers to develop their understanding and problem-solving skills.

**Q 7.1** What was one of the major reasons for the increased need for grid-scale energy storage between 2010 and 2017?

**A** Decreasing costs of fossil fuels.
**B** Increased integration of intermittent renewable energy sources like wind and solar.
**C** Reduced demand for electricity during peak hours.
**D** Improved grid infrastructure capable of handling variable generation.

**Q 7.2** What was a significant technical advantage of the Tesla Powerpack system used in the Hornsdale Power Reserve?

**A** Use of solid-state batteries for higher energy density.
**B** Lack of need for thermal management systems.
**C** Nickel-Manganese-Cobalt (NMC) chemistry for long cycle life and energy density.
**D** Ability to operate without inverters for grid synchronization.

**Q 7.3** What economic benefit did the Hornsdale Power Reserve provide within its first year of operation?

**A** Reduced South Australia's FCAS costs by 90%.
**B** Eliminated the need for coal-based power plants.
**C** Generated revenue solely through energy arbitrage.
**D** Reduced battery installation costs by 50%.

**Q 7.4** Which emerging technology could address some of the technical challenges faced by lithium-ion batteries in grid-scale applications?

**A** Lead-acid batteries for higher thermal stability.
**B** Solid-state batteries for improved energy density and safety.
**C** Nickel-Cadmium batteries for lower costs.
**D** Use of silicon-based inverters for reduced energy throughput.

**Q 7.5** Calculate the annual energy loss due to thermal management in a 150 MW/194 MWh battery system like the Hornsdale Power Reserve. Assume, thermal management consumes 4% of the total energy throughput annually. Thermal management is critical for maintaining the safety and efficiency of lithium-ion batteries. However, it requires energy to operate. The Hornsdale Power Reserve operates at a capacity of 150 MW with a total energy storage of 194 MWh. Also consider that the system operates at full capacity for 365 days a year.

**Formula:**

$$\text{Annual Energy Loss} = \text{Total Energy Throughput per Year} \times \text{Thermal Management Percentage}$$

$$\text{Total Energy Throughput per Year} = \text{MWh/day} \times \text{days/year}$$

## 7.4 Conclusion and Key Takeaways

This case study is an example of how grid-scale BESS can revolutionize today's power grids for renewable integration as well as grid resiliency. The project not only created a baseline for fast response times, operational excellence, and economic scalability but also exposed the difficulties and opportunities of grid-scale battery deployments. The takeaways from the case study are as follows:

1. **Technical Excellence and Performance:** HPR gave insight into the fast frequency response capability of lithium-ion batteries, especially NMC-based chemistry, as well as providing grid stability. Its modular design and innovative inverter, designed for ease of integration into the existing grid infrastructure guarantees proper functioning even in harshest of conditions.
2. **Economic Impact:** The project demonstrated the financial benefits of grid-scale battery storage. It reduced South Australia's FCAS costs by over 90% in its first year, and achieved payback within three years. Multi-stream revenue models, including energy arbitrage and ancillary services, underline the economic feasibility of such systems.
3. **Challenges in Scaling and Long-Term Viability:** Thermal management, battery degradation and grid harmonization are some of the technical challenges that must be overcome to make the BESS viable in the long-term. Also, maintenance, advanced control algorithms and regulatory requirements are necessary for sustained performance reliability.
4. **Future Innovations and Opportunities:** Essentially, new breakthroughs in battery chemistries, power electronics and grid integration paradigms are required to scale energy storage systems globally.

Finally, the Hornsdale Power Reserve shows that grid-scale battery storage can be a key enabler to the roll-out of renewable energy. It has economic and technical benefits to boost. Success on a global scale, however, relies on how well it overcomes scalability, cost and longevity challenges.

## 7.5 Discussion Points

The Hornsdale Power Reserve case study raises several critical points that highlight the complexities, opportunities, and challenges of implementing large-scale battery energy storage systems. These points encourage deeper analysis and debate around the technical, economic, and regulatory aspects of grid-scale storage and its role in the transition to renewable energy.

1. **The Role of BESS in Grid Stability:** How can battery energy storage systems balance the intermittency of renewable energy sources such as wind and solar?
2. **Economic Feasibility and Regulatory Considerations:** What financial incentives or regulatory frameworks are needed to encourage more private and public investments in battery storage projects?
   How can governments and policymakers support the adoption of battery storage systems while addressing safety, cybersecurity, and environmental concerns?
3. **Technical Challenges and Environmental Impacts:** How can emerging technologies, such as solid-state batteries and silicon carbide inverters, address issues like thermal management, battery degradation, and efficiency?
   What are the environmental implications of large-scale lithium-ion battery production, usage, and disposal?
   How can the lifecycle of batteries be extended, and what role does recycle or second-life applications play in reducing waste?

These discussion points are designed to help readers critically consider the implications of widespread grid-scale battery deployment. The Hornsdale Power Reserve project exemplifies the type of large-scale battery systems that could transform energy infrastructure globally. They also provide an early framework for building upon and discussing potential innovations. It highlights both technological advances and policy decisions that are necessary to meet sustainable energy future challenges.

## 7.6 Limitations of the Case Study

Although the Hornsdale Power Reserve case study offers important lessons in the promise of grid-scale battery energy storage systems, it is crucial to consider the study constraints in order not to over-interpret this understanding. These limitations suggest important directions for future research. They also help in providing a better evaluation of the broader implications of larger projects.

1. **Geographic and Contextual Specificity:** The success of the HPR is heavily influenced by South Australia's unique energy landscape, characterized by high renewable energy penetration and frequent grid instability. The findings

may not be directly applicable to regions with different energy mixes, grid conditions, or regulatory frameworks.

The case study does not fully explore how BESS could perform in regions with lower renewable energy integration or more stable grid infrastructures.

2. **Focus on Short-Term Outcomes:** Much of the analysis in the case study revolves around the immediate benefits of the HPR, such as reduced FCAS costs and rapid response times. There is limited information on the long-term performance, degradation rates, and operational costs of the system over its lifecycle.
3. **Limited Exploration of Alternative Technologies:** The case study focuses primarily on lithium-ion batteries, particularly those with nickel-manganese-cobalt chemistry. It does not compare the performance of alternative storage technologies, which could offer different advantages and trade-offs.
4. **Economic and Financial Assumptions:** The financial analysis relies on specific market conditions, such as high FCAS costs in South Australia, which may not be replicable in other markets.
5. **Environmental Considerations:** While the case study highlights the benefits of renewable energy integration, it does not delve into the environmental impacts of lithium-ion battery production, including resource extraction, manufacturing, and end-of-life disposal.
6. **Scalability and Global Relevance:** The case study primarily focuses on a single large-scale installation. It does not explore the challenges of scaling similar projects globally, especially in regions with limited infrastructure, funding, or technical expertise.

By recognizing these limitations, readers can better contextualize the findings of the Hornsdale Power Reserve case study and identify areas that require further research and analysis. Addressing these gaps can be crucial for understanding the full potential and challenges of grid-scale battery energy storage systems in diverse energy landscapes.

## References

1 Matthews, M.D. and Brown, S.A. "We are all gonna die: how the weak points of the power grid leave the united states with an unacceptable risk," PhD Thesis, Naval Postgraduate School, Monterey, CA, 2023.

2 International Energy Agency. "Batteries and secure energy transitions. World Energy Outlook special report," 2024, https://iea.blob.core.windows.net/assets/cb39c1bf-d2b3-446d-8c35-aae6b1f3a4a0/BatteriesandSecureEnergyTransitions.pdf (accessed February 25, 2026).

3 UNFCCC. "The Paris Agreement," https://unfccc.int/process-and-meetings/the-paris-agreement (accessed February 25, 2026).
4 UNFCCC. "COP28: Governments pledge to triple global renewable energy capacity by 2030," 2024, https://unfccc.int/cop28 (accessed February 25, 2026).
5 International Energy Agency. "Net zero by 2050: a roadmap for the global energy sector," 2024, https://www.iea.org/reports/net-zero-by-2050 (accessed February 25, 2026).
6 International Energy Agency. "Energy storage outlook 2024," 2024, https://www.iea.org/reports/world-energy-outlook-2024 (accessed February 25, 2026).
7 BloombergNEF. "Battery pack prices fall to an average of $132/kWh, but rising commodity prices start to bite," 2021, https://about.bnef.com/blog/battery-pack-prices-fall-to-an-average-of-132-kwh-but-rising-commodity-prices-start-to-bite (accessed February 25, 2026).
8 Hackernoon. "What is a BESS (battery energy storage system) and how does it work?," 2021, https://hackernoon.com/what-is-a-bess-battery-energy-storage-system-and-how-does-it-work. (accessed February 25, 2026).
9 Mexis, I. and Todeschini, G. "Battery energy storage systems in the United Kingdom: a review of current state-of-the-art and future applications," *MDPI Energies*, 13(14), 3616, 2020.
10 Denholm, P. and Margolis, R. "Energy storage requirements for achieving 50% solar photovoltaic energy penetration in California," *National Renewable Energy Lab. (NREL)*, 2016.
11 Miller, N.W., Shao, M., D'aquila, R., Pajic, S. and Clark, K. "Frequency response of the US eastern interconnection under conditions of high wind and solar generation," *In 2015 Seventh Annual IEEE Green Technologies Conference IEEE*, 21–28, IEEE, 2015.
12 Divya, K.C. and Østergaard, J., "Battery energy storage technology for power systems—an overview," *Electric Power Systems Research*, 79(4), 511–520, 2009.
13 Castillo, A. and Gayme, D.F. "Grid-scale energy storage applications in renewable energy integration: a survey," *Energy Conversion and Management*, 87, 885–894, 2014.
14 Fitzgerald, G., Mandel, J., Morris, J. and Touati, H. "The Economics of Battery Energy Storage," Rocky Mountain Institute, 2015.
15 BloombergNEF, "Energy storage market outlook 2H 2021," 2021.
16 Cunha, Á., Martins, J., Rodrigues, N. and Brito, F.P. "Vanadium redox flow batteries: a technology review," *International Journal of Energy Research*, 39(7), 889–918, 2015.
17 Gil, A., Medrano, M., Martorell, I., Lázaro, A., Dolado, P., Zalba, B. and Cabeza, L.F. "State of the art on high temperature thermal energy storage for power generation. Part 1—Concepts, materials and modellization," *Renewable and Sustainable Energy Reviews*, 14(1), 31–55, 2010.

**18** IRENA, "Hydrogen: a renewable energy perspective," 2019. https://www.france-hydrogene.org/app/uploads/sites/4/2021/11/IRENA_Hydrogen_2019.pdf (accessed February 25, 2026).

**19** Chernyakhovskiy, I., Bowen, T., Gokhale-Welch, C. and Zinaman, O. "USAID energy storage decision guide for policymakers (No. NREL/TP-6A20-78815)," *National Renewable Energy Lab (NREL)*, Golden, CO, United States, 2021.

**20** International Hydropower Association, "Pumped Storage hydropower," 2021, https://www.hydropower.org (accessed February 25, 2026).

**21** Weber, A.Z., Mench, M.M., Meyers, J.P., Ross, P.N., Gostick, J.T. and Liu, Q. "Redox flow batteries: a review," *Journal of Applied Electrochemistry*, 41, 1137–1164, 2011.

**22** Soloveichik, G.L. "Flow batteries: current status and trends," *ACS Publications-Chemical Reviews*, 115(20), 11533–11558, 2015.

**23** Glenk, G. and Reichelstein, S. "Economics of converting renewable power to hydrogen," *Nature Energy*, 4(3), 216–222, 2019.

**24** IRENA- International Renewable Energy Agency, "Hydrogen: a renewable energy perspective,"2024. https://www.irena.org/-/media/Files/IRENA/Agency/Publication/2024/Sep/IRENA_Shaping_sustainable_hydrogen_value_chains_2024.pdf (accessed March 23, 2026).

**25** U.S. Energy Information Administration, "Annual energy outlook 2017," Washington, DC, 2017, https://www.eia.gov/pressroom/presentations/sieminski_01052017.pdf (accessed February 25, 2026).

**26** California ISO, "What the duck curve tells us about managing a green grid," *CAISO*, 2016, https://www.caiso.com/documents/flexibleresourceshelprenewables_fastfacts.pdf (accessed February 25, 2026).

**27** Wingren, L. and Johnsson, J. "Battery energy storage systems as an alternative to gas turbines for the fast active disturbance reserve," Master's Thesis, Lund University Libraries, 2018.

**28** Wehner, J., Mohler, D., Gibson, S., Clanin, J., Faris, D., Hooker, K. and Rowand, M. "Technology performance report: duke energy Notrees wind storage demonstration project (No. DE-OE0000195)," Duke Energy Renewable, Charlotte, NC, 2015.

**29** Bryans, D., Jiminez, M.R., Maxwell, J.M., Mitxelena, J.M., Kerr, D. and Berlouis, L.E. "Standalone batteries for power backup and energy storage," *Advances in Energy Storage: Latest Developments from R&D to the Market*, 31–60, 2022.

**30** Rantamäki, A., "Carbon footprint of an electric car in comparison to a fossil fuel car," Bachelor's Thesis, Lappeenranta–Lahti University of Technology LUT, 2022.

**31** Li, X., Hui, D. and Lai, X., "Battery energy storage station (BESS)-based smoothing control of photovoltaic (PV) and wind power generation fluctuations," *IEEE Transactions on Sustainable Energy*, 4(2), 464–473, 2013.

**32** Araiza, J., Hambrick, J., Moon, J., Starke, M. and Vartanian, C. "Grid energy-storage projects: engineers building and using knowledge in emerging projects," *IEEE Electrification Magazine*, 6(3), 14–19, 2018.

**33** SBS News, "SA blackout cost business $367 million," 9 September 2016, https://www.sbs.com.au/news/article/sa-blackout-cost-business-367-million/eqh6oyil8 (accessed February 25, 2026).

**34** AEMC, Australian Energy Market Commission, "Review of the system black event in South Australia," October 2019, https://www.aemc.gov.au/ (accessed March 15, 2025).

**35** Department of Treasury and Finance, Government of South Australia, "2017-18 budget statement," 2017, https://www.treasury.sa.gov.au/__data/assets/pdf_file/0007/516625/2017-18_budget_statement.pdf (accessed February 25, 2026).

**36** Rodriguez, A. "A controversy analysis of Tesla's (big) battery in Australia," Doctoral Dissertation, Queensland University of Technology, 2022.

**37** International Energy Agency, "Renewables 2020: analysis and forecast to 2025," 2020, https://www.iea.org/reports/renewables-2020 (accessed 25 February 2026).

**38** Denholm, P., Ela, E., Kirby, B. and Milligan, M. "The role of energy storage with renewable electricity generation," (No. Tec. Rep NREL/TP-6A2-47187). National Renewable Energy Laboratory, 2010.

**39** Bayram, I.S. and Devetsikiotis, M. "Analytical Models for Emerging Energy Storage Applications," in *Advanced Data Analytics for Power Systems*, ed. A. Tajer, S.M. Perlaza, H.V. Poor, 455–480. Cambridge University Press, 2021.

**40** Gong, X., Wang, X. and Cao, B. "On data-driven modeling and control in modern power grids stability: survey and perspective," *Applied Energy*, 350, 121740, 2023.

**41** International Renewable Energy Agency (IRENA), "Renewable capacity statistics 2021," Abu Dhabi, 2021, https://www.irena.org/%20publications/2021/March/Renewable-Capacity-Statistics-2021 (accessed February 25, 2026).

**42** Yang, Y., Bremner, S., Menictas, C. and Kay, M. "Battery energy storage system size determination in renewable energy systems: a review," *Renewable and Sustainable Energy Reviews*, 91, 109–125, https://doi.org/10.1016/j.rser.2018.03.047, 2018.

**43** Lazard, L., "Lazard's levelized cost of energy analysis," 2020.

**44** Xu, B., Oudalov, A., Poland, J., Ulbig, A. and Andersson, G. "BESS control strategies for participating in grid frequency regulation," *IFAC Proceedings Volumes*, 47(3), 4024, 2014.

**45** AEMO, "Initial operation of the Hornsdale Power Reserve," 2018, https://www.aemo.com.au/-/media/Files/Media_Centre/2018/Initial-operation-of-the-Hornsdale-Power-Reserve.pdf (accessed February 25, 2026).

**46** Saleem, A., Shaw, L.L., Iqbal, R., Hussain, A., Akbar, A.R., Jabar, B., Rauf, S. and Majeed, M.K. "Ni-rich cathode evolution: exploring electrochemical dynamics and strategic modifications to combat degradation," *Energy Storage Materials*, 69, 103440, 2024.

**47** Matthews, T., Hirve, M., Pan, Y., Dang, D., Rawar, E. and Daim, T.U. “Tesla Energy,” in *Innovation Management in the Intelligent World: Cases and Tools*, ed. T. U. Daim and D. Meissner, Springer, 233–249, 2020.

**48** Aurecon Group, “Hornsdale Power Reserve impact study,” 2018, https://www.aurecongroup.com/-/media/files/downloads-library/thought-leadership/aurecon-hornsdale-power-reserve-impact-study-2018.pdf (accessed February 25, 2026).

**49** Hornsdale Power Reserve, “Hornsdale Power Reserve,” https://hornsdalepowerreserve.com.au (accessed February 25, 2026).

**50** Hornsdale Power Reserve, “Consumer Benefits,” https://hornsdalepowerreserve.com.au/consumer-benefits (accessed February 25, 2026).

**51** “Hornsdale Wind Farm – World’s largest Tesla battery storage,” 2025, https://www.ghd.com/en/projects/hornsdale-wind-farm---worlds-largest-tesla-battery-storage (accessed February 25, 2026).

**52** CEFC – Clean Energy Finance Corporation, “SA big battery: a game changer,” *Clean Energy Finance Corporation*, 2025, https://www.cefc.com.au/case-studies/sa-big-battery-a-game-changer (accessed February 25, 2026).

**53** Dockrill, P., “Elon Musk’s South Australian battery responded in just 140 milliseconds after a coal-fired power plant failed,” December 2017, https://www.sciencealert.com/elon-musk-s-south-australian-battery-responded-in-just-140-milliseconds-after-a-coal-fired-power-plant-failed (accessed February 25, 2026).

**54** Clean Energy Council, “Clean energy Australia report 2020,” 2020, https://cleanenergycouncil.org.au/Cec/media/Background/Resources/clean-energy-australia-report-2020.pdf (accessed February 25, 2026).

**55** Widding, T. and Haugland, T.H. “Mulige løsninger for distribusjon og lagring av energi for Utsira Nord,” Bachelor’s Thesis, Høgskolen på Vestlandet, 2021.

**56** Ma, S., Jiang, M., Tao, P., Song, C., Wu, J., Wang, J., Deng, T. and Shang, W. “Temperature effect and thermal impact in lithium-ion batteries: a review,” *Progress in Natural Science: Materials International*, 28(6), 653–666, 2018.

**57** Hesse, H.C., Schimpe, M., Kucevic, D. and Jossen, A. “Lithium-ion battery storage for the grid—a review of stationary battery storage system design tailored for applications in modern power grids,” *Energies*, 10(12), 2107, https://doi.org/10.3390/en10122107, 2017.

**58** Smith, K., Saxon, A., Keyser, M., Lundstrom, B., Cao, Z. and Roc, A., “Life prediction model for grid-connected Li-ion battery energy storage system,” in *2017 American Control Conference (ACC)*, 4062–4068, 2017.

**59** IEEE Power and Energy Society, “Grid-forming inverters: a critical asset for a reliable and resilient power grid,” IEEE PES Technical Report, 2021.

**60** BloombergNEF, “Lithium-ion battery pack prices hit record low of $139/kWh,” December 2023, https://about.bnef.com/blog/lithium-ion-battery-pack-prices-hit-record-low-of-139-kwh (accessed February 25, 2026).

**61** Kim, Y.-J., Lee, H., Hwang, S., Kim, W., Kim, S., Kim, S.-Y., and Bae, S. "Research needs for realization of zero-carbon power grids with selected case studies," *MDPI Applied Sciences*, 12(5), 2533, https://doi.org/10.3390/app12052533, 2022.

**62** Lazard, "Lazard's levelized cost of storage analysis – version 7.0," 2021.

**63** NFPA, "NFPA 855: standard for the installation of stationary energy storage systems," 2020.

**64** Alam, M., Kumar, K. and Dutta, V. "Comparative efficiency analysis for silicon, silicon carbide MOSFETs and IGBT device for DC–DC boost converter," *Springer SN Applied Sciences*, 1, 1–11, 2019.

**65** Wolfspeed, "The role of SiC in modern inverter design," 2022, www.wolfspeed.com (accessed February 25, 2026).

**66** Tesla, "Tesla Energy virtual power plant in South Australia," 2021, https://www.tesla.com/en_au/virtual-power-plant (accessed February 25, 2026).

**67** Australian Energy Market Operator, "Black system South Australia 28 September 2016," March 2017, https://www.aemo.com.au/-/media/Files/Electricity/NEM/Market_Notices_and_Events/Power_System_Incident_Reports/2017/Integrated-Final-Report-SA-Black-System-28-September-2016.pdf (accessed February 25, 2026).

# Part 2

# Semiconductors, Circuits, and Connectivity

## Relevant Subjects

Electrical Engineering principles, low power circuit design

## Relevant Learning Objectives

Analyze power efficiency, explore semiconductor innovations

# 8

# TSMC's N3 Breakthrough: Beyond Moore's Law

300 Million Transistors per Square Millimeter and the Future of Nanoscale Innovation

## 8.1 Breaking Barriers at 3 nm

What does it take to pack billions of transistors onto a single chip the size of a fingernail? And how does a company tackle the headaches of power leakage and manufacturing defects while staying at the forefront of semiconductor technology? The answer is immense scientific rigor and precision engineering.

Taiwan Semiconductor Manufacturing Company (TSMC) faced these challenges head-on while crafting its 3 nm (N3) process, a breakthrough that redefined the limits. This new process integrates an impressive 300 million transistors per square millimeter, delivering up to 15% speed improvements and 30% power efficiency compared to its 5 nm predecessor. As industry approaches the end of Moore's Law, the ever-shrinking transistors have brought new obstacles. Power-hunger is an issue in always-on systems that rapidly deplete battery life, with some devices losing as much as 20% of their charge overnight due to leakage.

In the last two decades, semiconductor technology has shrunk at an incredible pace with each transistor being reduced in size to deliver more powerful and energy-efficient integrated circuits. This scaling has been the key to technological progress, but at smaller and smaller dimensions transistors do not scale forever. TSMC has developed advanced techniques to overcome this challenge, including the use of high-k metal gate technology, 3D transistor architecture, and new manufacturing processes to control defects [1].

Scaling transistors to smaller nodes results in many technical challenges [2]. These include but are not limited to:

- Increased electrical leakage current due to the thinner gate dielectrics and lower transistor threshold voltages [3].
- Difficulty in maintaining sufficient on-state drive current and transconductance.

- Increased device variability due to statistical fluctuations in the number of dopant atoms and worsening short-channel effects.
- Increased manufacturing complexity and cost to pattern and fabricate the smaller feature sizes.

TSMC has tackled these challenges with innovations in transistor design, materials, and manufacturing processes. The industry is reaching the limits of how small transistors can shrunk, with the 3 nm node being a major milestone. One of the biggest issues is the complexity of manufacturing chips that follows with wafer size scaling in order to preserve a good cost–performance trade-off. As feature sizes are getting smaller defect inspection strategies will have to be developed that can deal with the difficulty of small defect detection. Power consumption and heat dissipation are other big bottlenecks especially in portable or always-on systems, where leakage power can quickly drain the battery [4]. In this perspective, analog signal processing can be a power-efficient alternative to the dominant digital signal processing in some applications [5].

## 8.2 Creating the Leading Edge

TSMC, founded in 1987, has set new levels for semiconductor technology and modern electronics. The company is the world's largest producer of sophisticated computer chips and with more than 50% of advanced semiconductor sales globally along with about 90% of the most advanced 5 nm and 3 nm chips [6]. As TSMC tackles another finishing filler generation of process nodes, it's been increasingly difficult to both shrink the transistor size and power efficiency while also keeping performance [7]. The N3, which signifies the breakthrough of 3 nm technology, is a clear example of these problems.

The N3 process is designed to deliver up to a 70% logic density gain, 15% performance improvement, and 30% power reduction than 5 nm technology. However, these benefits come with an unrivaled level of technical difficulties. Power leakage, induced by the quantum tunneling effect at such low scales rendered the chips inefficient [8]. Furthermore, to deliver acceptable manufacturing yields, majority of manufactured chips need to meet all the quality requirements, especially when it must address defects from the insanely complex fabrication process. But before explaining further, it is crucial to clarify the basics of semiconduction technology.

### 8.2.1 What Is a 3 nm Process Wafer and How Does It Work?

Have you ever wondered how the tiny devices in your pocket, like smartphones, manage to juggle complex tasks while staying small? The secret lies in the wizardry of semiconductor technology, especially the 3 nm process wafer – basically the superhero of chip manufacturing.

In essence, a process node is the secret sauce for cooking up semiconductor chips. So, the "3 nm" in 3 nm process is not a random digit, it's the feature size on the chip, the gate length of transistor where the unit is nanometer (nm). For context, 1 nanometer is so tiny it makes a human hair look like a giant redwood tree – it's about 80,000 times smaller.

Smaller process nodes are directly proportional to the integration of the number of transistors within a single chip, while making a chip leaner, faster, and more efficient. Manufacturers can cram more transistors onto a single chip. This means faster processing and lower power consumption and smaller chip size but with more bells and whistles like an iPhone. 3 nm provides better performance and lower power consumption than its previous generation of 5 nm process node. And this is almost as useful as trading a bicycle for a fueled rocket.

The "wafer" is a thin, circular slice of silicon that serves as the foundation for manufacturing semiconductors. Advanced lithography creates ever more complicated patterns onto these wafers for a billion or so electronic switches called transistors (each transistor functions as a simple on and off electronic switch). 3 nm process allows transistors to shrink down and stop the local flow with great precision, controlling electrical path of those transistors up to few electrons above threshold, which is vital for improving leakage and saving.

A concrete analogy of this is like packing a luggage for a trip. Packing a 3 nm process wafer is like packing for a vacation when you are a chronic over-packer. Imagine that your suitcase is the "chip" and it is already small, but you need to pile in all the necessary items for travel (transistors). The 5 nm process is like trying to pack bulky sweaters and random toiletries – there's so much wasted space in between. But for 3 nm, it is like we are moving from messy clothes to rolled up clean laundry. There, all the right pieces come together and there is space for that extra pair of shoes you told yourself you wouldn't bring. The 3 nm chip is equivalent to making your suitcase a bag that flies light, neat, and organized with more in it.

Table 8.1 illustrates transistor density relative to node process size and its equivalent TSMC wafers and pricing. As the node process advances, transistor size decreases and its density grows.

### 8.2.2 Moore's Law and Its Influence on TSMC's N3 Technology

Since its inception, Moore's Law has been the GPS for semiconductor innovation, guiding engineers like a relentless backseat driver since 1965. First proposed by Gordon Moore, it boldly claims that the number of transistors in a chip doubles approximately every two years – basically the tech world's version of "go big or go home." This prediction has fueled a nonstop race for better performance and efficiency, pushing technology to shrink, flex, and dazzle. But now, as we venture below 5 nm, it is like getting into extreme territories. Transistors are so tiny they are practically playing hide-and-seek, and engineers are battling power leakage,

**Table 8.1** Evolution of TSMC's transistor density and equivalent process wafer.

| Pricing | Node Process | TSMC Wafers | Year introduced | Transistor Density | Increase in Density |
|---|---|---|---|---|---|
| \$20,000 USD | 7 nm FinFET | 7 nm FinFET Wafer | 2018 | 90–102 million transistors/mm$^2$ | 2× increase: 8 nm to 7 nm |
| \$16,000 USD | 5 nm FinFET | 5 nm FinFET Wafer | 2020 | 130–230 million transistors/mm$^2$ | 1.8× increase: 7 nm to 5 nm |
| \$10,000 USD | 3 nm FinFET | 3 nm FinFET Wafer | 2022 | 300 million transistors/mm$^2$ | 1.3 increase: 5 nm to 3 nm |

*Source:* Adapted from [9].

quantum effects, and manufacturing complexity like they are in a sci-fi thriller. Moore's Law becomes harder for engineers in terms of power leakage, quantum effects, and manufacturing complexity.

TSMC's 3 nm (N3) technology represents one of the most significant efforts to uphold Moore's Law despite these obstacles. With transistor densities reaching approximately 300 million transistors per square millimeter, the N3 process delivers a 15% speed improvement at the same power or up to 30% power reduction at the same speed compared to its 5 nm predecessor. These improvements are enabled by several key innovations, including FinFlex™ technology, Gate-All-Around (GAA) transistor structures, and advancements in extreme ultraviolet (EUV) lithography.

The impact of Moore's Law on TSMC's N3 technology can be understood by examining its influence on transistor structure. Traditional FinFET transistors, which have powered chips for over a decade, struggle with leakage currents as gate lengths shrink. To address this, TSMC introduced FinFlex™, which allows chip designers to customize fin configurations for specific performance or efficiency needs. Additionally, GAA transistors, which fully surround the channel with gate material, provide superior control over current flow, reducing power leakage and improving efficiency. This transition represents a major shift in transistor architecture, marking one of the most significant advancements since the introduction of FinFETs.

Producing a 3 nm wafer is like trying to build a sandcastle with grains of sugar while wearing oven mitts. Part of this involves tackling quantum tunneling, where mischievous electrons decide to "leak" through thin layers like uninvited party crashers. Moreover, TSMC's aggressive adoption of EUV lithography plays a crucial role in maintaining transistor scaling. Using extreme ultraviolet lithograph, the EUV lithography enables finer patterning at the nanometer scale of ultra-short-wavelength light (13.5 nm). Basically, the closest thing to needle-point in pitch darkness and working at scales thousands of times smaller than a red blood cell. With this, engineers are able to build transistors with an accuracy few could have imagined decades ago. However, the implementation of EUV at such an advanced node introduces new challenges, including the need for highly reflective mirrors, defect control, and throughput optimization.

The iPhone 15 Pro serves as a prime example of the capabilities of TSMC's 3 nm process. The A17 Pro chip, powered by the N3 process, delivers significant improvements in both processing speed and energy efficiency. Thanks to this compact genius, a phone can effortlessly juggle demanding tasks without compromising battery life or portability.

Ultimately, TSMC's ability to uphold Moore's Law through its N3 process underscores the ongoing ingenuity of semiconductor engineers. Scaling could slow down because of physical and quantum limitations, meanwhile transistor architecture innovations but also material science and lithography push progress forward. Companies like TSMC are pushing new limits as we approach the angstrom era and will need to keep Moore's Law alive for the further evolution of computing.

While the relentless pursuit of miniaturization has brought us incredible computing power, the end of linear scaling is on the horizon, and forcing the semiconductor industry to chart a new course. Indeed, the gradual end of Moore's Law is ushering in a new era of information technology, where the focus shifts from mere shrinkage to coordinating the introduction of novel devices, integration techniques, and computing architectures [10].

### 8.2.3 Process Roadmap of Top Semiconductor Foundries, 2014–2026

In the face of the ever-looming power and yield challenges, the semiconductor industry must continue to innovate at a breakneck pace to deliver the capabilities that our digital world demands. To maintain its position as the industry leader and keep up with major players like Samsung and Intel, TSMC has had to tackle technological challenges that might make even rocket scientists break a sweat. The company has consistently pushed the boundaries of innovation – because apparently, "impossible" is just a suggestion.

Figure 8.1 shows the process roadmap of top foundries in the world. The rapid evolution of transistor technologies, driven by Moore's Law and increasing demand for high-performance, and energy-efficient chips.

The semiconductor race has been a high-stakes game, with TSMC, Samsung, and Intel battling for process leadership. In 2014, the industry was very enthusiastic about 16 nm and 14 nm FinFETs. It was the future, like exchanging flip phones for smartphones. Fast forward to 2018, TSMC's 7 nm chips were racing ahead

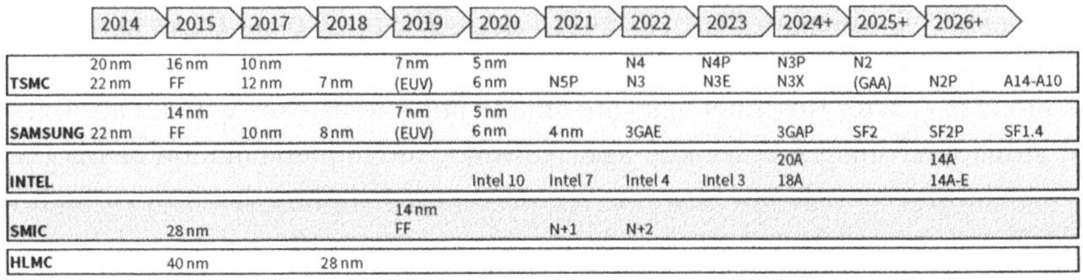


**Figure 8.1** Process roadmap of top foundries [11]. *Source:* Author's own work.

and Intel's 10 nm were at the starting line. By 2020, it was supposed to be 5 nm, with Apple's A14 plus variants essentially being the norm. Enter the 3 nm era, defined by extreme ultraviolet lithography and innovations like TSMC's FinFlex™, which allows designers to optimize the balance between performance and power efficiency. Looking ahead, the real match is in the 2 nm territory, where GAA transistors reshape the most fundamental elements. It is an unrelenting sprint, and so in this race a single nanometer can mean billions in revenue or billions of headaches.

### 8.2.4 Strategic Intellectual Property and Patent Leadership

Production capacity and process road maps are tangible measures of success, but a robust intellectual property (IP) protection strategy is just as critical. This was the factor that enabled TSMC to dominate the IP landscape. In today's highly competitive semiconductor environment, a "moat" of patents is needed to justify the multi-billion-dollar R&D investments for nodes like N3 and N2.

TSMC has long been recognized as one of the top patent holders in the United States, exemplifying its evolution from being primarily a manufacturer, to becoming a key enabler of global innovation. According to IFI CLAIMS Patent Services, TSMC's rise has been meteoric: the company was not even ranked in the Top 50 in 2011. By 2012, it broke into the rankings at #48 and has since surged among tech giants to secure the #2 spot globally in 2024.

Beyond the annual US rankings, the sheer scale of TSMC's total portfolio reflects an accelerating pace of innovation. In 2021, the number of patents held by TSMC increased to over 60,000. Their total patent count surged to more than 78,000 patents in 2022, showing a strong uptrend in innovation activity. TSMC's granted patents further increased to over 109,000 patents by 2023, up by around 39.5% in two years [12].

Table 8.2 highlights this decade-long trajectory, highlighting the correlation between their patent grants and their rise to technology leadership.

However, the significance of this expanding portfolio extends far beyond statistical milestones. This growth in patents is not only bureaucratic but also strategic, as they "wall off" their proprietary technologies. TSMC also makes it difficult for competitors to match its yields and energy-efficiency figures by holding thousands

**Table 8.2** TSMC US patent ranking and grants (2001–2024).

| Year | Rank (US Patent) | Status | Strategic Focus |
|---|---|---|---|
| 2024 | # 2 | Historic High | Surpassed major US tech firms; heavy focus on 2 nm and GAA IP. |
| 2023 | # 3 | Rapid Growth | ~3,600+ grants; cementing leadership in 3 nm FinFlex™ technology. |
| 2022 | # 3 | Steady | Launch of N3 (3 nm) and Nanosheet (GAA) research. |
| 2021 | # 4 | Steady | Key patents are filed for extreme ultraviolet (EUV) integration. |
| 2020 | # 6 | Rank Surge (+6 ranks) | Major jump in rankings; focus on 5 nm maturity. |
| 2019 | # 12 | 2,331 Grants | Key patents are filed for extreme ultraviolet (EUV) integration. |
| 2018 | # 6 | 2,465 Grants | Early consolidation of EUV leadership. |
| 2017 | # 9 | 2,425 Grants | Solidified Top 10 status. |
| 2016 | # 9 | 2,288 Grants | First entry into Top 10. |
| 2015 | # 13 | 1,774 Grants | Rapid accumulation of FinFET patents. |
| 2014 | # 23 | 1,460 Grants | Scaling operations and IP simultaneously. |
| 2013 | # 35 | 941 Grants | Approaching 1,000 annual grants milestone. |
| 2012 | # 48 | Market Entry | First entry into Top 50 (650 grants). |
| 2011 | – | Unranked | Not listed in Top 50. |

*Source:* Adapted from IFI CLAIMS Patent Services data [13].

of patents, particularly in FinFET structures, advanced packaging (CoWoS), and lithography techniques. TSMC ensures that competitors cannot easily replicate its yield rates or power-efficiency metrics without infringing on protected IP. This is the legal fortress to complement their physical foundries. TSMC is a dual-threat leader when it comes to producing hardware and intellectual property.

## 8.3 Context for Analysis – Mitigating Power Leakage in Transistors

TSMC faced monumental technical hurdles in developing its 3 nm semiconductor process, mostly due to power leakage and manufacturing yield issues. Basically, the tech equivalent of trying to build a sandcastle during high tide. These problems meant the performance, dependability, and cost-effectiveness of 3 nm chips were, well, still unpacking their suitcases at home.

When transistors shrink to the nanoscale, quantum tunneling effects become impossible to ignore. Electrons start sneaking through insulating barriers, now thinner than a sheet of tissue paper, like rebellious teenagers past curfew. Beyond quantum tunneling, achieving high yield rates becomes increasingly difficult. As manufacturing tolerances grow ever tighter at the 3 nm node, even microscopic defects can render chips non-functional. This means fewer working chips per wafer, driving up costs. Both issues threaten TSMC's ability to mass-produce cost-competitive, cutting-edge chips at scale.

The TSMC case study dives into highly complex semiconductor manufacturing processes, where trade secrets are guarded tighter than grandma's cookie recipe, and innovations are as cutting-edge as a freshly sharpened sword. The real-world engineering challenges behind TSMC's 3 nm process are so complex that even Einstein might've needed a coffee break to figure them out. These are tackled with advanced calculations, proprietary data, and foundry processes with futuristic vision. It is simply too complicated for general academic analysis. By exploring how FinFET technology tackles power leakage at the 3 nm scale, students can get a sneak peek into the genius of semiconductor design – without needing to access classified industry data or wear a lab coat.

In order to mitigate power leakage in transistors, students can analyze how FinFET technology addresses power leakage at the 3 nm scale. We can start by examining the relationship between transistor geometry and leakage current.

### 8.3.1 Transistor Geometry and Leakage Current

Transistors are the most basic electrical components in semiconductor devices. Fin Field-Effect Transistors (FinFETs) are state-of-the-art transistors in contemporary semiconductor design, as they provide a significant improvement on the control over current leakage with respect to planar transistors. FinFETs do this via a 3D design in which the conducting channel is literally under a "fin" and therefore have better gate control current. As illustrated in Figure 8.2, the most important geometric parameters for FinFET are fin height, fin width and gate length. These dimensions directly influence transistor performance, power consumption, and leakage current.

In the 3 nm process at TSMC, FinFlex™ also enables a single transistor design to have multiple fin configuration. Then, the company announced a 3-2 FinFlex™ that would be three fins for performance throughput and two for the long run of energy efficiency. These configurations can have an impact on leakage current and manufacturing complexity with the following relationships:

- Leakage current increases exponentially with reduced gate length and poorly controlled fin geometry. Often doubling for every 10 nm reduction in gate length.
- Improved control (taller fins or optimized width) mitigates leakage by up to 30% but can complicate manufacturing or reduce operational speed by approximately 15%.

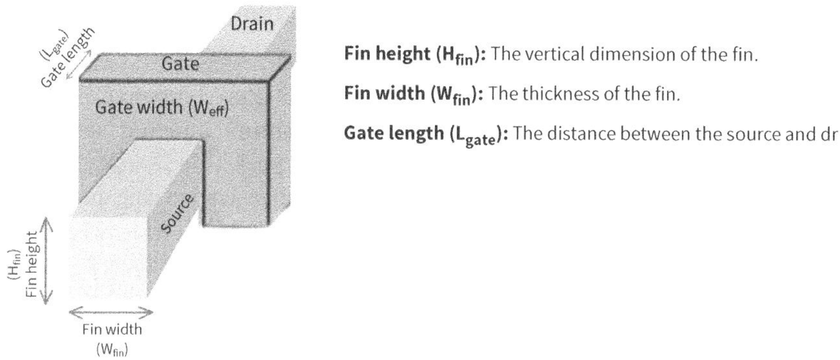


**Figure 8.2** Key dimensions of a FinFET. *Source:* Adapted from [14].

Leakage current, or the unintended flow of electrons when the transistor is in the "off" state, is a significant issue at nanoscale dimensions. The relationship between transistor geometry and leakage current can be understood through the following principles:

1. **Fin Width ($W_{fin}$):**
   - A narrower fin width enhances electrostatic gate control over the channel. It decreases the potential of leakage current due to the inhibition of integrated interfering electron flow.
   - Whilst very thin fins can increase resistance and drive current (performance). It means there is a trade-off in leakage current reduction with a cost in performance.
2. **Fin Height ($H_{fin}$):**
   - A taller fin increases the effective channel width, enabling more electrons to flow in the "on" state. It makes the drive current of transistor better.
   - Taller fin also means improved gate control over more surface area that lowers the leakage. But scaling down to nanoscales pose challenges of mechanical instability, especially with taller fins.
3. **Gate Length ($L_{gate}$):**
   - Shorter gate lengths result in faster speeds, but come with shorten-channel effects (SCE). Thus, leakage increases and gates no longer have control on the channel for higher end transistors.
   - To counteract SCEs for gate control, the fin height and width need to be tuned.
4. **Gate Thickness and Material:**
   - The thickness and dielectric material of the gate can cause leakage. High-k dielectrics are often used to reduce tunneling leakage without requiring excessively thick gates.

### 8.3.2 Applying Moore's Law to Power Leakage in Transistors

Computer chips keep getting tinier and moreefficient, thanks to Moore's Law, which claims that the number of transistors – and thus computing power – doubles about every two years. But as these chips shrink, quantum tunneling causes power leaks and turns your sleek device into a pocket-sized sauna.

Imagine a semiconductor company is designing a next-generation chip using 3 nm FinFET technology to reduce power leakage. Consider, 5 nm chip with 125 million transistors/$mm^2$ and average power leakage of 0.5 watts/$mm^2$. If they do follow Moore's Law, doubling transistor density into 3 nm, the new transistor density would be close to around ~250 million transistors/$mm^2$; and how about leakage current or power leak?

Here, FinFET technology is the backbone again to lower the power leakage at this scale of 3 nm. FinFET structures are a 3-dimensional high side-transistor design with a thin silicon fin. It grows vertically over the substrate and, resulting in superior control at source-drain channel. Thus, the leakage current is much lower than the process with traditional planar transistors. For example, the power leakage of 5 nm chips as 0.5 watts/$mm^2$, that got potentially improved to about only between 0.05–0.1 watts/$mm^2$ with the state of 3 nm process technology. It makes N3 an energy-efficient process. It reduces the leakage and leads to an increase in the density of transistors without having to higher levels, excessive power consumption or heat generation.

## 8.4 Think and Solve

**Q 8.1** If the company follows Moore's Law and doubles the transistor density at 3 nm, estimate the new transistor density and discuss how FinFET technology helps mitigate power leakage at this scale.

**Q 8.2** As transistors shrink to smaller nodes such as 3 nm, what is a key challenge engineers face?

**A** The cost of transistors decreases significantly with each node reduction.
**B** Smaller transistors always lead to higher performance with no drawbacks.
**C** Managing short-channel effects and leakage current becomes increasingly difficult.
**D** Larger transistors are needed to compensate for power leakage at smaller nodes.

**Q 8.3** Why are FinFET transistors preferred over traditional MOSFETs for power-efficient semiconductor design?

**A** They allow higher current flow, increasing overall power consumption.
**B** They provide better electrostatic control, reducing leakage current and power loss.
**C** They eliminate the need for low-power modes in IoT devices.
**D** They increase transistor size, making chips more power-efficient.

## 8.5 Conclusion and Key Takeaways

The relentless march of Moore's Law is taking the semiconductor industry toward smaller transistor dimensions and breaking all physical and construction rules. At TSMC's 3 nm process, miniaturization is an incredible opportunity, and huge challenges lie in addressing power leakage solutions, as well as pushing efficiency higher. Innovations like FinFET and the latest generation Gate-All-Around transistors have been critical for solving these problems, in that higher density transistors do not necessarily mean power consumption spikes.

The discussion on transistor geometry and leakage current demonstrates how engineering ingenuity allows us to keep improving performance despite physical limitations. Although the intricacy of semiconductor fabrication is one of those secrets best not known, studying these problems from a basic standpoint is a key to invoking this intuition between power, performance, and efficiency for students as well as engineers.

1. **Moore's Law and Scaling Challenges:** As transistors shrink to sizes so tiny they would make ants feel like giants, quantum effects like power leakage start sneaking in, as if the universe itself is trolling engineers. Since its inception, Moore's Law has been the semiconductor industry's trusty fortune cookie, predicting the doubling of transistors on a chip every two years. But now, as we venture into the 3 nm era and beyond, we are hitting physical barriers that feel like nature's way of saying, "Nice try, humans." Shrinking transistors to these ridiculously small scales invites quantum mischief – power leakage, increased resistance, and possibly a few electrons laughing at us. Yet, despite these microscopic headaches, companies like TSMC, Intel, and Samsung are charging ahead, proving that while Moore's Law might be limping a bit, innovation is still alive.
2. **FinFET and GAA Transistors:** Advanced transistor designs are like the overachievers of the tech world – they improve current control, reduce

leakage, and still manage to stay efficient at nanoscale nodes. Traditional planar transistors? Well, they had a bit of a leakage problem as they shrank, like trying to keep water in a leaky bucket. Enter FinFET (Fin Field-Effect Transistor) technology, the superhero of the transistor saga, swooping in to tighten electrical control, reduce leakage, and boost efficiency. But wait, as we dive below 3 nm, the GAA transistors take the stage, hugging the channel on all sides like a clingy friend – except this friend makes everything faster and more efficient. These constant design upgrades show that transistors are the tech world's version of "never settle," keeping chips speedy and energy-smart while leaving leakage behind like a bad habit.

3. **The Role of EUV Lithography:** Precision at the atomic level is critical for maintaining manufacturing accuracy and yield at 3 nm and beyond. Manufacturing chips at the 3 nm node demands extreme precision. Enter Extreme Ultraviolet (EUV) Lithography, a technique that uses light with a wavelength of just 13.5 nm — short enough to carve out structures smaller than a virus. This cutting-edge process allows manufacturers to print incredibly small and intricate transistor patterns with high accuracy. However, EUV comes with its own challenges, such as high costs and the need for near-perfect defect detection techniques. Still, it remains a cornerstone of next-generation semiconductor fabrication.
4. **Industry-Driven Innovation:** Companies like TSMC just love to push the limits of semiconductor technology, they regularly set new records on performance and power efficiency. The Race of Semiconductors is like the war between the foremost semiconductor foundries – TSMC vs. Samsung vs. Intel has turned into an arms race, and battling each other on chip technology. TSMC is the early front runner, getting incredibly high yields for its 3 nm process and Samsung/Intel are both trying. For these companies to keep ahead, they need to constantly be perfecting materials alternatives then tweaking the manufacturing process and transistor designs. There are plenty of lessons to be learned in the world of semiconductors about the nature of progress – it does not stop; it merely escalates in difficulty.
5. **The Future of Semiconductor Technology – Beyond 3 nm:** As transistors approach atomic-scale dimensions, the industry is exploring new frontiers, such as 2D materials like graphene, new chip architectures, and 3D stacking techniques. These breakthroughs will be essential for overcoming the limitations of traditional silicon scaling. We also need to consider that alternative computing paradigms such as quantum computing and neuromorphic chips could soon become the norm and completely reshape information processing. The 3 nm process is a state-of-the-art in semiconductor manufacturing, but developing even smaller, more energy-efficient and powerful chips is a long way yet to go.

This case study demonstrates that the semiconductor breakthrough is a dance of physics, advanced engineering, and state-of-the-art manufacturing. Recognizing these challenges gives future engineers the knowledge to participate in the next wave of computing technology innovations. The 3 nm process is much more than just a technical achievement – it is an unmatched testament to willful energy. Humans must keep pushing the impossible. As engineers and researchers continue to defy the limits of physics, the future of semiconductor technology remains as exciting as ever.

## 8.6 Discussion Points

The 3 nm process isn't just a technological milestone, it's a testament to human ingenuity, an engineering marvel akin to threading a needle while riding a roller-coaster at full speed. The relentless pursuit of smaller, faster, and more efficient chips is what keeps the semiconductor industry moving forward, despite the ever-mounting physical and economic challenges.

It is important to remember an analogy for simple understanding. Think of semiconductor scaling like packing for a vacation: at 10 nm, it is a decent-sized suitcase. At 5 nm, you downsize your suitcase to a smaller hand baggage but stuff even more into it. Now at 3 nm, you decide to use even more tiny hand baggage, but double your stuffing into that tiny bag, and hoping the zipper doesn't burst. And yet, engineers continue to make it work, fitting more transistors into a chip than ever before while keeping power consumption and performance under better control.

While skeptics have long declared the impending "end of Moore's Law," the industry has repeatedly found ways to prove them wrong. From FinFETs to GAA transistors, from high-k metal gates to EUV lithography, innovation has kept the law alive. The future will likely bring even more radical solutions, from 3D-stacked architectures to quantum and neuromorphic computing, ensuring that technological progress continues at an exhilarating pace.

But let's not forget the bigger picture: these advancements aren't just for the sake of pushing boundaries. They enable the devices that power our daily lives – from the smartphone in your pocket to the data centers running AI models that can predict the future. Without these innovations, we'd still be dealing with sluggish computers, power-hungry devices, and a digital world that moves at a snail's pace.

So, the next time you effortlessly stream high-resolution videos, marvel at the speed of your smartphone, or see AI generate a near-perfect cat meme in seconds, remember the hidden world of semiconductor engineering that makes it all possible. And if history has taught us anything, it's that the engineers who brought us 3 nm will soon be losing sleep over 2 nm, then 1.4 nm, and eventually, chips so small that measuring them requires a PhD in quantum mechanics.

In the race to push technology further, one thing remains certain: "impossible" is just another challenge waiting to be engineered away. Here are few more points to discuss.

1. **The Limits of Scaling:** Given the increasing difficulty of transistor miniaturization, what alternative approaches beyond Moore's Law might drive future semiconductor advancements?
2. **Power Efficiency Trade-offs:** With power leakage becoming a bigger concern at smaller nodes, what are some emerging solutions beyond FinFET and GAA transistors?
3. **Economic and Manufacturing Challenges:** As nodes shrink, the cost of semiconductor fabrication increases exponentially. What are the economic implications for companies and consumers?
4. **Real-World Applications:** Consider the impact of 3 nm chips on emerging technologies such as AI, 5G, and edge computing. How do advancements at the semiconductor level shape innovation in these fields?
5. **Quantum Tunneling and Engineering Solutions:** How does quantum tunneling affect transistor performance at the nanoscale, and what materials or architectural changes might help mitigate its effects?
6. **Ethical and Environmental Considerations:** With the rapid pace of chip development, what are the environmental and ethical concerns regarding e-waste and energy consumption in semiconductor manufacturing?
7. **The Role of AI in Chip Design:** AI is increasingly used to optimize semiconductor layouts and designs. What are the benefits and potential risks of relying on AI-driven design processes?

These discussion points encourage a deeper exploration of the technological, economic, and ethical aspects of semiconductor innovation while reinforcing the critical concepts covered in this chapter.

## 8.7 Limitations of the Case Study

Despite providing valuable insights into cutting-edge semiconductor manufacturing, this case study has inherent limitations:

1. **Proprietary and Confidential Data:** The semiconductor industry, particularly at the 3 nm scale, is heavily guarded by trade secrets. While this case study presents publicly available information and engineering principles, many crucial technical details, such as exact process optimizations, defect reduction techniques, and proprietary FinFlex™ technology specifics, remain undisclosed.

2. **Complexity Beyond Undergraduate Scope:** The intricacies of 3 nm transistor behavior, including quantum effects, material interactions, and advanced lithography challenges, often require a deep understanding of quantum mechanics, solid-state physics, and nanoscale engineering. These topics go beyond standard undergraduate coursework, making it difficult to present problems that are both realistic and solvable at this level.
3. **Rapid Technological Advancements:** Semiconductor technology progresses at an extraordinary pace, meaning that findings from this case study may become outdated within a few years. The industry is already researching 2 nm and sub-2 nm nodes, which could introduce entirely new paradigms in transistor design, power efficiency, and process integration.
4. **Manufacturing and Yield Variability:** While TSMC has optimized its 3 nm process, real-world fabrication experiences significant challenges. Yield rates, defect densities, and power-efficiency improvements can differ between production batches, influenced by factors like wafer material quality, environmental conditions, and equipment calibration. These nuances are difficult to model without access to real fabrication data.
5. **Focus on Power Leakage:** This case study emphasizes transistor geometry and leakage current mitigation, but the 3 nm process involves numerous other challenges, including heat dissipation, supply chain constraints, cost per wafer, and integration with chipset architectures. A holistic study would require a broader scope beyond transistor design alone.

By acknowledging these limitations, I ensure a realistic approach to studying advanced semiconductor manufacturing while keeping the analysis accessible for students and researchers. Despite these constraints, the core engineering principles discussed – transistor scaling, power leakage mitigation, and the impact of Moore's Law – remain highly relevant to the future of semiconductor technology.

## References


1 Dabbas, R.M., Huling, B. and Chen, H.N. "Factory shutdown: a managerial approach," *International Journal of Services Technology and Management*, 4(4–6), 408–423, 2003.

2 Kang, S.M. "Elements of low power design for integrated systems," in *Proceedings of the 2003 International Symposium on Low Power Electronics and Design*, 205–210. Association for Computing Machinery, 2003.

3 Buengener, R. "Defect inspection strategies for 14 nm semiconductor technology," in *Instrumentation, Metrology, and Standards for Nanomanufacturing, Optics, and Semiconductors VI*, vol. 8466, 45–51. SPIE, 2012.

**4** Pramanik, P.K.D., Sinhababu, N., Mukherjee, B., Padmanaban, S., Maity, A., Upadhyaya, B.K., Holm-Nielsen, J.B. and Choudhury, P. "Power consumption analysis, measurement, management, and issues: a state-of-the-art review of smartphone battery and energy usage," *IEEE Access*, 7, 182113–182172, 2019.

**5** Gregorio, F., González, G., Schmidt, C. and Cousseau, J. *Signal Processing Techniques for Power Efficient Wireless Communication Systems: Practical Approaches for RF Impairments Reduction*, Springer, 2020.

**6** Purohit, P.A. "Chip wars: the struggle for semiconductors supremacy," *Air Power Journal*, 19(2), 21–42, 2024.

**7** Bosch, D. "Simulation, fabrication and electrical characterization of advanced silicon MOS transistors for 3D-monolithic integration," Doctoral Dissertation, Université Grenoble Alpes, 2020.

**8** Hepp, S., Jetter, M., Portalupi, S.L. and Michler, P. "Semiconductor quantum dots for integrated quantum photonics," *Advanced Quantum Technologies*, 2(9), 1900020, 2019.

**9** "TSMC website," www.tsmc.com (accessed February 26, 2026).

**10** Theis, T.N. and Wong, H.S.P. "The end of Moore's law: a new beginning for information technology," *Computing in Science and Engineering*, 19(2), 41–50, 2017.

**11** "TrendForce Research Institute," www.trendforce.com (accessed February 26, 2026).

**12** Taiwan Semiconductor Manufacturing Company, "2023 sustainability report," TSMC ESG, June 2024, https://esg.tsmc.com/file/public/2024-TSMC-Sustainability-Report-e.pdf (accessed February 26, 2026).

**13** IFI, "IFI Claims Patent Services," 2024, https://www.ificlaims.com/rankings/top-50-2024 (accessed February 26, 2026).

**14** Swinnen, M. and Duncan, R. "Physical verification of FinFET and FD-SOI Devices," 2013, www.techdesignforums.com/practice/technique/physical-verification-design-finfet-fd-soi (accessed February 26, 2026).

# 9

# Nordic Semiconductor's Ultra-Low-Power Revolution: IoT Energy Trilemma

BLE Innovation, Dynamic Power Management, and the Quest for Ultra-Low Power Wireless

## 9.1 The Challenge of IoT Power Consumption

Back in 2010, as video calls were becoming a staple but still felt novel and clunky, many of us grew accustomed to the frustrations of technology still finding its footing [1]. Some phones got overheat during video calls for more than a few minutes, batteries depleted fast, range loss or audio drop were often experienced by users. Users tried to keep asking themselves, how come wireless communication consumes so much power? Perhaps somewhere in the world, engineers and researchers likely pondered about this question too and were seeking ways to optimize efficiency without sacrificing performance. Innovative advanced algorithms to redefine energy delivery and transmitting the most essential information with no redundant signals were just starting to emerge as part of the smarter energy usage [2]. The initial strides are not only focused on escape from overheating devices but also toward seamless connectivity, energy efficiency, and better streamline tools for the development [3].

Can a small chip solve massive battery life problems? Imagine your smartwatch running out of battery just shortly after full charge. Or think of a medical sensor failing in a critical moment because of power constraints [4]. Wireless connectivity is extremely important in the tech world, and power consumption is its ever-present partner, whether that device lives for years or operates just for a few days or months.

This case study offers a deep dive into the intricate trade-offs of power-efficient wireless communication. By applying concepts from power electronics, signal processing, and low-power circuit design, readers can explore:

- How advanced power management techniques, such as dynamic voltage scaling and duty cycling, maximize battery efficiency without compromising performance. Dynamic Voltage Scaling and Duty Cycling make use of the fact that some devices can actually lower their power consumption based on workload, thereby maximizing battery life. In wearable health monitors

(Oura Ring, Fitbit etc.) dynamic power scaling means the sensor is live at full power only when needed, for example heartrate measurements and is sleeping from low-power modes in-between.
- How FinFET transistor technology, with its ability to minimize leakage currents, meets the growing demand for ultra-low-power wireless chips. Traditional planar MOSFETs suffer from high leakage currents, leading to unnecessary energy loss, especially in always-on IoT devices. On the other hand, FinFET transistors with almost no leakage means they are well suited for energy sensitive applications. For example, smart home devices (like Philips Hue, Amazon Echo etc.) run on Nordic Semiconductor's ultra-low-power chipsets that integrate FinFETs to ensure a responsive device even in standby mode.

By tackling these challenges, readers can gain valuable insights into how cutting-edge engineering innovations drive the next generation of IoT connectivity and push efficiency to unprecedented levels.

## 9.2 Powering IoT with Ultra-Low Energy

One chilly morning in the early 2000s, in a modest lab in Trondheim, Norway there were some engineers at Nordic Semiconductor looking at the latest wireless prototypes the company had just developed – heavy, and battery hungry chips aimed at niche industrial customers. They faced a challenge that could make even the most stoic engineer break into a nervous laugh: how to stretch battery life without turning their high-performance connectivity into a glorified game of charades?

Frustrated by the limitations of existing technologies, they wondered: What if wireless connectivity could be reimagined not just for factories, but for everyday life such as sensors, wearables, and smart homes, without burning up batteries? At that time, Bluetooth and similar standards were too energy intensive. They were also still years away from supporting the always-on and constantly listening devices that the future demanded.

Nordic Semiconductor's journey began in 1983 as a small Norwegian company focused on developing integrated circuits. Operating initially under the name Nordic VLSI, the company focused on Application Specific Integrated Circuits (ASICs) for various fields. In the 1990s, it strategically expanded into wireless communication positioning itself for the rise of short-range radio. One of the company's defining moments came in the early 2000s when it turned its attention to ultra-low-power wireless solutions. The company was probably best known for focusing on ultra-low-power wireless solutions back in the early 2000s. Having realized that the base class of wireless tech (Bluetooth) was too power hungry for small battery-operated things, Nordic Semiconductor created new ways to save power: Bluetooth Low Energy (BLE), a technology that changed the rules for wearables and health tech, but also smart home devices.

To understand Nordic Semiconductor's dominance in this market, it's important to understand the origin of the technology itself. Bluetooth Low Energy did not emerge from thin air; it's beginning can be traced back to a technology known as "Wibree," developed by mobile maker Nokia back in 2001 [5]. Nokia had a vision for a low-power radio technology that would enable coin-cell-powered accessories to connect to mobile phones. But they needed a partner to test the physics behind this vision, to see if the idea could work on a silicon chip.

While Nokia had the protocol stack solution, it was Nordic Semiconductor who had the essential hardware capability to implement the Wibree technology in an ultra-low power radio [6]. This partnership between them laid the technical foundation for modern BLE. When Wibree was later incorporated into the Bluetooth SIG as Bluetooth 4.0 in 2010, Nordic Semiconductor was not merely an adopter of the standard. Instead, they helped to shape it, designing the underlying silicon that made the technology commercially viable.

Under the leadership of key figures like Svein-Egil Nielsen, the company's CTO, Nordic developed groundbreaking solutions in short-range radio technology. Nordic was instrumental in the 2010 launch of Bluetooth 4.0 introducing BLE to the world. The nRF series of wireless chips from the company became the industry low-power connectivity standard, driving everything from health sensors to fitness trackers.

Figure 9.1 shows the evolution of Bluetooth technology and Nordic Semiconductor's BLE product series from 1999 to 2024. Highlighting the rapid advancements in

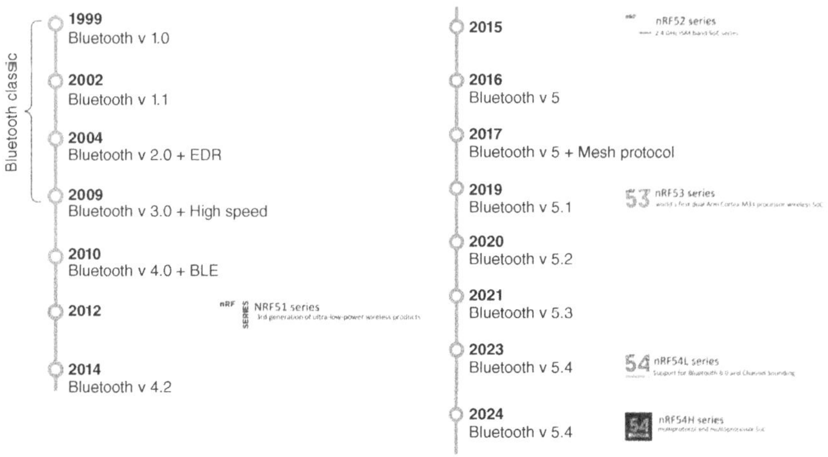


**Figure 9.1** Bluetooth technology timeline and Nordic Semiconductor's BLE product series (1999–2024). *Source:* Adapted from [7].

low-power wireless communication. Classic Bluetooth dates back to the 1990s when pagers were still widely used, and mobile technology was just beginning to take off. As small devices grew more sophisticated, power limitations turned into an important problem. To solve this, engineers desperately needed a communications standard that focused on energy efficiency. The introduction of BLE v4.0 in 2010 was the answer, enabling the development of battery-powered wearables, medical devices, and smart home technology.

One key actor in this shift was Nordic Semiconductor, who introduced its first BLE System-on-Chip (SoC), the nRF51 series, in 2012. The architecture transformed the landscape of the IoT by allowing small size battery-powered devices that could run for months or years without recharging. With each subsequent release, including the industry-standard nRF52 series and the high-performance nRF54 series, Nordic has brought more processing power, greater security, and multi-protocol support. Their roadmap demonstrates a consistent ability to anticipate market needs and driving the adoption of ultra-low-power wireless technology.

With the maturing of the Internet of Things (IoT) market, Nordic Semiconductor adapted. BLE was not all it had, but it also opened to cellular IoT, and LTE-band NB-IoT for reliable low-power wide-area connectivity over greater distances. It was this move that enabled Nordic to go from a niche player to the global wireless semiconductor market leader, challenging the titans like Texas Instruments and Qualcomm.

Today, Nordic Semiconductor continues to push the boundaries of low-power wireless innovation, having research centers around the world and being backed by major tech corporations. From smart lighting to asset tracking, the technology is embedded in millions of devices and defining what the future might hold for connected electronics.

As the world clamored for compact and energy-efficient IoT applications, Nordic had to tackle this issue head-on because no one wants their smart device to act like it's on a permanent coffee break. Battery life is still a key design criterion for IoT devices in health care, as well as logistics and wearables – all of which need regular maintenance. These sectors collectively make up for low power IoT adoption.

Figure 9.2 is a hypothetical breakdown of IoT device categories based on their reliance on long battery life. The medical (30%) + industrial (25%) + wearables (20%) categories sum to 75%, reflecting applications where frequent battery replacement is impractical or costly [8]. The suggested percentages reflect the proportional representation of device categories where battery longevity is prioritized. These are the percentage of IoT device categories where extended battery life is a core design requirement. It is based on operational needs (e.g., medical safety), accessibility (e.g., industrial devices in the field) and user expectations (wearables).

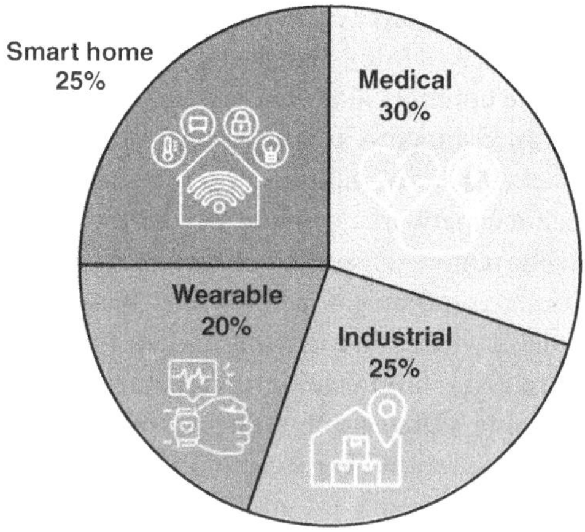


**Figure 9.2** Hypothetical grouping of IoT device categories based on their reliance on long battery life. *Source:* Author's own work.

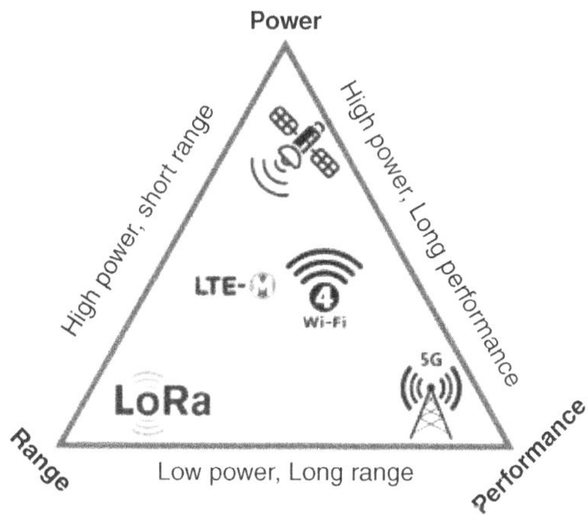


**Figure 9.3** Power efficiency trilemma. *Source:* Author's own work.

### 9.2.1 Trade-Offs in Wireless Connectivity

The Power Efficiency Trilemma in wireless communication represents a fundamental trade-off between three competing factors: power consumption, data rate, and communication range. As Figure 9.3 illustrates, optimizing one parameter often necessitates compromises in the others due to inherent physical and technical limitations. For instance, 5G mmWave achieves blazing-fast data rates but consumes significant power. It also suffers from short transmission ranges due to high-frequency signal attenuation. Conversely, LoRaWAN is the tortoise in this race; slow and steady, with low data rates but unbeatable energy efficiency, long range, and perfect for IoT devices that transmit small data packets intermittently. Meanwhile, satellite communication systems require extended range and higher power capabilities to overcome long-distance transmission challenges while maintaining high performance and achieving low latency.

The optimal balance between power consumption, performance, and communication range is fundamentally application dependent. Wearables and health trackers prioritize ultra-low power consumption to achieve extended battery life, cellular networks demand maximum performance for high-speed connectivity, while remote sensors in industries require robust long-range transmission capabilities to communicate across vast distances. The trilemma reminds us of the golden rule in wireless design: you cannot have it all. So, pick your priority, because in wireless communication, compromise is not just expected, it is mandatory.

Nordic's strategically redirected its focus to ultra-low-power chips, stripping down protocols, optimizing sleep modes, and pioneering BLE. This radical shift enabled devices to operate on minimal energy, drastically extending battery life. By 2006, the nRF24L01 chip they introduced was a cornerstone of IoT, engraving efficiency into the DNA of IoT technologies from fitness trackers to smart thermostats [9, 10].

But then, a new challenge emerged: How do you connect battery-powered IoT devices to cellular networks without rapidly depleting their energy reserves? Nordic Semiconductor addressed this in 2018 with the nRF9160 System-in-Package (SiP), a low-power cellular IoT solution supporting LTE-M and NB-IoT standards [11]. This device integrated advanced power management, edge computing, and dynamic sleep modes. The result was a 90% reduction in energy usage when compared to those traditional, power-hungry cellular modules.

This innovation was not just groundbreaking; it became the necessity in the IoT revolution. Devices like smart medical sensors and industrial trackers now boast battery lifespans for over a decade on a single charge. Real-world deployments have already shown asset trackers living their best lives, transmitting data daily while consuming less than 5 microamps in idle mode and just 1.4 μA/kB for data prioritization. That is like a marathon runner surviving on a single granola bar through a long race. Table 9.1 compares the power consumption of early wireless devices (2010) to current IoT devices powered with Nordic Semiconductor chips. The comparison highlights improvements in power efficiency for wireless devices with innovations like low-power circuit design, BLE, and Nordic's nRF9160 SiP.

Certified for global cellular networks, the nRF9160 has put the spotlight on Nordic Semiconductor for IoT energy efficiency. With over 50 million units deployed worldwide as of 2023, this little SiP has proven that connectivity and sustainability can be compatible [13]. To put that number into perspective: it is more than the entire population of Spain. If each unit were a person, they could fill over 600 soccer stadiums (each holding 80,000 people) to capacity. In the IoT world, hitting 50 million deployments is not just a milestone; it's a seismic shift, proving Nordic Semiconductor's technology is not just mainstream but the actual backbone of our connected future.

**Table 9.1** Power consumption comparison (2010 vs. 2024).

| | 2010 (Early Wireless Devices) | 2024 (IoT Devices with Nordic Chips) |
|---|---|---|
| **Device Type** | Smartphones, early IoT prototypes | Smart medical sensors, industrial trackers |
| **Battery Life** | 4–6 hours (video calling) | 10+ years (e.g., Nordic nRF9160 SiP) |
| **Heat Generation** | 45–50 °C (during video calls) | <30 °C (idle/active modes) |
| **Data Loss Rate** | 15–20% (due to range/dropouts) | <1% (BLE/optimized protocols) |
| **Power Consumption** | 500–1000 mW (active mode) | 1–5 mW (active mode, e.g., nRF9160) |
| **Standby Current** | 10–50 mA | 1–5 μA (ultra-low idle modes) |

*Source:* Adapted from [12].

## 9.3 Optimizing IoT Connectivity: FinFET Technology and Dynamic Power Management

The rapid growth of the Internet of Things, with 17.08 billion connected IoT devices in 2024, and this figure is expected to almost double to 30 billion by 2030 [14]. This means we are heading toward a future where even your toaster might judge your breakfast choices. Well, since more than 80% of these IoT devices are battery-powered by default, and therefore their energy is a concern. The desperate begging of millions of AA batteries around the world for a little bit of compassion. This requires new thinking on how we build networks and manage energy before our homes become battery graveyards [15]. Figure 9.4 is an illustration of how far IoT connectivity has come. The global proportion of IoT devices shows that as of 2020, IoT devices already outnumbered non-IoT devices (encircled in Figure 9.4). And by 2025, IoT devices are expected to outnumber non-IoT by a proportion of 3:1.

IoT is full of battery-operated sensors and devices that are used in many smart homes, and in healthcare, industrial automation, and environmental monitoring [17]. Research shows that IoT devices operate on batteries with an average lifespan of 1–2 years or less, depending on the application and power optimization strategy. The main objective is to achieve power-efficient wireless connectivity in remote or hard-to-reach areas, where battery-operated IoT devices are often left uncharged or unmaintained. To solve this daunting challenge, Nordic Semiconductor created a low-power solution optimized for LTE-M/NB-IoT, which is built to deliver

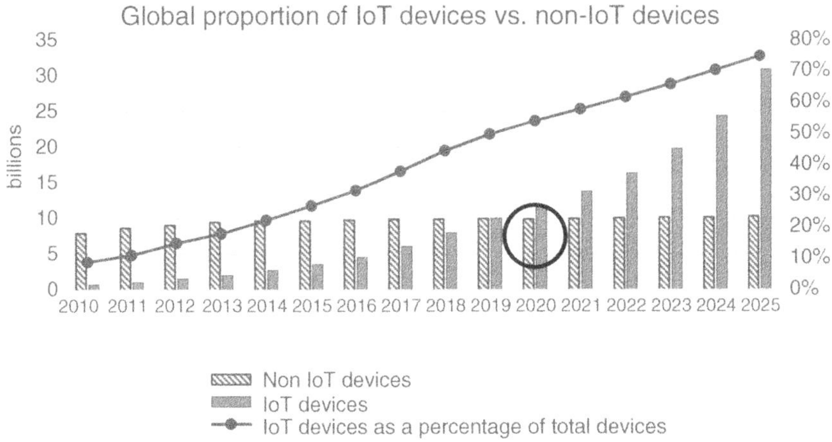


**Figure 9.4** Global proportion of IoT devices vs. non-IoT devices. *Source:* Adapted from [16].

low-bandwidth, and power-efficient IoT communication. Their solution is power-balancing between battery consumption and network performance that provide back-end days of battery life of up to 10 years in some use cases without compromising the reliability and stability of wireless connections.

## 9.4 Engineering Challenges of Power-Efficient Wireless Communication

Creating energy-efficient wireless devices demands profound insight into power leakage through transistor geometry and how it affects the overall power consumption of the device. In this case, students can explore the impact of FinFET and circuit optimizations on lower power for IoT devices. Unfortunately, getting efficiencies in the real world is not just about having better transistors networked, but signal level interference and battery caps all come into play. By analyzing these challenges, one can gain insight into the trade-offs between balancing power efficiency, power performance, and a reliable connection range.

To reinforce the key concepts explored in this case study, the following multiple-choice questions may help to test the understanding of power efficiency, semiconductor design, and IoT connectivity. These questions push the reader to reason about power management techniques, how FinFETs help in leakage reduction and what the future is for energy-efficient wireless communication in general. Answering the questions in this manner allows readers to be mentally more engaged with practical and technical perceivable low-power IoT practice.

## 9.5 Think and Solve

**Q 9.1** Power Efficiency vs. Connectivity: You are designing a battery-operated IoT device that must last at least five years while maintaining reliable connectivity. What is the best strategy to maximize battery life without sacrificing performance?

**A** Transmit data continuously in real-time to ensure constant connectivity.
**B** Increase transmission frequency while always maintaining full signal strength.
**C** Implement dynamic sleep modes, adjust signal strength based on network conditions, and optimize transmission intervals.
**D** Use the highest available power setting to always ensure strong network coverage.

**Q 9.2** Transistor Geometry and Leakage: A conventional MOSFET-based IoT device design has a leakage current of 100 nA per transistor. If a FinFET-based alternative reduces leakage by 90%, what is the new leakage current per transistor?

**A** 100 nA.
**B** 50 nA.
**C** 10 nA.
**D** 1 nA.

**Q 9.3** IoT devices, such as smart meters and health monitors, require long battery life while maintaining reliable connectivity. What is one of the biggest challenges in designing these devices?

**A** Increasing the physical size of batteries to store more power.
**B** Reducing power consumption while ensuring effective wireless communication.
**C** Using high-power transmission at all times to maintain a strong connection.
**D** Eliminating all forms of power leakage, regardless of performance trade-offs.

**Q 9.4** Which of the following best describes the role of dynamic sleep modes in an IoT device?

**A** They allow devices to continuously transmit data without any interruptions.
**B** They reduce power consumption by putting the device into a low-power state when it is not actively transmitting or receiving data.
**C** They amplify the device's signal strength to improve network coverage.
**D** They increase power usage to compensate for weak network signals.

## 9.6 Achieving Power-Efficient Wireless Communication

The growing demands of IoT applications, ranging from smart homes to industrial automation, have driven the need for more energy-efficient and reliable wireless connectivity solutions. From an engineering perspective, the problem can be broken down into two key areas: hardware optimization and software-based power management.

Striking a balance between hardware improvements, software optimizations, and the constraint of implementation is the challenge of achieving power-efficient wireless communication. With how quickly various industries are adapting to IoT, engineers have to constantly focus on performance optimization with energy consumption, more reliability, and cost efficiency.

Wireless communication technology has seen dramatic change in the last 20 years. Back in the early 2000s, when Nordic Semiconductor first launched ultra-low-power wireless products in industry, the major semiconductor players of the time (Texas Instruments, Qualcomm, Broadcom, STMicroelectronics) were basically focused on wireless chip on high performance for mobile, computers and industrial consumer electronics. Low-power wireless communication represented less than 5% of the worldwide wireless chip market, which was approximately $20 billion in 2001 at that time [18]. Other than that, short-range communication connectivity was limited to classic Bluetooth (pre-Bluetooth Low Energy), Zigbee or other RF solutions and they were power hungry. Classic Bluetooth devices had current draw from 30 mA during active transmission, and not suitable for battery-powered IoT devices.

In contrast, Nordic Semiconductor's approach was somewhat different, focusing on ultra-low power wireless solutions, targeting small devices with batteries. The development of Bluetooth Low Energy was their first breakthrough, and it was optimized for commercialization during the 2010s. Although standard Bluetooth and Wi-Fi both had robust portfolios from companies such as Texas Instruments and Qualcomm, at the time they were generally more focused on performance rather than power consumption. Seeing the demand for low-powered, "always connected" and "always on" devices growing, Nordic became one of the pioneers in the making of BLE from sparse support to commonplace across IoT applications.

Although the contributions of Nordic Semiconductor are notable, what was more important was that they solved a problem that many competitors where not really prioritizing: energy-efficient wireless connectivity that could double or triple the battery life from months to years. The SoC series nRF51, nRF52, and nRF53 totally changed the world together with ARM Cortex processors in some applications. They work through Wi-Fi, ultra-low-power radio, and dynamic power control. This was the base to build tiny wireless devices that could stay online for days without the need to be recharged.

Apart from BLE, Nordic has been leading the way to cellular IoT (LTE-M and NB-IoT), and its nRF91 series are probably well known. Cellular connectivity prior to this innovation was dead space for the power-hungry 2G, 3G and LTE that could not be used in IoT devices. Modem market was controlled by companies like Qualcomm and MediaTek, they were looking for high-speed mobile communications instead. Nordic Semiconductor flipped the switch with cellular IoT capabilities that could reliably deliver ultra-low power capabilities for remote sensors, asset tracking, and smart meters. What makes Nordic stand out is its engineering-first philosophy. Unlike its larger competitors, where they focus on mass-market consumer electronics, Nordic has specialized in wireless solutions optimized for ultra-low-power applications. Their open-source software ecosystem, developer-friendly tools, and strong customer support have made them a preferred choice for engineers designing IoT products.

The innovations of Nordic have had major impact on wearables to smart home devices and eventually even medical/industrial IoT sensors. Things have drastically changed with BLE and cellular connected IoT. Before having Nordic's solutions, most small connected IoT products were powered by a cheap battery, and had short range and/or high-power costs. With progress from Nordic, we are now able to design smarter products with battery-powered devices that will work for years from one charge. For example, empirical studies indicate that typical IoT devices based on traditional networking often had battery lives of less than six months whereas Nordic's advancements have extended this up to 10 years on a single battery [19, 20]. Also, their BLE technology has improved 50% increase in range from previous standards while lowering power consumption by up to 40%. As reported in the industry, over five billion devices are powered by BLE technology today and the IoT market is expected to reach $1.1 trillion in 2028 with the majority relying on innovations such as these from Nordic [21].

Expanding into the bioscience area with their ultra-low-power wireless technology now supporting medical wearables, remote patient monitoring, and envisioned next-generation biosensors. BLE and cellular IoT platforms are being broadly adopted in connected health devices such as glucose monitors, ECG patches, smart inhalers and biosensors for continuous monitoring of vital signs. The company has been a critical provider of secure, trustworthy and energy-efficient real-time health monitoring support. It enables patients and their healthcare service providers to monitor the life-critical parameters.

A good example would be the nRF52 and nRF53 series from Nordic Semiconductor, which are widely used in biomedical research and wearable biosensing tech. These advanced chips provide seamless above-ground connectivity between medical sensors and mobile devices. Ensuring efficient data transmission while maintaining a focus on real-time and live driven health diagnostics. Such capability makes those very well suited for applications in precision medicine where the AI algorithms can be used to identify personalized treatment strategies from health data. As the need escalates for tools that are non-invasive and have capabilities of continuous monitoring, Nordic Semiconductor stands at the center of the bioscience and digital health solutions. Power- and energy-efficient wireless connectivity enables development of cutting-edge solutions to support sustainable health monitoring and benefit patient care.

Looking ahead, as wireless connectivity evolves with Matter, Wi-Fi 6, and AI-powered IoT applications, the demand for power-efficient solutions will only keep skyrocketing. Nordic Semiconductor's pioneering work in BLE, ultra-low-power RF, and cellular IoT lays the groundwork for a future where IoT devices are not only efficient, scalable, and sustainable but also unlikely to leave you often hunting for a charger.

## 9.7 Discussion Points

Nordic Semiconductor's journey from a niche Norwegian company to a global leader in ultra-low-power wireless technology presents valuable insights into the evolving semiconductor industry. Their new ideas and innovations in terms of communication centered around FinFET design, dynamic power management, and IoT has shifted the paradigm for power-efficient computation. The following are some important discussion points of this case study. This is to challenge reflection on the competitive environment and future considerations for wireless technology placing these notes in a broader context from Nordic's distinct market standpoint up to its potential contribution in bioscience and sustainable electronics and the implications way beyond the engineering or technology space.

1. **Balancing Power Efficiency and Connectivity:** Nordic Semiconductor had to tackle power consumption versus reliable wireless connectivity that is achievable with the Nordic. Can you compare FinFET-based designs and dynamic power management with what their low-power semiconductor approaches may have looked like?
2. **Nordic's Competitive Edge:** In a market crowded with giants such as Qualcomm, Texas Instruments and STMicroelectronics, what is the one or two things that makes Nordic Semiconductor different? What gave them a distinct advantage in the market by their laser-like focus on ultra-low power wireless?
3. **The Role of IoT in Semiconductor Evolution:** From smart homes to industrial automation, the IoT applications are growing, and Nordic Semiconductor has played a strong role. What are some of the toughest engineering challenges in creating wireless chips for IoT, and how has Nordic addressed these?

4. **Bioscience Applications and the Future of Health Tech:** As Nordic branches out into bioscience and connected healthcare, there are implications of ultra-low power wireless in medical wearables, patient monitoring etc. Why does real-time data transmission beneficial for health care outcomes?
5. **Impact on Sustainable Technology:** Sustainable electronics, and power efficiency are the key parameters. What does Nordic's contribution bring to the table in terms of larger sustainability achievements like lowering e-waste and encouraging IoT longevity? With future semiconductor designs, what could be the other moves that can be taken to increase energy efficiency further?

## 9.8 Limitations of the Case Study

While this case study provides an in-depth analysis of Nordic Semiconductor's innovations in low-power wireless communication, it is important to acknowledge its limitations. Case studies, by nature, focus on specific scenarios, and variables of real-life situations do not always represent those exemplified through the pages of a case by case.

1. **Technological Advancements:** One key limitation is the scope of technological advancements considered. Nordic Semiconductor has made substantial progress in transition to FinFET and switching to dynamic power management. However, the exploration of new semiconductor technologies such as quantum computing or other materials (e.g., gallium nitride or GaN) is not fully covered. This LTE-M and NB-IoT centric review overlooks the other wireless communication technologies (LoraWAN or 5G mmWave) which could complement these networks but also have different benefits and use cases.
2. **Market and Competitive Analysis:** Another limitation concerns market and competitive analysis. Though the case study flag some of the major competition to Nordic Semiconductor Innovation, the semiconductor world moves so fast that new entrants could change market dynamics. The analysis does not take into breakthroughs by others or a completely new wireless standard that could change the wireless landscape for power-efficient IoT connectivity.
3. **Technical and Engineering Perspectives:** Furthermore, the study primarily emphasizes technical and engineering perspectives, leaving out broader economic and geopolitical influences. Factors such as semiconductor supply chain volatility, global chip scarcity and trade regulations are all major factors in the industry landscape but not discussed thoroughly within this case.

   In addition, while the case touches on bioscience applications but the breadth of Nordic Semiconductor's involvement in this field could be better addressed with more case examples.

4. **Real-World Implementation:** Finally, real-world implementation often presents unforeseen challenges that laboratory simulations cannot fully capture. Although Nordic solutions have undergone large-scale deployment in different environments, newly discovered limitations have yet to surface.

This case study provides valuable insights into the challenges of ultra-low power wireless technology development while showcasing Nordic Semiconductor's key contribution to IoT advancement. Further work on complementary technologies, economic rationality and large-scale deployment case studies may enable a more holistic understanding.

## References


1 Shah, A., Tuggle, C., Clune, J.E., Steinbacher, D. and Patel, A. "FaceTime with Apple and information technology," *Plastic and Reconstructive Surgery*, 129(3), 562e–563e, 2012.

2 Budagavi, M., Sze, V., Demircin, M.U., Dikbas, S., Zhou, M. and Chandrakasan, A.P. "Description of video coding technology proposal by Texas Instruments Inc," *JCTVC-A101*, 16, 45, 2010.

3 Wunder, G., Boche, H., Strohmer, T. and Jung, P. " Sparse signal processing concepts for efficient 5G system design," *IEEE Access*, 3, 195–208, 2015.

4 Bohorquez, J.L., Yip, M., Chandrakasan, A.P. and Dawson, J.L. "A digitally-assisted sensor interface for biomedical applications," in *2010 Symposium on VLSI Circuits*, 217–218. IEEE, 2010.

5 Lappeteläinen, A. "How Nokia triggered the global rise of Bluetooth LE," Nokia, 9 April 2024, https://www.nokia.com/blog/how-nokia-triggered-the-global-rise-of-bluetooth-le (accessed February 26, 2026).

6 Shandle, J. "Nordic Semiconductor is key player in Wibree," *EE Time*, October 2006, https://www.eetimes.com/nordic-semiconductor-is-key-player-in-wibree (accessed February 26, 2026).

7 Nordic Semiconductor. "Bluetooth low energy," https://www.nordicsemi.com/Products/Wireless/Bluetooth-Low-Energy (accessed February 26, 2026).

8 Hercog, D., Lerher, T., Truntič, M. and Težak, O. "Design and implementation of ESP32-based IoT devices," *Sensors*, 23(15), 6739, 2023.

9 Sonavane, S.S., Kumar, V. and Patil, B.P. "MSP430 and nRF24L01 based wireless sensor network design with adaptive power control," *ICGST International Journal on Computer Network and Internet Research (CNIR)*, 8(II), 11–15, 2008.

10 Basu, D., Gupta, G.S., Moretti, G. and Gui, X. "Protocol for improved energy efficiency in wireless sensor networks to support mobile robots," in *2015 6th International Conference on Automation, Robotics and Applications (ICARA)*, 230–237, 2015.

**11** Al-Sammak, K.A., Al-Gburi, S.H., Marghescu, I., Drăgulinescu, A.M., Marghescu, C. and Al Sammak, N.A.H. "An Experimental Study of Power Consumption in Narrowband IoT Devices," in *2024 15th International Conference on Communications (COMM)*, 1–6. IEEE, 2024.

**12** Nordic Semiconductor, "nRF9160," https://www.nordicsemi.com/Products/nRF9160 (accessed February 2026).

**13** Nordic Semiconductor, "Q4 2023 financial report, 3," 2023, https://www.nordicsemi.com/-/media/Investor-Relations-and-QA/Quartely-Reports/2023/Q4_Quarterly_Report_2023.pdf (accessed February 26, 2026).

**14** Singh, S., Sharma, M. and Hossain, S.A. "Navigating the threat landscape of IoT: an analysis of attacks," in *International Conference On Innovative Computing And Communication*, 25–48. Springer Nature, 2024.

**15** Shirvanimoghaddam, M., Shirvanimoghaddam, K., Abolhasani, M.M., Farhangi, M., Barsari, V.Z., Liu, H., Dohler, M. and Naebe, M. "Towards a green and self-powered internet of things using piezoelectric energy harvesting," *IEEE Access*, 7, 94533–94556, 2019.

**16** Vailshery L.S. "IoT and non-IoT connections worldwide 2010-2025," *Statista*, 2022.

**17** Shah, S.W.H., Mian, A.N., Aijaz, A., Qadir, J. and Crowcroft, J. "Energy-efficient mac for cellular IoT: state-of-the-art, challenges, and standardization," *IEEE Transactions on Green Communications and Networking*, 5(2), 587–599, 2021.

**18** Andrea, K., "Little tigers' big surprise: The evolution of telecommunication based on Gartner's hype cycle theory," *International Journal of Management Cases*, 3(1), 5–26, 2010. https://www.researchgate.net/publication/296750927 (accessed March 23, 2026).

**19** Liberg, O., Sundberg, M., Wang, E., Bergman, J. and Sachs, J. *Cellular Internet of Things: Technologies, Standards, and Performance*, Academic Press, 2017.

**20** Ergun, K., Ayoub, R., Mercati, P. and Rosing, T. "Dynamic optimization of battery health in IoT networks," in *2019 IEEE 37th International Conference on Computer Design (ICCD)*, 648–655. IEEE, 2019.

**21** das Neves Duarte, I.M.C. "Nordic Semiconductor-Bluetooth Low Energy leader facing difficulties in the post-pandemic era," Master's Thesis, Universidade NOVA de Lisboa, Portugal, 2023.

# 10

# Intel's 3D Transistor Stacking: Vertical Scaling Breakthrough

Pioneering 3D Chip Architecture, and the Future of High-Performance Computing

## 10.1 Synopsis

What if the key to unlocking the next era of computing power wasn't just about making transistors smaller but about thinking vertically? Parallel to the way sky-scrapers dominate cityscapes by stacking upwards into space, Intel 3D stacking, and advanced transistor technology from high performance, have redefined chip architecture by expanding upwards rather than simply scaling down.

Moore's Law, the statement that the number of transistors on a chip would double about every two years, has guided the semiconductor industry for decades. But when traditional scaling was running out, the question chipmakers faced is how to continue improving performance without drastically increasing power consumption or increasing production costs?

As a pioneer in semiconductor innovation, Intel addressed the challenge with an audacious idea that combined vertical integration of 3D stacking. Thus, the next-generation transistor phenomenon is like FinFET, as well as its successor, the RibbonFET. This freed Intel from the two-dimensional (2D) layout and enabled performance possibilities via the new three-dimensional stacking of chip elements for high-performance computing, artificial intelligence and data-intensive workloads. This case study examines Intel's strategic approach to 3D packaging and cutting-edge transistor technology. The analysis explores how this strategy has reshaped the semiconductor industry to achieve faster, and more power-efficient chips while overcoming traditional transistor scaling limits.

What exactly led to this shift? And what were the struggles for Intel on the way? Also, what does this mean going forward in the evolution of computing?

## 10.2 Background and Case Scenario

The fundamental building block at the heart of every modern electronic device is the transistor. This component is an electrical switch that controls the flow of current in a circuit. From simple calculators to advanced AI systems, transistors enable everything by turning on and off at incredible speeds. Transistors are built on semiconductors. They are ideal for fabricating complex circuits because materials like silicon can alternate between conducting and insulating electricity. Figure 10.1 shows the different types of traditional transistors and their major applications in our daily lives.

The Sankey diagram illustrates the flow of transistor types (BJT, MOSFET, IGBT) into primary applications such as digital logic, switching, amplification, LED driving, and temperature sensing. These applications power our daily technologies from smartphones and laptops to electric vehicles, home automation systems, and industrial equipment and sensors. The thickness of each flow represents the relative significance of that application and highlights the role transistors play in shaping today's digital efficient world.

For decades, the semiconductor industry has been riding the express train, assuming chips would just keep getting faster, smaller, and more efficient, like some overachieving student who delivers outstanding performance year after year without

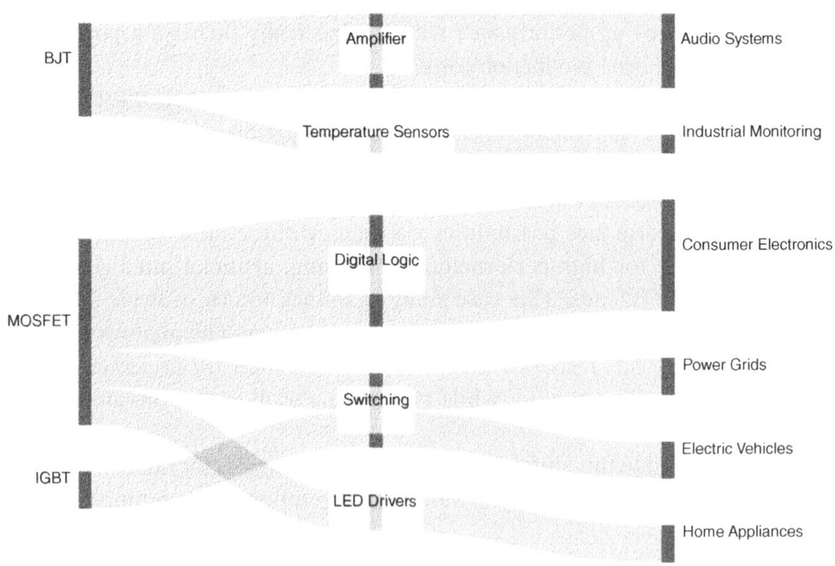


**Figure 10.1** Sankey diagram of transistor applications in modern electronics. *Source:* Author's own work.

fail. But now, the chipmakers are hitting a wall – literally. Imagine trying to cram one more couch into an already packed studio apartment; at some point, physics will just throw up its hands.

This problem revolves Moore's Law, a projection done by Intel's co-founder Gordon Moore in 1965. He observed that the number of transistors on a chip was doubling approximately every two years and performance, and efficiency improvements were not going to stop. For decades, this continued to work and drive continuous advancement in computers, smartphones, cloud and artificial intelligence. But as transistors shrunk down to atomic scales, maintaining this exponential growth was getting harder.

### 10.2.1 The Roadblocks of Traditional Scaling

In the early days of semiconductors, increasing the number of transistors on a chip was the primary method to achieve performance improvements, a principle known as Dennard scaling. Smaller transistors enabled faster switching speeds, lower power consumption and higher computational density per square millimeter of silicon. This means a 35% performance improvement and 65% power reduction on the move from 90 nm technologies in the early 2000s to 7 nm in 2018. This scaling enabled exponential growth in terms of processing power and efficiency, leading to the technological advancements of traditional computing. But once transistors shrank to the single-digit nanometer regime, problems like quantum tunneling, heat dissipation, and variations in manufacturing became true roadblocks for continued scaling. At the smaller dimensions – like 5 nm for example, leakage currents from quantum tunneling can increase as high as 1000 times larger than in bigger nodes, stalling the traditional scaling and calling for new ways to keep pushing chip performance.

As the transistor approached the limits of miniaturization, engineers were left with more and more problems that threatened the semiconductor scaling of future. Power leakage and heat dissipation were one of the biggest issues at the time. At extremely small scales, quantum mechanical effects such as electron tunneling caused little currents to leak unintentionally through transistor gates. This leakage caused a larger power draw and excessive heat that resulted in even more complex cooling and power subsystems.

But as time goes by, transistor shrinkage is going to be less important. While earlier generations of chips saw doubling in speed and efficiency gains with each node shrink, the benefits became less significant. Beyond a certain threshold, making the transistors smaller was no longer going to necessarily result in faster or more power-efficient processors. In addition to this the manufacturing complexity and costs were going up.

The move from 90 nm to ~7 nm and beyond needed more advanced fabrication techniques like Extreme Ultraviolet Lithography (EUV). It drove enormous costs into semiconductor production. The increased cost associated with research and development as well as fabrication at these advanced nodes, thereby constraining companies on continued scaling within the traditional model. This meant the industry no longer had a choice but to reflect on its current path under these growing pressures. With transistors unable to be shrunk anymore in 2D, could vertical scaling be the answer?

### 10.2.2 A Small Switch That Changed the World

Back in 1947, William Shockley, John Bardeen, and Walter Brattain at Bell Labs made the first point-contact transistor, which was a significant innovation that eventually led vacuum tubes to die out in electronic devices. That tiny semiconductor device could both amplify and switch electronic signals, ushering in the golden age of transistors. By the 1950s, millions of transistors were being made per year, already outsizing vacuum tubes in both size and power consumption. But the real transformation occurred in 1959, with Mohamed Nasiruddin Atalla and Dawon Kahng at Bell Labs introducing the Metal-Oxide-Semiconductor Field-Effect Transistor (MOSFET) – a compact, low energy device.

Unlike earlier bipolar junction transistors, MOSFETs have completely changed the world with their low power consumption and high operating speed, along with the capability for dense integration on a silicon wafer. This advancement resulted in the development of planar MOSFET, where the transistors were produced on a flattened silicon surface and gate controlled the electrons from the source to meet drain.

By the early 1960s, Fairchild Semiconductor, founded by Gordon Moore and Robert Noyce with a focus on MOSFET technology to drive so much of the electronics industry. The company began to lead an era of development for silicon transistors and paving the way for current day computing. Moore and Noyce both left to form Intel Corporation in 1968 to continue their quest for advancing the limits of innovation. They wanted to change semiconductor technology and build things that have never been built before. Figure 10.2 shows this evolution from the early 70s to the modern FINFET era.

By 1971, the first major success for Intel was with the introduction of the world's first microprocessor, Intel 4004. This chip, made up of 2,300 transistors (each with 10-micron-wide process) could do 92,000 instructions per second. Intel 4004 revolutionized the computing process because the central processing unit (CPU) was integrated into it. It put Intel in the forefront of the microprocessor era. This invention paved the way for creating personal computers and started the technological age.

Gordon Moore predicted the exponential growth of computing power in 1965, and it became known as Moore's Law. This prediction was driven by

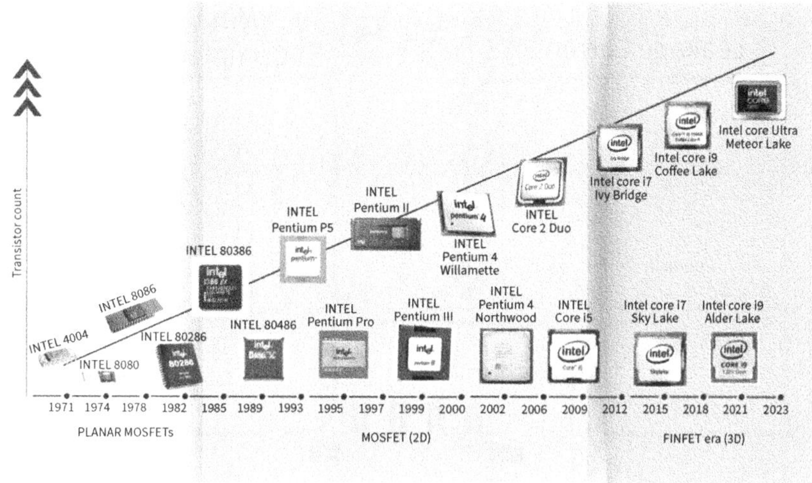


**Figure 10.2** Intel's microprocessor evolution and transistor count. *Source:* Author's own work with data from [1, 2].

continuous improvements in fabrication node technologies. Fabrication node represents the minimum transistor dimensions achievable with the available manufacturing process in each era. Early microprocessors used 10-micron technology, and it progressed to 1 μm fabrication nodes during the 1980s and 1990s. The move into sub-micron dimensions enabled outstanding advances in transistor density and computing performance. Intel was instrumental in this development.

Intel's persistent innovation has made the company maintain its market leadership until now. Transistor technology has driven the relentless scaling of transistors since its first invention. The Intel 80386 based on 1.5 μm node was launched in 1985 with the 32 bit computing power while the Pentium in 1993 followed 0.8 μm node process. Pentium provided a quantum increase in computational power [3]. By 1999, Intel Pentium III shrunk down to 180 nm process [4]. Then followed by Pentium 4 in 2000, where planar MOSFETs became as small as they could ever get at 130 nm [5].

However, by the early 2000s, it became clear that physical and electrical limitations threatened to slow down progress. As illustrated in Figure 10.3, issues such as gate leakage, thermal hotspots, and escalating manufacturing costs meant that traditional planar MOSFET scaling had reached a breaking point. Intel, along with other semiconductor leaders, needed new transistor architecture. If transistors could no longer be efficiently scaled in two dimensions, could they be reimagined in three?

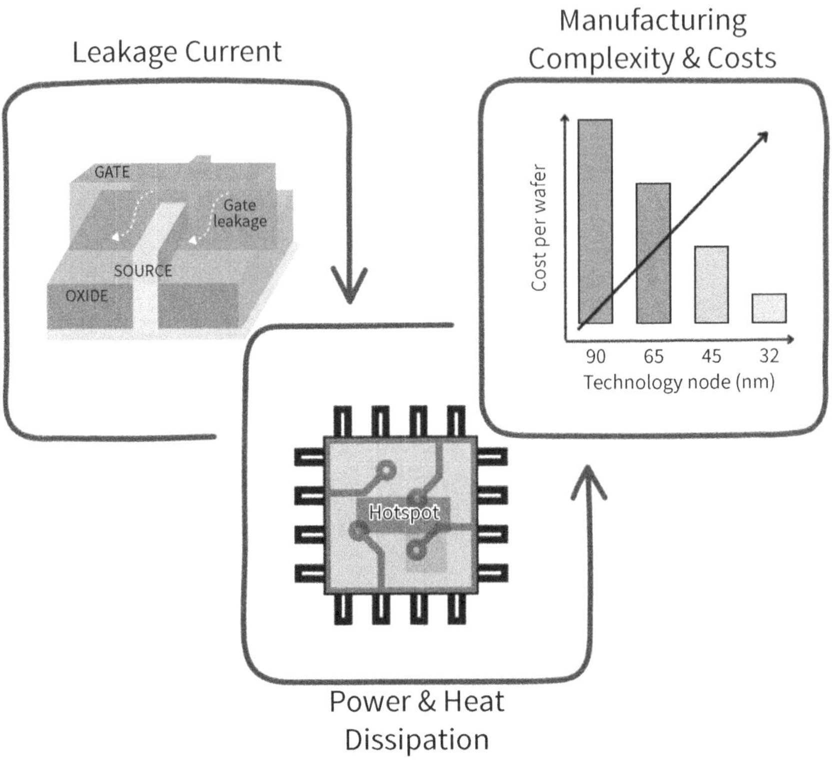


**Figure 10.3** Challenges of planar scaling: leakage, power, and cost. *Source:* Author's own work.

### 10.2.3 From Shrinking to Stacking

The relentless quest for transistor performance and energy efficiency has led to a seismic shift in transistor technology, from conventional planar MOSFET to Fin Field-Effect Transistors (FinFET) and other three-dimensional transistor architectures [6]. Planar MOSFETs, eventually became unable to overcome their own limitations. To address these challenges and maintain the pace of Moore's Law, novel device architectures in the form of FinFET have been conceived as feasible alternatives [7].

FinFETs were first conceptualized in the 1990s by researchers from University of California, Berkeley. They showed that a possibility to improve transistor quality and were able to avoid leakage power dissipation. FinFETs were first commercialized at the 22 nm node by Intel in 2011 with the introduction of Ivy Bridge [8]. This transformation enabled one of the most substantial transistor density and power consumption improvements in history.

As Intel introduced FinFET technology, this also marks a shift from 2D to partial 3D structures. FinFETs used a raised silicon "fin" to make the current control more efficient than planar transistors. FinFETs replaced the traditional planar transistor as the industry standard for leading edge chips, which later paved the way to higher performing and power efficient processors [9].

Following Intel's breakthrough, the semiconductor industry widely adopted FinFETs for advanced nodes, including 14 nm, 10 nm, and 7 nm technologies. Refining the fin structure in each new iteration, increased transistor density and increased power efficiency. FinFETs may have fabrication problems at first (lithography and etching get harder with the additional fin infrastructure) but manufacturing has significantly improved over the last decade. As a result, this type is the current standard for high-performance chips [10].

Other semiconductor manufacturers, including TSMC, Samsung, and GlobalFoundries, have also integrated FinFET technology into their process nodes. Figure 10.4 shows the difference between FinFETs and planar MOSFETs and defines how the FinFET structure represents a significant change from planar technology in terms of better gate control over the channel, reduced short-channel effects, and higher current drive [11].

The FinFET's complex topography thus requires state-of-the-art lithographic technology and simulation methods for accurate modeling of the architecture as well as implementation of the fabrication [12]. Table 10.1 summarizes the reasons why FinFETs outperform planar MOSFETs in key aspects such as efficiency, power and performance.

This transition is often dubbed the "FinFET revolution" because it serves as a critical juncture in semiconductor technology. It allows for continued scaling in accordance with Moore's Law by architecturally innovating alternatives to mere dimensional shrinking. It became imperative to bridge this gap in order to tackle

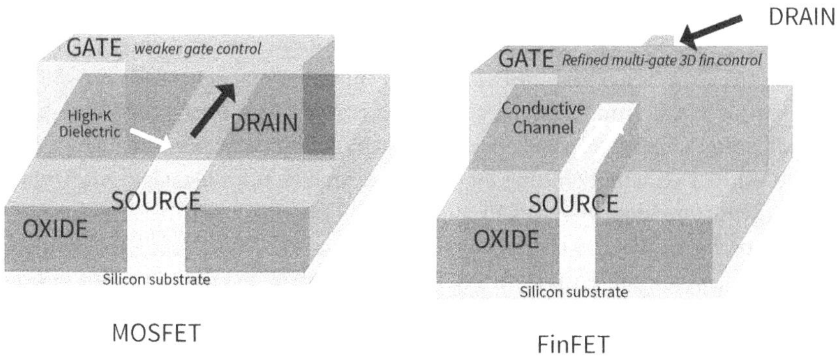


**Figure 10.4** Comparative diagram of planar MOSFETs and FinFET technology. *Source:* Author's own work.

**Table 10.1** Advantages of FinFETS over planar MOSFETs.

| Feature | Planar MOSFET | FinFET |
|---|---|---|
| **Electrostatic Control** | Weaker gate control, prone to short-channel effects | Multi-gate structure improves control, reducing leakage and short-channel effects |
| **Power Consumption** | Higher leakage currents lead to increased power usage | Operates at lower voltages, reducing power consumption while maintaining performance |
| **Drive Current** | Limited due to a single gate and narrow channel width | Higher drive current due to 3D fin structure, increasing effective channel width |
| **Performance and Speed** | Limited due to increased resistance and lower carrier mobility | Enhanced switching speeds and processing efficiency |
| **Scalability** | Scaling below 22 nm is challenging due to leakage and control issues | More scalable, enabling continued transistor miniaturization |
| **Manufacturing Complexity and Cost** | Simple fabrication process, lower costs | More complex fabrication but offers better efficiency and performance benefits |

*Source:* Author's compilation with data from [13–15].

the emerging issues of leakage current, short-channel effects and high-power consumption that drastically limited the scaling of conventional MOSFETs further.

The FinFET architecture, characterized by a fin-shaped channel, allows for electrostatic gate control from multiple sides and made it possible for more aggressive scaling of transistors. This also effectively suppresses short-channel effects and enabling higher performance at lower operating voltages. It has emerged as the dominant design in modern microprocessors, memory chips, and other integrated circuits [16]. However, even FinFETs are facing fundamental limits as device dimensions shrink to the 5 nm technology node and beyond. This is primarily due to the intensification of short-channel effects that degrade transistor performance [11].

With FinFET scalability hitting the wall, new architectures like Gate-All-Around Field-Effect Transistors (GAAFET) and Intel's RibbonFET will be introduced to improve performance and efficiency much further [17]. The GAAFETs enhanced gate control, which can be analytically explained only by the natural width or how far in the channel the electric field of source penetrates [18]. Researchers are looking for ways to circumvent the limitations of FinFET, alternative materials, strain engineering and 3-dimensional architecture for technological progression [19]. GAAFETs have greater gate controllability, which is expected to enhance performance and energy efficiency hence it is one of best options that would replace FinFETs [15]. Additionally, novel channel materials such as germanium, indium gallium arsenide, and graphene are being investigated to boost carrier mobility, enabling faster operation and lower power consumption.

### 10.2.4 A New Dimension in Semiconductor Design

As transistor architecture continued to evolve, another fundamental paradigm shift emerged. No longer would all the components be on a single layer as chipmakers started building them vertically. The breakthrough, called 3D stacking technology, was due to requirements for higher performance with lower power and more dense integration.

For decades, the semiconductor industry followed transistor scaling by reducing transistor sizes and increasing transistor density per unit area. However, as process nodes shrank below 10 nm, traditional planar scaling encountered fundamental physical and economic barriers as discussed in the previous section. 3D stacking emerged as a solution to extend Moore's Law by optimizing not just transistor size but also chip layout and connectivity.

Instead of relying on lateral scaling, 3D stacking lays out various things like CPU cores and memory with power delivery circuits in layers above each other [20]. It shrinks signal distances and enhances the communication from layer to layer. It facilitates an order of magnitude increase in computational capability by providing higher bandwidth, lower latency and more efficient power delivery.

Intel pioneered the practical application of 3D stacking with its Foveros technology, first introduced in 2018. Unlike conventional monolithic designs which can have all components etched onto a single silicon die, Foveros is designed for heterogeneous integration stacking of components of different logic and memory in series.

Intel's Foveros 3D stacking technology changed chip architecture. It allows heterogeneous integration and different process node combinations to trade-off power efficiency and performance. This method lowers power consumption and heat, providing faster data transfers between logic as well as memory cells. With a dense stack of these parts, Foveros shrinks the total chip size which makes it a better fit for mobile and high-performance computing.

2020 saw the first major product to hit the market that really highlighted Foveros. It was Intel's Lakefield processor [21]. Lakefield showed what 3D stacking can do in general by using it to mix high-performing cores and power efficient cores in a vertically stacked configuration. While Intel pioneered 3D stacking with Foveros, other semiconductor manufacturers have embraced similar innovations to push the boundaries of chip performance. Figure 10.5 highlights the advancements from other top semiconductor foundries. This widespread shifts toward 3D stacking, and it is becoming a cornerstone of most modern semiconductor technology today.

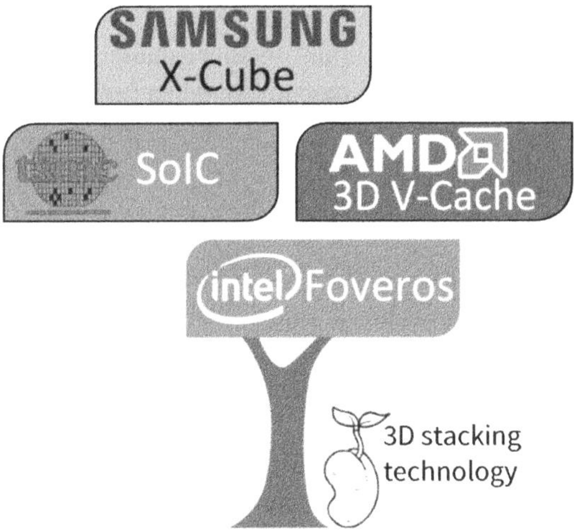


**Figure 10.5** Industry-wide adoption of 3D stacking. *Source:* Author's own work.

Despite its numerous benefits, 3D stacking brings challenges too, and engineers have to eventually find a solution to overcome those challenges. The active components are much closer together in compact stacking, and so thermal management gets increasingly difficult and demands some advanced cooling techniques. Aligning the stacked layers precisely and bonding them is a difficult manufacturing problem. It places constraints on bonding and interconnection technologies. The fabrication costs remain high and stacking enable yield rate is yet to be realized in high volume. But breakthrough innovations such as through-silicon vias (TSV) and hybrid bonding are addressing these challenges. These developments show that 3D stacking is a feasible reality rather than a theoretical concept.

The transition from traditional transistor scaling to 3D stacking, signals are a game changer in semiconductor design paradigm. This is not just an alternative to shrinking, but it is high-performance computing going forward. The refinement and scaling of 3D stacking technology by industry leaders can have broad implications beyond processors such as AI, edge computing, and advanced mobile devices. Through continued innovation, 3D stacking becomes a key to a third act in the evolution of computing. This technology promises chips that compute much faster than ever while using less power and potentially leading the next era of technological advancement.

## 10.3 Think and Solve

To better understand the challenges and key takeaways from this case study, the following questions summarize its critical aspects. These questions focus on the technical limitations, advantages, and implications of 3D stacking technology while evaluating its real-world application.

**Q 10.1** What is the primary reason for semiconductor manufacturers adopted 3D stacking technology?

**A** To reduce the cost of fabrication.
**B** To improve transistor density and performance while overcoming planar scaling limits.
**C** To eliminate the need for new materials in semiconductor design.
**D** To increase the physical size of microchips.

**Q 10.2** Which of the following is a key advantage of Intel's Foveros technology?

**A.** It allows heterogeneous integration of different process nodes.
**B** It eliminates all thermal challenges in semiconductor design
**C** It replaces the need for advanced lithography techniques.
**D** It solely focuses on improving memory performance.

**Q 10.3** One major challenge of 3D stacking technology is:

**A** Limited transistor scaling due to photolithography constraints.
**B** Decreased bandwidth compared to planar architectures.
**C** Difficulty in managing heat dissipation within stacked layers.
**D** The inability to integrate multiple logic components.

**Q 10.4** Why is 3D stacking considered a paradigm shift in semiconductor engineering?

**A** It replaces traditional transistors with optical computing components.
**B** It eliminates the need for Moore's Law by keeping chips two-dimensional.
**C** It redefines chip design by stacking components vertically rather than scaling them laterally.
**D** It allows chips to operate without requiring external power sources.

**Q 10.5** A semiconductor company is designing a 3D-stacked processor. One of the biggest challenges in the design is heat dissipation. The power consumption of a stacked processing unit is 120 W, and its total surface area available for heat dissipation is $0.01\,m^2$. The maximum allowable operating temperature is 85 °C, and the ambient temperature is 25 °C. The company plans to use a heat sink with a thermal resistance of 0.5 °C/W.

- Calculate the junction temperature of the processor with the given heat sink.
- Determine whether the selected heat sink is sufficient to keep the processor within the allowable operating temperature. If not, suggest an alternative thermal management strategy.

**Given:**

- Power dissipation, $P = 120\,W$
- Heat sink thermal resistance, $R_{th} = 0.5\,°C/W$
- Maximum operating temperature, $T_{max} = 85\,°C$
- Ambient temperature, $T_{ambient} = 25\,°C$

The junction temperature can be estimated using the thermal resistance equation:

$$T_{junction} = T_{ambient} + (P \times R_{th})$$

## 10.4 Conclusion and Key Takeaways

The evolution of semiconductor technology has reached a pivotal moment with the adoption of 3D stacking. Intel's Foveros technology (and similar moves from TSMC, AMD, and Samsung) show that transistor scaling vertical integration is not the end of high-performance computing, but a future. 3D stacking paved the way for more advanced computing architecture. It provides heterogeneous integration with better power efficiency under same footprint.

However, as with all breakthroughs, challenges remain. Thermal management, complexity in manufacturing and cost concerns stall out for wide scale usage. Through-silicon vias (TSV), hybrid bonding and new cooling solutions can address these obstacles. It seems challenges can be ironed out with every other innovation, and the trend toward 3D stacking is steadily viable.

Ultimately, the impact of 3D stacking extends beyond processors. The realm of emerging technologies like AI, edge computing and high-performance applications demand more power-efficient packaging, cost-effective and compact designs. To supply such demands, the semiconductor industry continues its innovative journey with advanced architecture and keeps pushing the computing capabilities.

These are the key takeaways from this study:

**3D stacking:** 3D stacking is a paradigm shift in semiconductor design, moving away from the traditional Moore scaling of transistors.

**Leading Innovations:** Intel's Foveros, TSMC's SoIC, AMD's 3D V-Cache, and Samsung's X-Cube are leading innovations in this space.

**Heterogeneous Integration:** Allows different process nodes to be combined for optimal performance and efficiency.

**Key benefits of 3D stacking:** Include reduced energy consumption, improved bandwidth, and a smaller chip footprint.

**Challenges:** Heat dissipation, manufacturing precision, and cost remain challenging, but ongoing advancements in TSVs and hybrid bonding are addressing these issues.

**The Future:** The future of 3D stacking lies in enabling more powerful, efficient chips for AI, mobile devices, and edge computing.

As industry continues to innovate, 3D stacking plays a crucial role in defining the next era of semiconductor advancements and shaping the future of computing for years to come.

## 10.5 Discussion Points

The adoption of 3D stacking technology has introduced new opportunities as well as challenges in semiconductor design. Although it increases performance and efficiency, it also throws important questions of its viability in long term scaling and its impact on computing architecture. In the following discussion points, the reader can dive deeper into different facets of 3D stacking, and steer future semiconductor innovation.

1. **The Shift from Planar Scaling to 3D Stacking:** How does 3D stacking compare to traditional transistor scaling in terms of power efficiency, performance, and cost? Can 3D stacking fully replace planar transistor scaling, or will they continue to coexist?
2. **Industry Adoption:** Intel, TSMC, AMD, and Samsung have all developed their own 3D stacking solutions. Will 3D stacking become a standard in semiconductor manufacturing, or will its adoption remain limited to high-performance applications?
3. **Economic and Manufacturing Considerations:** The cost of implementing 3D stacking remains high. How can economies of scale help to reduce these expenses over time?
4. **Future of 3D Stacking in Computing:** How 3D stacking influence the design of processors for AI, edge computing, and mobile devices?

## 10.6 Limitations of the Case Study

Although this case study discusses 3D stacking technology, several limitations must be acknowledged. This is partly due to the proprietary restrictions, scope limitations and the dynamism of semiconductor research in general.

**Lack of Proprietary Data and Industry-Specific Insights:** Most of the 3D stacking advancements are created by leading semiconductor companies (Intel, TSMC, Samsung) that hold very tightly to their inhouse proprietary technologies. Consequently, this case study is based mainly on open literature and corporate disclosures. The specific processes, materials and performance benchmarks are not available for the case study. Therefore, further technical analysis is limited.

**Limited Scope of Comparative Analysis:** Intel's Foveros technology and more broadly the industry at large is the main subject of this case study. A more in-depth performance comparison between separate 3D stacking technologies (e.g., SoIC at TSMC or AMD – 3D V-Cache) is beyond the scope of this discussion. Broad industry trends have been stated but a closer look at the comparison would necessitate proprietary design information and possibly benchmark data.

**Ongoing Technological Developments:** The semiconductor industry is rapidly developing with developments on the through-silicon vias, hybrid bonding, and cooling landscape. Newer and further breakthroughs may appear. Conclusions in this case study are derived from current knowledge which may be outdated by new and better ways of stacking methods in future.

**Economic and Market Factors:** Although technical feasibility is central to this case study, it does not offer a detailed economic assessment of the cost–benefit trade-offs for 3D stacking. Several key factors significantly influence the market viability of this technology. Manufacturing costs, yield rates, and supply chain bottlenecks play critical roles in determining economic feasibility. The confidential nature of semiconductor manufacturer cost data prevents a complete financial assessment, as this information is not publicly disclosed.

This case study touches broadly on 3D stacking technology but with limitations such as lack of proprietary data access and no direct competitive comparisons. Eventually, further research may bridge these gaps through industry collaborations and empirical testing. It could utilize first-hand economic modeling to provide a more complete picture of 3D stacking capabilities.

## References


1 "Intel and other manufacturers' 8080 CPU images and descriptions," cpu-collection.de (accessed February 26, 2026).

2 Intel, "Intel Microprocessor datasheet," www.datasheetarchive.com (accessed February 26, 2026).

3 Shankar, R. and Fernandez, E.B. *VLSI and Computer Architecture*, vol. 20, 3. Academic Press, 2014.

4 El-Atab, N. and Hussain, M.M. "The future of CMOS: more Moore or a new disruptive technology?" *Advanced Nanoelectronics: Post-Silicon Materials and Devices*, 1–31, 2018.

5 Auth, C. and Shankar, A., "Evolution of transistors: humble beginnings to the ubiquitous present," *IEEE Solid-State Circuits Magazine*, 15(3), 20–28, 2023.

6 Tan, C., Yu, M., Tang, J., Gao, X., Yin, Y., Zhang, Y., Wang, J., Gao, X., Zhang, C., Zhou, X. and Zheng, L. "2D fin field-effect transistors integrated with epitaxial high-k gate oxide," *Nature*, 616(7955), 66–72, 2023.

7 Karimi, K., Fardoost, A. and Javanmard, M. "Comprehensive review of FinFET technology: history, structure, challenges, innovations, and emerging sensing applications," *Micromachines*, 15(10), 1187, 2024.

**8** Haq, S.U. and Sharma, V.K., "Review of the nanoscale FinFET device for the applications in nano-regime," *Current Nanoscience*, 19(5), 651–662, 2023.

**9** Guo, X., Verma, V., Gonzalez-Guerrero, P., Mosanu, S. and Stan, M.R. "Back to the future: digital circuit design in the FinFET era," *Journal of Low Power Electronics*, 13(3), 338–355, 2017.

**10** Zimpeck, A., Meinhardt, C., Artola, L. and Reis, R., "FinFET technology," in *Mitigating Process Variability and Soft Errors at Circuit-Level for FinFETs*, 7–27. Springer, 2021.

**11** Reddy, M.N. and Panda, D.K. "A comprehensive review on FinFET in terms of its device structure and performance matrices," *Silicon*, 14(18), 12015–12030, 2022.

**12** Radamson, H.H., Zhu, H., Wu, Z., He, X., Lin, H., Liu, J., Xiang, J., Kong, Z., Xiong, W., Li, J. and Cui, H. "State of the art and future perspectives in advanced CMOS technology," *Nanomaterials*, 10(8), 1555, 2020.

**13** Singh, M., Chaudhary, T., Raj, B., Wadhwa, G. and Tripathi, S.L. "Comparative analysis of MOSFET and FinFET," in *Integrated Devices for Artificial Intelligence and VLSI*, ed. B. Raj, S.L. Tripathi, T. Chaudhary, K. Srinivasa Rao, and M. Singh, 1–24. Wiley, 2024.

**14** Shan, T. "Advantages of FinFET over traditional CMOS: reasons and implications," *Theoretical and Natural Science*, 14, 219–223, 2023.

**15** Duan, H. "From MOSFET to FinFET to GAAFET: the evolution, challenges, and future prospects," *Applied Computer Engineering*, 50(1), 113–120, 2024.

**16** Karimi, K., Fardoost, A. and Javanmard, M., "Comprehensive Review of FinFET Technology: History, Structure, Challenges, Innovations, and Emerging Sensing Applications," *Micromachines*, 15(10), 1187, 2024.

**17** Srivastava, S. and Acharya, A. "Challenges and future scope of gate-all-around (GAA) transistors: physical insights of device-circuit interactions," in *Device Circuit Co-Design Issues in FETs*, ed. S. Tayal, B. Smaani, S.B. Rahi, S. Labiod, and Z. Ramezani, 231–258. CRC Press, 2023.

**18** Suman, J.V., Swathi, Y. and Kumar, R.S. "Fabrication techniques and material selection for gate-all-around FET (GAAFET) in advanced nanoelectronics," in *2024 Second International Conference on Advances in Information Technology (ICAIT)*, vol. 1, 2024.

**19** Kandpal, J. and Goel, E. "Transition from conventional FETs to novel FETs, SOI, double gate, triple gate, and GAA FETs," in *Nanoscale Field Effect Transistors: Emerging Applications*, 25–46. Bentham Science Publishers, 2023.

**20** Xie, Y. and Zhao, J. *Die-stacking architecture*, Springer Nature, 2022.

**21** Noor, R., Kottur, H.R., Craig, P.J., Biswas, L.K., Khan, M.S.M., Varshney, N., Dalir, H., Akçalı, E., Motlagh, B.G., Woychik, C. and Yoon, Y.K. "US microelectronics packaging ecosystem: challenges and opportunities," arXiv preprint: arXiv:2310.11651, 2023.

# 11

# AMD–NVIDIA GPU Revolution: Power and Performance Dominance

Gaming, Data Centers, and AI - Engineering the Future of Graphics Computing

## 11.1 Synopsis

What happens when two biggest technological companies fight to become not only better in terms of performance, but also more economical to an energy conscious world? Imagine two elite marathon runners' records, each striving to finish first but at the same time using the least amount of energy possible. This analogy mirrors the raging rivalry between AMD and NVIDIA in the arena of the graphics processing unit (GPU). They are locked in a continuous battle to deliver power-efficient computing solutions to consumers and businesses alike [1].

In the fast-paced electrical engineering and semiconductor technology world, GPUs have transitioned from being a vogue component for the gaming community as well as critical hardware powering datacenters, AI, and cryptocurrency mining [2]. As the applications continue expanding, the need for energy-efficient GPUs that can work at high speed has exploded. The industry leaders AMD and NVIDIA, have decades of innovation under their belts. They never stop exploring both the frontiers of engineering design, semiconductor materials, and power management techniques to satisfy the new demands [3].

This case study unravels some of the central engineering decisions, strategic shifts, and game changing tech breakthroughs that are driving the current battle between AMD vs. NVIDIA. It identifies the key stakeholders, ranging from individual consumers and gamers up to data center operators and regulatory agencies. It highlights how each company's decisions on power efficiency have shaped the landscape of the market, industry standards, and the environment.

## 11.2 Background and Case Scenario

Have you ever wondered how the graphics on your screen evolved from simple pixelated images to ultra-realistic visuals that we know today? At the heart of this transformation lies the Graphics Processing Unit, a specialized electronic circuit designed for hard real-time processing and rendering images, animations, and video [4]. Imagine the GPU being the artist behind the canvas that draws first ambiguous images and gradually gets better until he's an ace. This journey of technologic artistry was written by fierce competition and forward-thinking engineers that dared to push boundaries.

### 11.2.1 Origins and Milestones of GPU Technology

How did GPUs move from being just an assistant handling basic tasks to being today's high-powered computation device? Just like vehicles evolved from simple design into sophisticated machines capable of navigating autonomously. GPUs have similarly expanded their role and become central to artificial intelligence [5]. Scientific workloads and general simulations run on GPUs rather than CPUs [6, 7].

Modern GPUs have evolved far beyond rendering the graphics for primitive screens, to dependable co-processors for doing hard number-crunching chores. Thanks to well-developed hardware and software stacks that provide IEEE-compliant and double precision support [7]. GPU technology at large has been driven by the requirements of the gaming industry. It has driven substantial investments and developments with consumer hardware that leads the way into inexpensive yet power-efficient solutions suitable for scientific high-performance computations [8]. I start with a brief history of GPU technology from early graphical accelerators to modern GPUs. Tracking the history and milestones of GPU technology is like exploring the evolution of transportation – from humble beginnings to cutting-edge innovations.

The idea of graphical processing, also known as computer graphics, can be traced back in the 1960s. Notably, Ivan Sutherland's first Sketchpad in 1963 [9]. Sketchpad is the first interactive graphical interface which established the fundamentals to all computer graphics stand today. It gave users a method of drawing and editing digital graphics directly. This was a huge leap forward to the rapid evolution of graphical user interfaces (GUI) and prepared the groundwork for even better graphics processing in years to come. Figure 11.1 shows the photo and illustration from Sutherland's (1964) MIT paper, and it is also part of his doctoral thesis. It shows the TX-2 operating area where the Sketchpad is in use. Sutherland is holding the light pen and the push buttons, "draw" and "move" are on the box in front of the author.

In the late 1970s and early 1980s, there was dedicated graphics hardware for the purpose of aiding the CPU in high-end graphical activities [10]. The advancements landed in the market due to the popularity of arcade games, and companies designed specialized chips to implement more complex visuals [11]. But, at that time, the chips were specifically meant for more basic functions such as drawing lines, filling polygons, and handling sprites; all of which helped the graphics render smoother and look better than a CPU probably could [12].

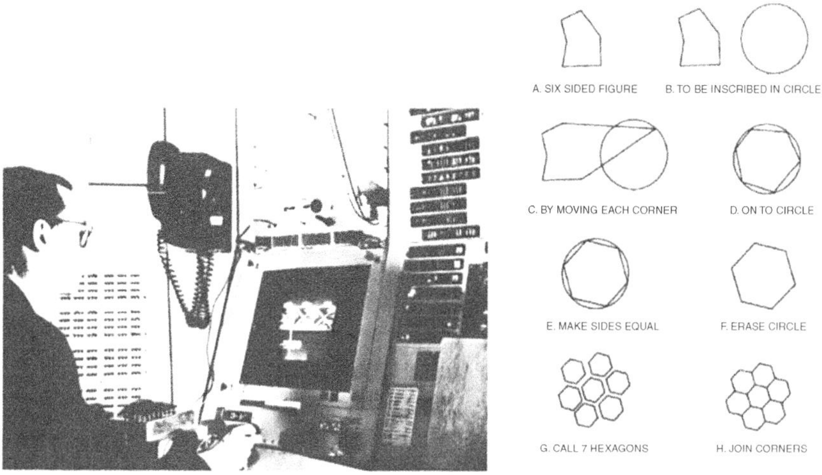


**Figure 11.1** Photo of the first Sketchpad by Ivan Sutherland in 1963 and the first illustrative examples. *Source:* [9] / with permission of ACM (The Association for Computing Machinery).

GPUs were only starting to take on recognizable forms in the 1980s and 1990s. The introduction of a geometry engine from Silicon Graphics in 1982 was a huge milestone in the history of GPU technology [13]. This pioneering graphics processor was intended to take the burden of difficult 3D rendering from the main CPU, which resulted in improved and high-quality graphic processing. In the next few years, the capacity of the geometry engine was improved to perform geometric transformations, lighting calculations, and primitive rendering, which are the predecessors for more advanced GPU designs.

As personal computers entered the gaming arena, graphic cards needed a step-up in resolution, color depth, and essentially performance. Historically, early manufacturers such as Hercules and CGA focused on addressing these deficiencies [10]. The graphic cards usually had very elementary features, only capable of driving a standard display but also lacked the 3D capabilities.

In the late 1990s, there was a time of rapid expansion of better graphics processing power in 3D gaming and personal computers. They were advancing quickly but early 2D graphical accelerators were used mostly for simple 2D rendering. The first GPUs had only fixed functions, which means they did exactly one thing based on fixed instructions, and this could not be easily modified or programmed otherwise [14].

A significant breakthrough happened in 1999 when NVIDIA launched GeForce 256, often considered as the world's first GPU [15]. This is when the term "GPU" was popularized [16]. It contained a single chip with the engines to transform lighting and rendering as well as the triangle setup and clipping of its early reconfigurable graphics processing hardware [17].

Later advances got the rise of dedicated graphics processing companies like ATI Technologies and NVIDIA. They became the leading players in the GPU market. They built increasingly powerful GPUs which enabled support for more complicated 3D graphics and computing workloads. It largely enabled the mass adoption of GPU technology in personal computers as well as gaming consoles. NVIDIA for instance developed itself in the GPU world by introducing creative chip and software solutions. These solutions led to leading-edge graphics as well as outstanding high-performance computing.

#### 11.2.1.1 Advancements and Technological Breakthroughs

The introduction of programmable shaders in the early 2000s significantly advanced GPU capabilities [18]. Shaders provided developers with the ability to programmatically set graphical rendering and eventually helped professionals to get realistic and interactive images [19]. That move from the fixed-function pipeline to programmable architecture paved the way for gaming, simulation, and visualization workloads. Programmable Shaders provided developers with a means to customize the graphics rendering pipeline, by tweaking and optimizing how 3D scenes were being rendered programmatically [20]. This flexibility created a fertile ground for new visual effects, more accurate lighting and materials as well as complex post-processing effects. Such advanced effects were impossible with the prior fixed-function graphics hardware. The shift toward programmable architecture was a huge turning point in the GPU evolution, which leveraged GPUs for massive amounts of use cases besides graphics rendering.

In recent years the GPU has matured beyond graphics rendering into the core hardware for artificial intelligence and machine learning. GPUs are also versatile and can be used for high-performance computing and deep-learning workloads thanks to architecture like modern NVIDIA Ampere (A100) GPUs. Figure 11.2 shows the key GPU technology pioneers who have inspired GPUs to what it is today. From Ivan Sutherland who laid the fundamental principles for interactive graphics, to David Kirk of NVIDIA and his programmable GPU architectures: all are remarkable. Then under Jensen Huang's leadership at NVIDIA, the company was placed at the epicenter of AI acceleration. There is also Lisa Su, who redefined both segments of AMD's GPUs and CPUs to revitalize competition and innovation.

Beyond NVIDIA and AMD, many other emerging companies are becoming key players in the development of GPU. In China, companies like Moore Threads, Illuvatar CoreX, and Jingjia Micro are becoming a major force in the GPU arena driving semiconductor innovation. In embedded systems, gaming consoles, and mobile GPUs for specialized markets, companies like Realtek, Sony, Arm, and Imagination Technologies contribute to the development of GPUs. For networking,

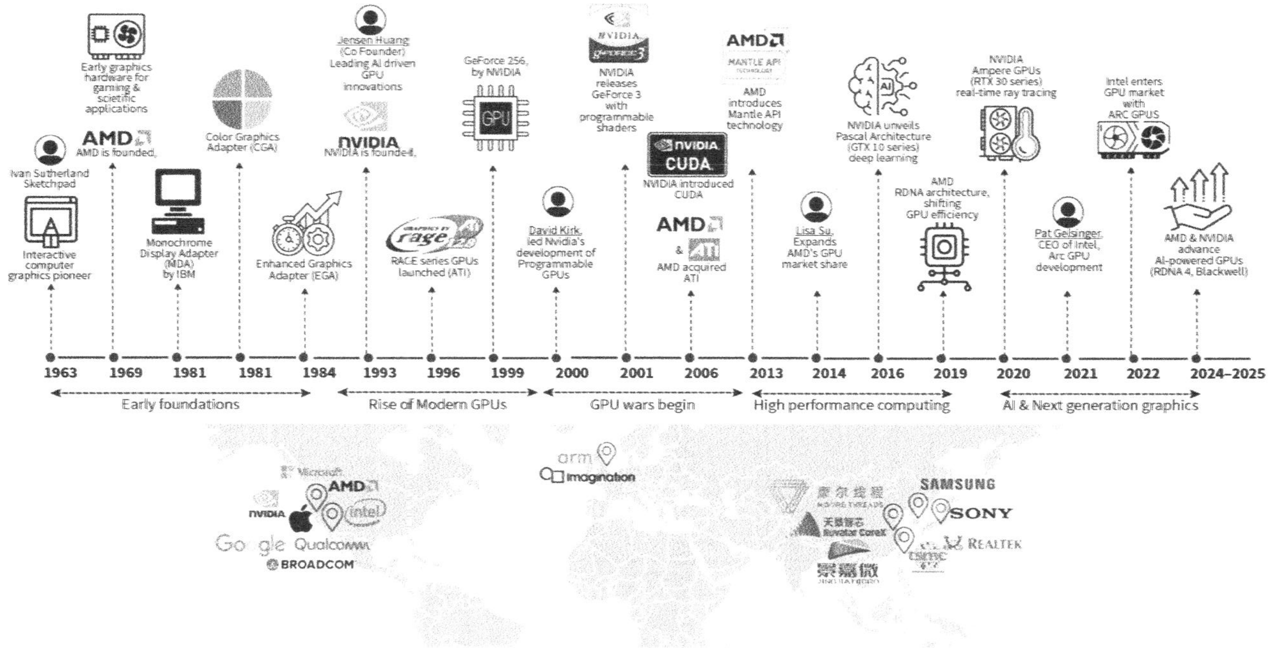


**Figure 11.2** Timeline of GPU innovation and global landscape featuring pioneers, prominent companies, and hubs.
*Source:* Author's own compilation.

AI and smartphone chipsets, Broadcom and Qualcomm are the most prominent among those using GPU technology in broad applications.

The geographical spots of GPU innovation, in other words, the global hot spots for GPU innovation are Silicon Valley and Taiwan. Silicon Valley being home to NVIDIA, AMD, and other high technology giants while TSMC in Taiwan is a leading semiconductor foundry that produces advanced GPU chips globally for other enterprises. As industry evolves, there are more such companies and regions at the epicenter of GPU innovation.

GPUs were originally very simple, and capable of ultra-low-level rendering with little computing power. But modern GPUs are much more advanced and even a toy powered by such GPU could execute trillions of simple computations per second. Today, there are GPUs available in the market that can do much more than just graphics. Many other applications are powered by this technology including artificial intelligence, simulations for science, and mining cryptocurrency. This multifaceted nature speaks a lot about its versatility. GPUs have opened entire new worlds of computing in the past few years, supporting researchers, developers, and enterprises to solve problems which previously did not have an answer.

### 11.2.2 AMD and NVIDIA: Shaping the GPU Landscape

The prominent companies who laid the foundation of today's GPU landscape are AMD and NVIDIA. Digging into their long rivalry, market games, and new technologies can help us to grasp how the GPU business has grown and where it is now. They have both established unique hardware, software, and architecture designs. It directly impacts the expansion of today's gaming and Computer-Generated Imagery (CGI) industries including professional visualization and high-performance computing.

The competition between AMD and NVIDIA has been a primary catalyst for innovation in the GPU market [21]. Advanced Micro Devices (AMD), originally founded in 1969, is primarily a semiconductor company specializing in microprocessors and integrated circuits. Over time, AMD expanded its line-up with some very aggressive moves when it acquired ATI Technologies for its GPU in 2006. That strategic move enabled AMD to compete in both CPU and GPU markets to offer solutions in data as well as graphics processing.

NVIDIA Corporation, AMD's main competitor, was founded much later, in 1993. Its initial products being purely graphical processors for gaming and multimedia. But NVIDIA quickly rose to prominence with the release of GeForce 256 in 1999, also known as the first GPU in world. Through its emphasis on innovation (graphics rendering and parallel computing in particular), the company was able to win a large share of the market and became an important player in GPU technology.

Matrox, a Canadian company founded in 1976, was another major player, especially in professional 2D graphics and multi-display solutions. Matrox had a successful track-record of producing unique graphics cards and solutions, with both medical imaging and industrial markets benefiting from this. While Matrox is still active, its consumer graphics market share grew extremely small as it focused more on niche professional applications.

Imagination Technologies, a British semiconductor company, specialized in developing and licensing GPU intellectual property. Their most notable development is the PowerVR graphics used in early mobile and embedded systems (Apple's first iPhones and iPads). Imagination Technologies has had a larger presence in mobile and embedded GPU markets over the desktop GPU market, but it has not been competing head-to-head with giants like NVIDIA and AMD.

Intel, traditionally dominant in the CPU market, has periodically ventured into the GPU market over the years. Most recently, Intel has aggressively re-entered the GPU arena with the launch of its Intel Arc series of graphics cards. For Intel, this is a strategic play to broaden its portfolio and future footprint in the rapidly growing GPU space where graphics processing power advancements are key to so many applications domains. The fierce competition among these firms caused the GPU market to move quickly in technological advancements and innovation.

The intense rivalry between AMD and NVIDIA only escalated after AMD bought ATI Technologies in 2006. Until the acquisition, NVIDIA was pretty much ATI's only competition. ATI was a Canadian semiconductor powerhouse known for their Radeon GPU array. This period was a time of high competition, as technology advanced quickly (from programmable shaders to the first real multi-GPU systems) and aggressive marketing strategies began to wear on gamers and PC fans. AMD's acquisition of ATI changed the competition character and hence AMD could now rival NVIDIA in CPU as well as GPU markets. The rivalry escalated as both companies poured a lot of money into R&D to battle for technological supremacy.

Figure 11.3 illustrates the primary factors of competition between AMD and NVIDIA. The competition is shaped by leadership in technology. NVIDIA has its own proprietary innovations (CUDA and ray tracing) while AMD emphasizes open standards, and competitiveness. Gaming is still where the big market share battles play out, along with data centers and professional visualization. NVIDIA dominates discrete GPUs and AMD excelling in integrated graphics. More strategic positioning further differentiates them as NVIDIA focuses on premium brand and cutting-edge technology while AMD prioritizes value and open platform collaboration. Additionally, branding wars and marketing strategies, including sponsorships and targeted campaigns, play a crucial role in shaping consumer loyalty and industry influence.

NVIDIA's CUDA (2007) is a closed-source programming model and software environment. The result was that GPUs could perform general purpose computing. This greatly expanded the arena in which to deploy GPUs beyond legacy

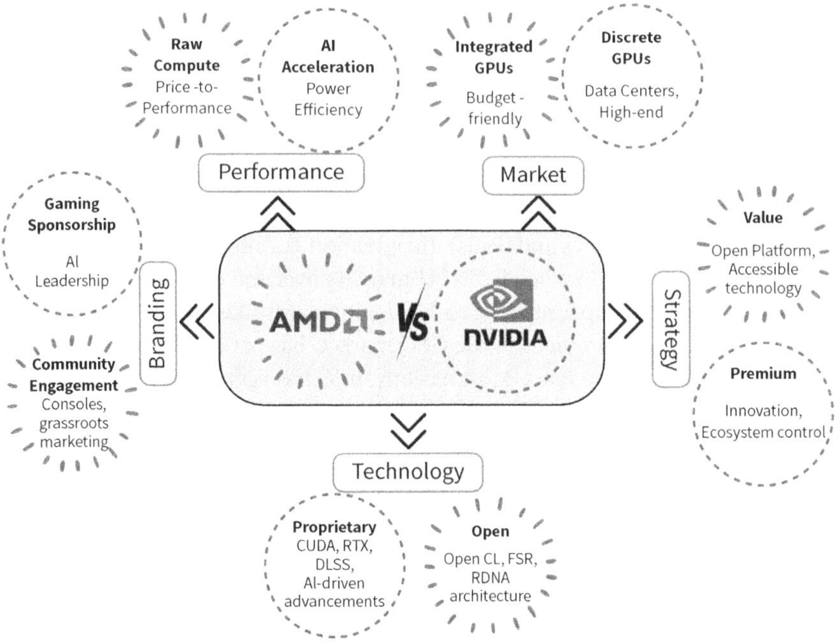


**Figure 11.3** Key factors driving the AMD–NVIDIA rivalry. *Source:* Author's own work.

graphics rendering. It sparked various areas of development such as scientific computing, machine learning algorithms, and parallel processing. Responding to this, AMD created its own standards (OpenCL) to create an extremely competitive environment that drove both of them to keep innovating in GPU computing. The establishment of CUDA, as well as subsequent competitive progress from both AMD and NVIDIA opened the door for the quick evolution and proliferation of GPUs across many industries and applications.

NVIDIA has been heavily influenced by the gaming market [22]. NVIDIA partnered with major game developers and computer makers to allow its chips to work with popular games. Despite a drop in the gaming business in the second fiscal quarter of 2023, NVIDIA's market share remained high at 79%, even increasing from the previous quarter, although still slightly less than the previous year [23].

The rivalry also went beyond technology and into branding and partnerships. Focusing on the GeForce brand and proprietary technology, NVIDIA differentiated themselves from AMD. But AMD focused on Free sync to compete as open initiatives plus working with strategic partnerships (Sony, Microsoft) in the area of games consoles. The ambition of each company was to dominate the market for advanced gaming graphics performance and attract emerging industries like artificial intelligence, data centers, and professional visualization.

The competition between AMD and NVIDIA helped the GPU market to reach new heights, and spurred advances in processing power and new applications.

In order to fully understand the separation of their dominance in the market, we need to examine the base technologies that underly their current strategies: NVIDIA's CUDA and AMD's RDNA.

NVIDIA's CUDA (Compute Unified Device Architecture) is a parallel computing platform, and this programming model was developed in 2006 [24]. Through its CUDA parallel computing platform, NVIDIA revolutionized the semiconductor industry and expanded GPU access beyond traditional graphics rendering and enabled general purpose parallel processing across diverse computational workloads. CUDA operates across the entire computing stack, from GPU virtual instruction sets and on-chip software to end-user applications. This positions CUDA as the central enabler of the AI and data center revolution. NVIDIA's powerful proprietary software ecosystem made its hardware the critical foundation for deep-learning, scientific simulations, and high-performance computing applications.

AMD's RDNA (Radeon DNA) was released in 2019 to replace GCN (Graphics Core Next) architecture and it focused heavily on improving gaming performance [25]. While GCN was a graphics/compute hybrid architecture, RDNA was designed from the bottom up with high performance gaming in mind. It is designed to have low latency, high power efficiency (performance-per-watt), and a highly optimized instructions pipeline. This design offered AMD the ability to compete again in the ever-growing gaming space. It was the technology that drives major gaming consoles such as PlayStation 5 and Xbox Series X/S.

### 11.2.3 The Race for Power Efficiency: Stakes and Significance

Power efficiency has emerged as a critical factor in GPU development. This is due to growing environmental concerns, rising energy costs, and increased reliance on GPUs for intensive applications. The power consumption of GPU-datacenters, for instance, can be up to 5–6 times more than typical commercial CPU-based servers. Let us consider older data centers that relied on CPUs, typically ran at roughly 150–200 watts per chip [24]. But, GPUs for AI use an average of 400 watts (2022), and by 2023, modern GPUs run at 700 watts with the expectation of this increasing to 1200 watts for the next generation chips [25]. This is because GPUs are around four times more power-dense than CPUs [26]. Cryptocurrency mining operations also consume a seemingly large amount of resources, around 0.5% of global electricity consumption [27]. As GPU becomes central to these energy-intensive processes, the ability to deliver high-performance and ultra-low energy efficiency becomes even more economically and ecologically pressing.

Achieving optimal power efficiency is crucial to deal with the environmental impact of GPU-based technologies. In order to address the increasing demand for energy-efficient computing solutions across various sectors, manufacturers must not stop innovating and optimizing their GPU designs. The challenge is to strike a balance between computing power and energy efficiency.

AMD and NVIDIA, the two leading GPU manufacturers, have adopted distinct strategies to enhance power efficiency. AMD has pushed architectural innovation and advanced semiconductor manufacturing nodes. Architectures (i.e., RDNA3 and RDNA4) made substantial improvements in terms of lower power consumption by up to 50% on top of the previous generation [28]. This was achieved through optimized performance-per-watt designs and a shift to smaller semiconductor nodes using TSMC's 5 nm processes and below, which offer up to 30% reduction in power draw [29].

NVIDIA, on the other hand, has also pursued architectural advancements, notably with its Ampere and Hopper GPU architecture. NVIDIA has targeted improving its GPU micro-architecture and power management technologies. This new architecture uses intricate AI-software based techniques that adds driving capabilities of dynamically adapting workloads. It results in a two times improvement of energy efficiency over the previous NVIDIA GPU generations. It also offers advanced power gating and clocking abstractions [30].

For example, the Ampere-based A100 GPU (a high-performance graphics processor designed for AI, data centers, and scientific computing) achieves up to 19.5 TFLOPS (Tera Floating-Point Operations Per Second) in computational speed from its high-performance graphics processor [31]. TFLOPS is the number of floating-point calculations per second that a processor can perform. It is a measure of how powerful the processor's GPU is. This power efficiency increase can be seen in the A100 that consumes approximately 400 W and achieves this result.

Similarly, the latest Hopper-based H100 GPU, optimized for AI workloads, offers up to 30 TFLOPS of AI performance with a huge power reduction of approximately four times the new benchmarks vs. previous NVIDIA GPUs [32]. These hardware and software breakthroughs have enabled NVIDIA to remain the market leader in power-competitive GPU computing for data centers, AI, and HPC (High-Performance Computing). The agility by which AMD and NVIDIA approach power efficiency is once again a testament to minimizing GPU technology's overall environmental impact as well as improving future energy sustainability [33].

Table 11.1 shows the comparative power consumption and performance efficiency of CPUs and GPUs from older data centers to (selected) modern AI-focused GPUs. The chart highlights NVIDIA's Ampere (A100) and Hopper (H100) GPUs, as well as AMD's RDNA3 improvements, showing significant TFLOPS per watt advancements over time.

Better power efficiency has wide consequences for major stakeholders. Consumers get to pay less on their utility bills, and get better performance in gaming and multimedia applications. For instance, this can include reduced operational costs for

**Table 11.1** Power consumption and efficiency improvements in GPUs.

| Year | Processor Type | Power Consumption (Watts) | Performance (TFLOPS) | TFLOPS per Watt |
|---|---|---|---|---|
| 2010 | CPU (Data Centers) | 150–200 W | ~0.5 TFLOPS | ~0.0025–0.0033 |
| 2015 | Early GPU for AI | 250 W | ~5 TFLOPS | ~0.02 |
| 2020 | NVIDIA Ampere A100 | 400 W | 19.5 TFLOPS | ~0.0487 |
| 2022 | AMD RDNA3 | 300 W | 61 TFLOPS | ~0.2033 |
| 2023 | NVIDIA Hopper H100 | 700 W | 30 TFLOPS | ~0.0428 |
| 2024+future projections | Next-Gen AI GPUs | 1200 W (Projected) | 100+ TFLOPS (Projected) | ~0.0833+ |

*Source:* [34, 35].

companies (especially data centers and cloud service providers) along with increased sustainability profiles. This allows developers to leverage more capable and optimal hardware, resulting in more complex applications being developed. Similarly, regulatory control incentivizes manufacturers to innovate for upcoming products in terms of power efficiency.

At a larger scale, the competition of AMD and NVIDIA in power efficiency is pushing the entire industry toward more sustainable development practices and influencing the direction of future technology. With increased awareness of climate change worldwide, it is likely to see a higher demand for energy-efficient computing solutions that drive industries toward more ambitious sustainable technologies. The ever-progressing efforts from these two companies not only creates benchmarks in the industry but also enable frameworks and environmental standards for new semiconductor technologies in full swing.

## 11.3 Think and Solve

The GPU evolution is a fast-moving train on computing power, energy throughput, and AI acceleration. Despite this progress however, it has not been without several challenges where the two heavy hitters, NVIDIA and AMD, battled back and forth. Competition is born from technology first, then market dominance, also strategic collaborations and branding. NVIDIA has been exclusively pushing premium and proprietary tech while AMD has been pushing open standards and cost-effective solutions. This competitive environment is forcing every company to face the challenges of the perpetual innovation cycles, pricing, and differentiation in the market.

To assess the reader's understanding of the case study, the following questions summarize key aspects of the AMD–NVIDIA rivalry and the GPU industry.

**Q 11.1** What is a key differentiator in NVIDIA's approach to GPU technology compared to AMD?

**A** Open-source software and collaboration.
**B** Focus on proprietary technologies like CUDA and ray tracing.
**C** Emphasis on low-power integrated graphics solutions.
**D** Manufacturing its own semiconductor chips.

**Q 11.2** Which market segment has traditionally been dominated by NVIDIA?

**A** Integrated graphics for budget laptops.
**B** Professional visualization and discrete gaming GPUs.
**C** Mobile GPU development for smartphones.
**D** Automotive infotainment systems.

**Q 11.3** Why is TSMC crucial to the GPU industry?

**A** It manufactures GPUs for multiple companies, including NVIDIA and AMD.
**B** It develops proprietary GPU architectures.
**C** It is the largest supplier of gaming consoles.
**D** It owns the patents for ray tracing technology.

**Q 11.4** What strategic factor allows AMD to compete with NVIDIA despite having a smaller market share?

**A** Exclusive partnerships with Intel.
**B** A monopoly over gaming consoles.
**C** A focus on value-driven GPUs and open-source initiatives.
**D** Developing GPUs specifically for cryptocurrency mining.

**Q 11.5** A GPU's performance can be estimated using the theoretical floating-point operations per second (FLOPS). Formula: FLOPS = cores × clock speed × operations per cycle

Using the specifications below, calculate the theoretical FLOPS performance of this GPU.

**Given:**

- A GPU has 6,000 cores
- A clock speed of 1.5 GHz
- Each core performs 2 operations per cycle

## 11.4 Conclusion and Key Takeaways

This rivalry between AMD and NVIDIA has been the spark that ignites GPU innovation, gaming breakthroughs, artificial intelligence, and high-performance computing. Despite NVIDIA having its own proprietary technologies (e.g., CUDA and ray tracing), AMD has taken hold of a meaningful market share by commoditizing costs and open-source collaboration. It should not be forgotten that semiconductor manufacturers such as TSMC play an important role too, because they provide the competition with their leading fab technologies.

Key takeaways from this case study include but not limited to the following:

**Technological Differentiation:** NVIDIA focuses on proprietary software and premium hardware, while AMD emphasizes affordability and open platforms.

**Market Positioning:** NVIDIA dominates discrete GPUs for gaming and professional applications, while AMD competes aggressively in integrated graphics and gaming consoles.

**Strategic Influence:** Branding, partnerships, and software ecosystems play a vital role in each company's success.

**Future Trends:** The evolution of AI, cloud computing, and power efficiency continues to shape the GPU landscape, making competition even more dynamic.

Ultimately, this competition is good for the consumers because of innovation, performance improvement, and to bring more powerful computing solutions to the hands of masses. The interaction of this rivalry gives us a great understanding of how the semiconductor industry keeps evolving.

## 11.5 Discussion Points

The competitive environment in GPU industry between NVIDIA and AMD exposes important themes like innovation, market standing and different technologies. As these companies keep bringing innovation and performance on GPUs, AI workloads, and energy efficiency, we see their efforts provide lessons for the struggle of high-tech competition. Here are some discussion points to dig deeper into the areas mentioned above.

1. How do proprietary and open-source approaches impact innovation and market competition in the GPU industry?
2. What factors contribute to NVIDIA's dominance in discrete GPUs, and what strategies could AMD adopt to challenge its lead?
3. How do GPU advancements influence AI, deep-learning, and scientific computing beyond gaming?
4. What role do semiconductor foundries like TSMC play in shaping the competitive landscape of the GPU market?
5. How do power efficiency and sustainability concerns affect GPU development and adoption in modern data centers?

## 11.6 Limitations of the Case Study

This case study provides a good summary of the AMD–NVIDIA rivalry and its implication to the GPU industry, but it also has some limitations. First, the analysis is very company specific on these two. It does not include the influence from other emerging players like Intel, Moore Threads, and Imagination Technologies. Second, it emphasizes historical and strategic perspectives but does not account for real-time market fluctuations or future technological disruptions. In addition, factors like supply chain tailwinds, geopolitical forces, and environmental constraints are not covered in detail. Lastly, this study is based on publicly available information, datasheets, industry reports, academic journals, and documents.

## References

1 Peddie, J. *The History of the GPU-New Developments*, Springer, 2022.
2 Huq, N., Lin, P., Reyes, R., and Perine, C. "A survey of cloud-based GPU threats and their impact on AI, HPC, and cloud computing," Trend Research, Technical Report, 2024.
3 Jain, R. *Advancements in AI and IoT for Chip Manufacturing and Defect Prevention*, CRC Press, 2024.
4 Wyrwas, E. "Body of knowledge for graphics processing units (GPUs)," (No. 2018-561-NEPP), 2018.
5 Chopra, B. "Enhancing machine learning performance: the role of GPU-based AI compute architectures," *Journal of Knowledge Learning and Science Technology*, 6386, 29–42, 2024.
6 Bridges, R.A., Imam, N. and Mintz, T.M. "Understanding GPU power: a survey of profiling, modeling, and simulation methods," *ACM Computing Surveys (CSUR)*, 49(3), 1–27, 2016.

**7** Georgescu, S., Chow, P. and Okuda, H. " GPU acceleration for FEM-based structural analysis," *Archives of Computational Methods in Engineering*, 20, 111–121, 2013.

**8** Volkema, G. and Khanna, G. "Scientific computing using consumer video-gaming hardware devices," arXiv preprint arXiv:1607.05537, 2016.

**9** Sutherland, I.E. "Sketch pad a man-machine graphical communication system," in *Proceedings of the SHARE Design Automation Workshop*, 6–329, 1964.

**10** Peddie, J. "1980–1989, graphics controllers on other platforms," in *The History of the GPU-Steps to Invention*, 31–97. Springer International Publishing, 2023.

**11** Peddie, J. "The first era of GPUs," in *The History of the GPU-Eras and Environment*, 105–150. Springer International Publishing, 2023.

**12** Silvano, C., Ielmini, D., Ferrandi, F., Fiorin, L., Curzel, S., Benini, L., Conti, F., Garofalo, A., Zambelli, C., Calore, E. and Schifano, S. "A survey on deep learning hardware accelerators for heterogeneous hpc platforms," *ACM Computing Surveys*, 57, 1–39, 2025.

**13** Sawicki, M., and Moody, J. "A brief history of computer graphics," in *Filming the Fantastic with Virtual Technology*, 21–58. Routledge, 2020.

**14** Owens, J.D., Houston, M., Luebke, D., Green, S., Stone, J.E. and Phillips, J.C. "GPU computing," *IEEE* 96(5), *879–899*, 2008.

**15** von Zitzewitz, V.L.C. "NVIDIA's bet on artificial intelligence," Master's Thesis, Universidade NOVA de Lisboa (Portugal), 2024.

**16** Islam, M.S., Subhani, M., Tareque, M., and Rafi, M. "Real-time 3D integral imaging system using a faster elemental image generation method using GPU parallel processing," BSc Thesis, BRAC University, 2017.

**17** Hu, D., Jia, A.L., Jien, F.Y., Ai, X. and Chen, C.Z. "Products of digital integrated circuit," in *Handbook of Integrated Circuit Industry*, ed. Y. Wang, M.-H. Chi, J.J.-C. Lou, and C.-Z. Chen), 187–232, 2023.

**18** Dally, W.J., Keckler, S.W. and Kirk, D.B. "Evolution of the graphics processing unit (GPU)," *IEEE Micro*, 41(6), 42–51, 2021.

**19** Parisi, T. *Programming 3D Applications with HTML5 and WebGL: 3D Animation and Visualization for Web Pages*, O'Reilly Media, Inc., 2014.

**20** Nair, R. "Using Raymarched shaders as environments in 3D video games," Drexel University, 2020.

**21** Sousa, B., Alves, C., Mendes, M. and Au-Yong-Oliveira, M. "Competing with Intel and Nvidia: the revival of advanced micro devices (AMD)," in *16th Iberian Conference on Information Systems and Technologies (CISTI)*, IEEE, 2021.

**22** Peng, C. "Nvidia Corporation: PC gaming industry strategic audit," Undergraduate Honors Thesis, University of Nebraska-Lincoln, 2023.

**23** Wang, Y. "NVIDIA's operating strategy analysis based on multiple valuation method and SWOT," *BCP Business & Management*, 45, 24–34, 2023.

**24** Ebrahimi, K., Jones, G.F. and Fleischer, A.S. "A review of data center cooling technology, operating conditions and the corresponding low-grade waste heat recovery opportunities," *Renewable and Sustainable Energy Reviews*, 31, 622–638, 2014.

**25** Shah, A." Generative AI to account for 1.5% of world's power consumption by 2029," *HPCWire*, 8 July 2024, www.hpcwire.com/2024/07/08/generative-ai-to-account-for-1-5-of-worlds-power-consumption-by-2029 (accessed March 2, 2026).

**26** Studer, M. "The energy challenge of powering AI chips," 6 November 2023, www.robeco.com/en-int/insights/2023/11/the-energy-challenge-of-powering-ai-chips (accessed March 2, 2026).

**27** Fadeyi, O., Krejcar, O., Maresova, P., Kuca, K., Brida, P. and Selamat, A. "Opinions on sustainability of smart cities in the context of energy challenges posed by cryptocurrency mining," *Sustainability*, 12(1), 169, 2019.

**28** "AMD RDNATM architecture," *AMD*, 2024, www.amd.com/en/technologies/rdna.html (accessed March 2, 2026).

**29** TSMC. "5nM technology," 2024, https://www.tsmc.com/english/dedicatedFoundry/technology/logic/l_5nm (accessed March 2, 2026).

**30** Luo, W., Fan, R., Li, Z., Du, D., Wang, Q. and Chu, X. "Benchmarking and Dissecting the Nvidia Hopper GPU Architecture," in *IEEE International Parallel and Distributed Processing Symposium (IPDPS)*, 656–667, 2024.

**31** Krashinsky, R., Giroux, O., Jones, S., Stam, N. and Ramaswamy, S. "Nvidia ampere architecture in-depth," *NVIDIA Technical Blog*, 14 May 2020, https://developer.nvidia.com/blog/nvidia-ampere-architecture-in-depth (accessed March 2, 2025).

**32** Nvidia. "NVIDIA H100 Tensor Core GPU datasheet," 2024, https://resources.nvidia.com/en-us-gpu-resources/h100-datasheet-24306 (accessed March 2, 2026).

**33** Kyung, C.M. and Yoo, S. "Energy-Aware System Design: Algorithms and Architectures," 6–7, Springer, 2011.

**34** NVIDIA, "NVIDIA data center GPU resource center," 2025, https://resources.nvidia.com/l/en-us-gpu (accessed March 2, 2025).

**35** AMD, "AMD graphics cards," 2025, https://www.amd.com/en/products/graphics/desktops/radeon.html (accessed March 2, 2026).

# 12

# Silicon Carbide Breakthrough: STMicroelectronics' EV Leadership

How a 200-Year-Old Crystal Discovery Became the Key to EV Efficiency

## 12.1 Synopsis

What if you could drive your electric car 60 kilometers further on a single charge, by simply changing one of the car's smallest electronic components? Imagine the function of the light switch that turns ON and OFF and if that function could control massive amounts of power while barely getting warm. That is what the cutting-edge silicon carbide technology from STMicroelectronics did.

Silicon carbide (SiC) is a high temp and power handling compound of silicon and carbon that has much higher temperature tolerance than common silicon. Imagine this as an ultra-formal super-silicon, that can work in above +200 degrees Celsius with less energy consumed. This is ideal for electric vehicles where efficiency and thermal management matters [1]. This remarkable material permits smaller, more effective and robust electronic components than traditional silicon [2].

STMicroelectronics, from 2019–2024, held the title for leader in SiC tech especially the latest advanced SiC MOSFETs [3]. Operating from their specialized facility in Catania, Italy, the company secured a landmark partnership with Tesla and other major automotive manufacturers [4]. This was the third generation SiC MOSFETs that reached unprecedented efficiencies of above 99%, enabling electric vehicle range increase to 10% more than with silicon-based solutions.

This case study involves the semiconductor industry, particularly power electronics in the automotive sector. The innovation demonstrates how advances in seemingly small electronic components can have substantial ramifications in making electric vehicles more practical and accessible to consumers [5].

## 12.2 From Discovery to Power Electronics Revolution

The story of SiC begins in 1824 when Swedish chemist Jöns Jacob Berzelius, while attempting to create artificial diamonds, accidentally synthesized a new compound by combining silica and carbon [6]. This mysterious new material, initially appearing as dark crystalline particles, would wait nearly two centuries before revealing its full potential in electronics.

Silicon carbide in its early history was known as carborundum. Its exceptional hardness property was largely used in industrial application for abrasive purposes [7]. This material's history of development is fascinating for the fact that it first emerged as a byproduct in the hunt for developing synthetic diamonds. Although its status as a byproduct is long gone, its modern-day applications in the electric vehicle industry mark a transition. It shows the remarkable advancements are possible when scientific exploration uncovers hidden potential of a material.

The electrical properties of silicon carbide remained largely unexplored until 1907, again, when the British electrical engineer Henry Joseph Round produced the first light emitting diode with a SiC crystal [8]. The discovery of the semiconductor property in silicon carbide occurred very early on, but this initial breakthrough remained dormant for several decades. Subsequently, Round's pioneering work laid the foundation to advance the field of SiC electronics.

The real breakthrough came in the 1990s when researchers at Cree Research (now Wolfspeed) developed the first commercially viable process for growing large SiC crystals [9]. This development opened the door for power electronic applications, though significant challenges remained in reducing defects and manufacturing costs. The advantages and disadvantages of each device are actively debated and compared, as researchers look to improve existing technologies and find new applications for silicon carbide.

In the late twentieth century, a revival of SiC occurred with better crystal growth techniques. There has been considerable research activity worldwide to bypass constraints and fully tap the silicon carbide for power electronics. This technological renaissance demanded the development of new semiconductor materials for electronic devices. The new materials needed to handle higher voltages, frequencies, and temperatures than silicon, with particular emphasis on aerospace, defense, and high-power industrial systems [10]. The intrinsic material characteristics are designed to optimize device performance and functionality, but the trade-offs between these objectives must be carefully considered [11].

The first commercial SiC diodes were introduced in the early 2000s, launching silicon carbide electronics on its initial step of development. Although, SiC MOSFETs are widely considered as the most promising advancement, they remained elusive

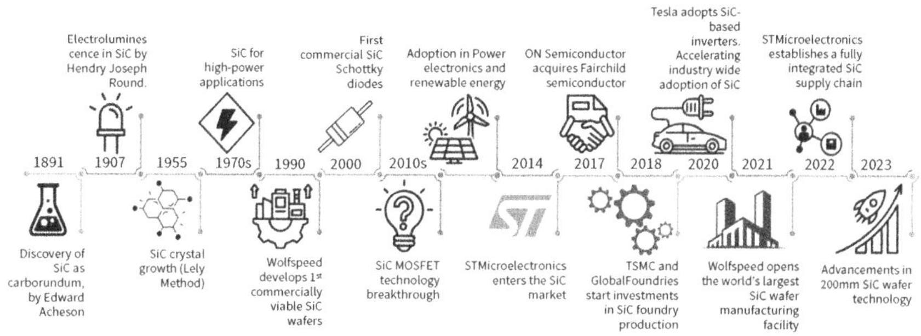


**Figure 12.1** Evolution of silicon carbide (SiC) from discovery to power electronics. *Source:* Author's own work.

due to their fundamental gate oxide interface problems. This material interface was particularly difficult to make, and a massive investment in research and development was required [12]. These stubborn hurdles were overcome only in the 2010s, marking a revolutionary moment for SiC in power electronics industry. Particularly, the advancement helped to achieve high-performance SiC-based MOSFETs to power more efficient converter systems that dramatically expanded its applications.

Figure 12.1 shows the complete evolution of silicon carbide from its discovery to its modern-day phenomenon. SiC is becoming a game-changing power electronics material. The timeline highlights major milestones including technological advancements of SiC, commercialization and semiconductor pioneers who are shaping SiC through leading companies such as STMicroelectronics, Wolfspeed (formerly Cree), ON Semiconductor, and TSMC. SiC has come a long way from an industrial abrasive to the backbone of emerging next-generation power electronics. Tesla pioneered the commercialization of SiC inverters in its Model 3 in 2020, bringing the technology into the mainstream electric vehicle market. Following this breakthrough, companies like Wolfspeed, STMicroelectronics, ON Semiconductor, and TSMC scaled up their SiC manufacturing capabilities to reduce costs and to achieve improved performance. The market is now transitioning to 200mm SiC wafers. It promises even greater power efficiency gains in critical power converter applications such as electric vehicles (EVs), renewable energy systems, and datacenters.

Silicon carbide is widely used in the production of construction materials and temperature-resistant products due to its ability to endure substantial mechanical and thermal stress [13]. Its increasing application in the electronics industry is largely attributed to its potential to replace metal components and also enabling the development of compact and high-performance SiC devices [14]. The selection of optimal sintering conditions is influenced by factors such as granulometric structure, particle shape, and surface properties of the initial powder [15]. In power electronics, high-voltage silicon carbide (HVSiC) devices are transforming

**Table 12.1** Comparison of efficiency and performance of silicon and silicon carbide.

| Aspect | Silicon (Si) | Silicon Carbide Improvements |
|---|---|---|
| Breakdown Voltage | Standard | 3× Higher |
| Power Losses in High Frequencies | High losses | 50% lower losses |
| Cooling Requirements | High thermal load | Up to 80% lower cooling needs |
| Energy Efficiency | Moderate efficiency | Higher system efficiency |
| Thermal Conductivity (W/cm · K) | 1.5 W/cm · K | 3.7 W/cm · K |
| Switching Speed | Limited | Faster (MHz range) |

*Source:* Author's own work.

modern variable-speed medium-voltage systems, including electric motor drives, renewable energy integration, energy storage, micro-grids, traction control, and compact pulsed power systems. This transition is feasible mainly due to their enhanced efficiency and high-temperature reliability [16].

Silicon carbide offers significant advantages over traditional silicon (Si) in power electronics, particularly in high-efficiency applications. As shown in table 12.1, SiC has three times higher breakdown voltage, that enabling devices to handle higher voltages with improved reliability. Additionally, SiC exhibits 50% lower power losses in high-frequency operations, and reducing energy wastage while improving overall system performance. The superior thermal conductivity of SiC (3.7 W/cm · K compared to 1.5 W/cm · K of silicon) allows for more efficient heat dissipation, as well as reducing cooling requirements by up to 80%. Furthermore, its faster switching speeds in the MHz range further enhances the efficiency in high-speed applications such as electric vehicles and renewable energy systems. These properties make SiC a game-changer in power semiconductor technology, offering higher efficiency, lower energy consumption, and improved thermal management.

The growing importance of silicon carbide in high power, high frequency, and high temperature applications is immense [17]. It has also been widely regarded as important for structural ceramics in heat engines where careful powder characterization and processing are needed to realize optimal material performance [18]. Wide bandgap, high breakdown field and excellent thermal conductivity of the material make it a natural choice for power devices for higher voltages, temperatures, and switching frequencies than normal silicon-based systems. Such benefits imply additional efficiency, smaller system size and weight as well as, most importantly, the overall lower cost. SiC devices at present are under development for more efficient power conversion in a suit of applications from the electric vehicle through renewable energy systems and industrial motor drives.

### 12.2.1 Market Evolution and Industry Drivers

What makes SiC so special for power electronics? The answer lies in its atomic structure and subsequent characteristics. As SiC becomes more prevalent in the realm of power electronics, its materially superior nature over conventional silicon-based semiconductors led to growing adoption. This wide bandgap of 3.3 eV (vs. 1.1 eV for silicon) enables SiC's capabilities to perform extremely well in higher voltages, extreme temperatures, and switching speeds, with higher efficiency and reliability [19]. Furthermore, its superior thermal conductivity (3.7 W/cm · K vs. silicon 1.5 W/cm · K) provides better heat dissipation at lower power gradients and enabling smaller system designs [20].

SiC has higher thermal conductivity, which means devices can transport heat away from them and dissipate more efficiently for better reliability and device life [21]. Thermal conductivity represents the ability of a material to transfer heat. Measured in watts per centimeter per kelvin (W/cm · K), this indicates how well heat is able to move through a specific material. This all comes down to thermal conductivity, the better that is, the more heat can be conducted away from sensitive electronic parts. Thus, it addresses the overheating issues in sensitive and critical electronic parts, where such overheating is a major concern. This brings a significant impact within the power electronics industry, in which the performance and reliability of a device directly depend on efficient heat management. SiC enables higher power conversion efficiency, smaller and lighter systems, and better performance in diverse applications. However, unlike silicon devices, SiC components also have unique failure mechanisms that require further characterization techniques for their reliable operations [22]. This has propelled the development of industry standards and guidelines significantly faster.

At a time when the industry is still developing its fabrications for SiC and increasing material quality across the stack, the market for these semiconductors is evolving quickly. Wafer manufacturing and processing have started to reduce the costs for running SiC technology. It is becoming more cost competitive. The global SiC market has been booming with claims expecting its compound annual growth rate (CAGR) to be over 30% despite uncertainties projected between 2019 and 2024 [22]. The growth is mostly made possible from power semiconductors being in high demand across a variety of sectors such as EVs, renewable energies, smart grids, aerospace, industrial automation, and data centers.

Figure 12.2 represents major advantages of SiC and their penetration in major industries. The six SiC properties are at the heart of developing several applications. The properties highlighted between the hexagon applies to the sectors directly. SiC is most beneficial to sectors in electric vehicles (efficiency of inverters, and longer run time on batteries), renewable energy (solar/wind: Efficient inverter), industrial automation and motor drives (motor power loss

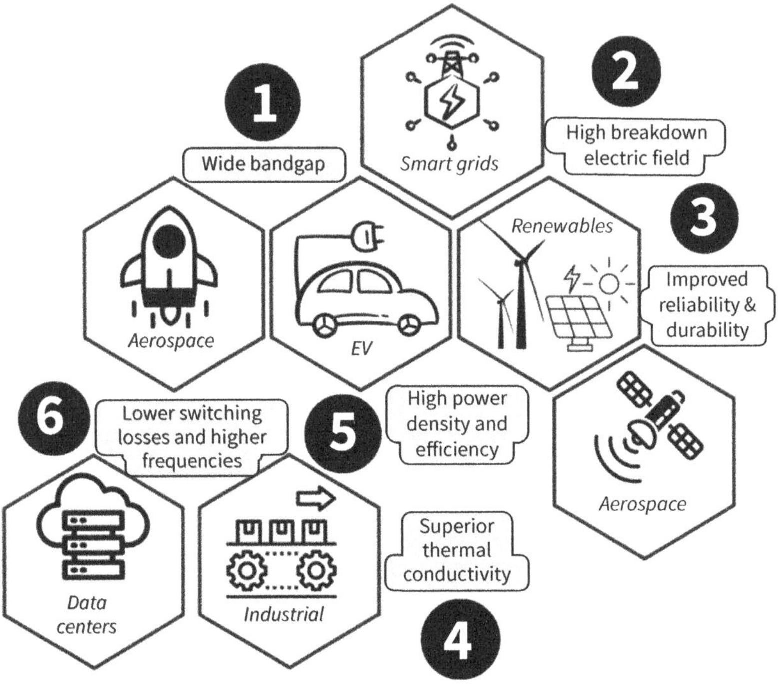


**Figure 12.2** Advantages and applications of silicon carbide (SiC) in electronics. *Source:* Author's own work.

reduction), data centers and telecommunications (power supply, and less energy lost), aerospace and defense (high-temperature operation, and high- radiation tolerant), and grid infrastructure and smart grids (an order of magnitude improvement via solid-state transformers in energy distribution).

The benefits of SiC are not limited to its application sectors but also affect the performance features of power electronics. While Figure 12.2 highlights how SiC's unique properties drive adoption in key industries, Figure 12.3 builds on this by illustrating how SiC compares to other semiconductor technologies in terms of power and frequency capabilities. This comparison from industry applications to metrics of performance explains why SiC is preferred for high-power, and high-efficiency systems over traditional silicon-based devices and other wide bandgap materials.

Figure 12.3 shows the power–frequency curve of some power semiconductor technologies such as silicon IGBT, SiC, and gallium nitride (GaN). Wide bandgap (WBG) materials, such as SiC and GaN, provide reduced power loss and increased thermal switching frequency leading to better overall system efficiency. The devices

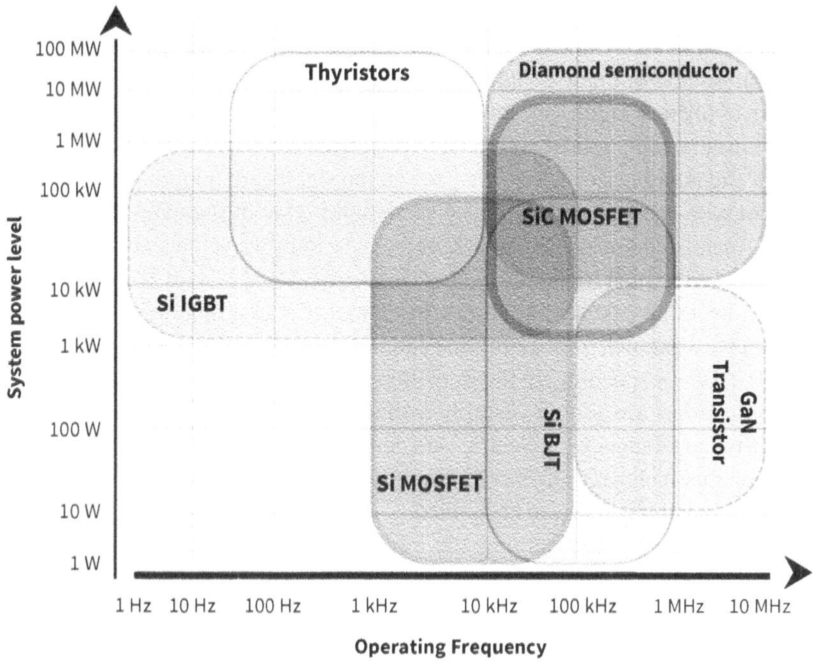


**Figure 12.3** System power level and operating frequency of different power semiconductor technologies. *Source:* Author's own work.

based on GaN, especially high electron mobility transistors (HEMTs), are excellent for high-frequency applications. It is pushing into MHz where power supplies are in the kW range such as DC-DC converters and on-board chargers (OBCs). They are also restricted to low-power (<100 kW) applications by their lateral structure. Instead, SiC has much higher power density (MW range), which makes it very interesting for electric vehicles, industrial motor drives, and grid applications. Although GaN electrons travel more easily, so that devices can be smaller and/or operate at much higher frequencies (>100 kHz to MHz). But its high-power density and relatively low thermal conductivity are challenges when it comes to cooling requirements. Figure 12.3 demonstrates the power and frequency operating trade-offs between Si, SiC, and GaN technologies across different power and frequency ranges.

Other applications of silicon carbide can be found in bioelectronics and biosensors, due to SiC's exceptional electrical properties, encompassing high electron and hole mobility, high breakdown electric field, and the capacity for high conductivity through doping [23]. SiC has the ability to grow a thermally stable native silicon dioxide layer. It makes it directly compatible with well-established micro- and nano-fabrication technologies for silicon-based devices and complementary metal-oxide semiconductor applications [24]. Additionally, the human body's harsh environment contains various salinized and ionic chemicals and proteins. It further highlights SiC's suitability for bioelectronic and biosensor applications due to its robust and reliable performance in such conditions.

One of the many places where we can see the significant adoption of SiC is in alignment with the global shift to electric mobility. One of the major EV market players, Tesla, is deploying SiC-based inverters to improve vehicle efficiency and battery range. Through higher switching frequency available with SiC MOSFETs, switching losses are reduced as shown in Figure 12.4, and therefore it improves drivetrain performance over traditional silicon-based insulated-gate bipolar transistors (IGBTs). By speeding up the transition to SiC power electronics in the EVs, car manufacturers are going to comply with energy efficiency regulations as well as longer range efficiency for electric cars demands [25]. These trends have made SiC market much bigger than ever and continue to develop innovative fabrication techniques and applications to be applied.

Beyond automotive applications, the integration of renewable energy sources into power grids has created another crucial market for SiC technology [5]. Solar and wind power systems require highly efficient power conversion technologies. Particularly, SiC-based power electronics excel in solar inverters due to their ability to operate at higher voltages and temperatures, improving energy conversion efficiency and reducing system size. Additionally, SiC devices enhance the performance of solid-state transformers and grid infrastructure, while facilitating the transition toward smart and decentralized energy systems.

This technical advancement has extended into industrial applications, where reliability and efficiency are paramount. Motor drives of high power UPS (uninterruptible power supplies) and high frequency switching power supplies gain benefits from the outstanding thermal and electrical properties of SiC, for example, in the heavy energy-draining data centers where SiC-based power converters

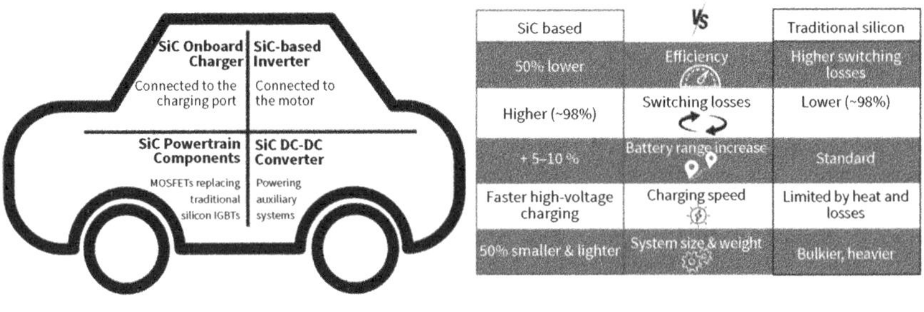


**Figure 12.4** SiC in EVs and a comparison table of SiC vs. traditional silicon. *Source:* Author's own work.

make for more efficient energy management. This can improve both operational expenditure and environmental impact. The shift toward cloud computing and artificial intelligence (AI) is simultaneously increasing the demand of SiC-based infrastructures [26].

While the advantages of SiC technology are clear, barriers to mass adoption have mainly centered on material availability, manufacturability and the cost of materials [6]. Large custom wafers of SiC are very specialized devices, hence much more expensive than an equivalent silicon wafer [27]. However, continuous advancements in crystal growth techniques, such as improved sublimation growth methods and wafer thinning technologies, have led to a steady decline in production costs. SiC fab investments by STMicroelectronics and Wolfspeed look like they are ramping up manufacturing to meet increasing demand with cost reductions via economies of scale [28].

In this rapidly evolving landscape, STMicroelectronics is one of the key players in SiC power electronics. As far as actual production of silicon carbide wafers, Wolfspeed has historically been the leader, but STMicroelectronics has deployed significant capital into vertical integration, driven by its new state-of-the-art manufacturing plant in Catania, Italy. Its strategic alliances with Tesla and wafer self-production initiative have given it a very interesting position among the major semiconductor companies.

Unlike competitors such as Infineon Technologies and ON Semiconductor, who primarily rely on external wafer supply, STMicroelectronics' approach to controlling the entire supply chain from substrate to finished device has proven, particularly advantageous during recent market uncertainties. Through these strategic initiatives and continuous technological innovation, STMicroelectronics has not only strengthened its market position but has also played a crucial role in accelerating the commercialization of SiC-based power solutions. As SiC technology matures and costs decline, STMicroelectronics' early investments and technological expertise position the company to capitalize on the expanding global power electronics market, particularly in automotive and industrial applications.

### 12.2.2 STMicroelectronics' Strategic Implementation

Based in Geneva, Switzerland, multinational electronics and semiconductor company STMicroelectronics has long been the market leader of power semiconductor technologies. Founded in 1987 as a merger of SGS Microelettronica and Thomson Semiconductors: both were leading European semiconductor firms [29]. With a visionary founder, Pasquale Pistorio firmly established STMicroelectronics as a powerhouse for the development and know-how of power electronics. The company has built numerous top-class production plants for SiC manufacturing in Europe. But its key location and so far, most expansive investment is in Catania, Italy.

STMicroelectronics' entry in silicon carbide market was a strategic choice due to increasing trend of high-efficiency power semiconductors across different industries. With its intrinsic material properties in mind, notably high bandgap, high breakdown voltage, and enhanced thermal conductivity, STMicroelectronics identified the opportunity to challenge the traditional silicon power electronics market. The foresight of this strategy has given STMicroelectronics a competitive advantage in the SiC power semiconductor market and offers its customers primarily automotive but also industrial applications, and renewable energies [30].

The company's strategic transformation began in 2018 with a decisive move to establish vertically integrated SiC manufacturing capabilities [31]. At the heart of this initiative lies in the company's 150,000 square meter facility in Catania, Sicily, where STMicroelectronics has invested more than €730 million to create a complete, in-house SiC manufacturing chain. It encompassed every step from the growth of high-quality SiC substrate crystals to the fabrication, and testing of advanced SiC power devices, including MOSFETs and diodes [32]. Figure 12.5 shows that by controlling the entire production process the company does not need to rely on external suppliers, ensuring a reliable and secure supply of SiC materials and components. This helped the company in optimizing quality, cost-efficiency, and scalability as it ramps up production to meet the growing global demand for SiC-based power electronics.

The Catania facility is one of the most sophisticated SiC plants in the world, equipped with state-of-the-art 150 mm wafer production lines [33], and it is staffed by specialists and highly skilled engineers. This allows STMicroelectronics to be

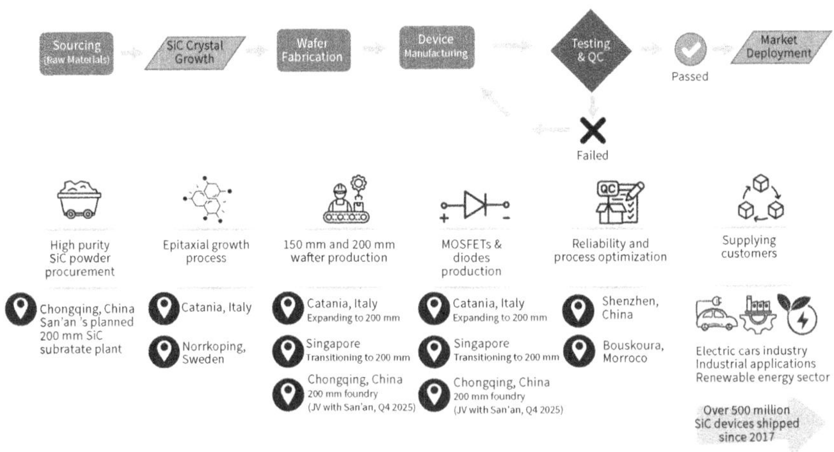


**Figure 12.5** STMicroelectronics' vertically integrated SiC manufacturing strategy. *Source:* Author's own work.

able to access the region's strength in semiconductor fabrication, and benefit from highly trained workforce and favorable climate for this industry.

Sicily may not be renowned as a major global semiconductor center but the region's uniquely mild Mediterranean climate gives it a favorable condition for SiC crystal growing, which is sensitive to thermal changes [34]. In May 2024, the company reported its upcoming new high volume 200 mm silicon carbide manufacturing facility for power devices, and models as part of efforts to ramp up manufacturing and decreasing costs [35]. That accomplishment was a major advancement in scaling up to larger wafer sizes that has long presented huge technical hurdles in SiC crystal growth. Table 12.2 provides a structured summary of STMicroelectronics' strategy, highlighting its strengths and challenges in the SiC market.

In addition to its internal manufacturing capabilities, STMicroelectronics has forged strategic partnerships to strengthen its position in the SiC ecosystem. A pivotal moment in STMicroelectronics' SiC strategy came with the 2020 announcement of a multi-year, multi-billion-dollar supply agreement with Tesla that STMicroelectronics secured through its technological expertise and manufacturing competence [31]. This partnership catalyzed further investments in manufacturing capacity, and

**Table 12.2** STMicroelectronics' SiC manufacturing strategy and key advantages.

| Aspect | STMicroelectronics' Approach | Key Benefits |
|---|---|---|
| Market Entry Strategy | Vertically integrated SiC manufacturing | Competitive advantages in quality, cost, and scalability |
| Manufacturing Facility | €730 M investment in 150,000 $m^2$ Catania plant | Controls full SiC production process, no reliance on suppliers |
| Wafer Production | 150 mm production, transitioned to 200 mm in 2024 | Larger wafers reduce production costs and increase output |
| Automotive Market Strategy | Long-term supply contract with Tesla (since 2020) | Stable revenue, leading role in EV SiC market |
| R&D and Innovation | Focus on MOSFET improvements, trench structures, and epitaxial growth | Lower on-resistance, improved reliability, and cost reductions |
| Supply Chain Management | In-house control from raw material to final testing | Ensures supply security and high-quality standards |
| Challenges | Scaling up SiC production while maintaining quality | Requires process optimization and yield improvement |

*Source:* Author's own work.

solidified the company's position in the automotive semiconductor market. The agreement specifically focused on supplying SiC power devices for Tesla's vehicle inverters, demonstrating the technology's crucial role in electric vehicle efficiency.

The significance of these long-term contracts cannot be over-emphasized. These agreements establish a stable environment for companies to undertake major long-term investments; ramp up manufacturing capacity and engage deeply with industry leaders in order to improve their technology, However, academic research has also identified both the advantages and disadvantages of such partnerships [36]. On one hand, long-term contracts help firms in supply chain risk mitigation, having a pipeline of predictable cash flows and access to actual customer insights to persist in innovation. On the other hand, long-term contracts can also create potential lock-in effects and limit a company's flexibility to adapt to market changes. Nevertheless, by establishing these long-term commitments, companies can make bolder decisions, commit resources, and focus on delivering the advanced solutions demanded by their key customers. In the case of STMicroelectronics, its strategy of securing long-term deals with industry leaders like Tesla appears to be creating a new trend in the SiC power electronics market. It shows that the vertically integrated manufacturers are better positioned to meet the growing demand for high-performance and energy-efficient power devices.

In addition to automotive applications, STMicroelectronics has expanded its SiC product by offering to encompass MOSFETs and diodes for industrial and power conversion applications. Research and development have focused on increasing device performance, while driving cost reduction in the manufacturing process. Main innovations are improved trench in metal-oxide-silicon (MOS) field effect transistors with better epitaxial growth process that led to low on-resistance and reliability improvements.

The corporate manufacturing strategy involves in-house control of the full supply chain (from raw material processing to testing and final output). Though capital intensive, this method offers some major improvements in quality control and supply reliability. Over the next year, STMicroelectronics grew its SiC revenue to more than USD$ 1 billion a year – which is truly monumental relative to the initial investments.

However, significant challenges remain in scaling SiC production while maintaining high quality and reducing manufacturing costs. The company continues to invest heavily in process optimization and yield improvement efforts, particularly in the critical areas of substrate crystal growth and epitaxial layer formation. These technical challenges are essential for STMicroelectronics to meet the rapidly increasing global demand from automotive, industrial, and other key customers, while also competing effectively with other major SiC device manufacturers.

Growing large SiC crystals and depositing high-quality epitaxial layers at scale involves significant complexities. Overcoming these challenges is very crucial for companies to achieve the desired cost reductions and production ramp-up needed to satisfy the market's increasing demand for SiC-based power electronics. Moreover, the company has had to address critical reliability issues, such as threshold voltage stability and gate oxide reliability, which have recently been investigated by different research groups [37, 38].

On a global scale, the market share of SiC power belongs to six main companies. Figure 12.6 shows that in 2023, the global SiC device market experienced tremendous growth led mainly by electric vehicle applications, where STMicroelectronics held 32.6% market share as the leader. In total, the top five manufacturers controlled 91.9% of the market. Major players are aggressively expanding their manufacturing capabilities, particularly in 8-inch wafer production, and STMicroelectronics is developing its Catania facility and forming a strategic joint venture with Sanan Optoelectronics in China. It is indicative of a crucial shift from horizontal integration and economies of scale, in that ONsemi (US) has steadily grown its self-sufficiency of substrate. Substrate self-sufficiency refers to a manufacturer's capability to produce its own silicon carbide wafer base materials in-house, rather than relying on external suppliers.

This approach provides greater control over costs, quality, and supply chain security. Infineon commits to its supplier diversification strategy and Wolfspeed has material leadership despite execution issues and ROHM committing to produce through the acquisition of the Kunitomi plant in Japan. ROHM's acquisition of Solar Frontier's Kunitomi plant represents a strategic expansion of their SiC manufacturing capabilities. They convert a former solar panel facility into a fourth SiC production site to boost their 8-inch wafer production capacity. The competitive environment is marked by significant capital investments in 8-inch wafer facilities. This underscores the industry's transition from conservative growth to aggressive expansion strategies.

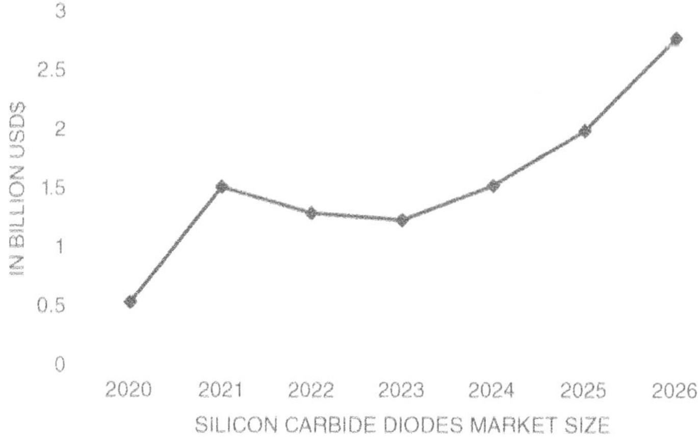


**Figure 12.6** Silicon carbide diodes market size, 2020–2026 (in billions USD $). *Source:* [39, 40]. *Note:* Figures for 2025 and 2026 are forecasts.

In the highly competitive market, STMicroelectronics has positioned itself as an original device manufacturer (ODM) with the full chain of SiC under its control. This strategy is unique, and in general, they buy their own wafers independently, which is good for some of the cost control. It is also a benefit for the supply security for STMicroelectronics. As the demand for SiC-based power solutions continues to surge across various applications, the ability to maintain a reliable and cost-effective supply chain is a critical factor in determining market leadership and long-term success.

## 12.3 Think and Solve

The following questions are designed to test the reader's understanding of the key concepts and strategic decisions discussed in the STMicroelectronics SiC case study. These questions focus on critical aspects of manufacturing strategy, market positioning, and technical challenges in SiC production. The problem-solving question specifically addresses the economic aspects of SiC wafer production, while helping readers to understand the relationship between manufacturing efficiency and cost implications.

**Q 12.1** What was STMicroelectronics' primary strategic advantage in SiC production compared to its competitors?

**A** Lower manufacturing costs.
**B** Vertically integrated supply chain.
**C** Superior product quality.
**D** Larger market share.

**Q 12.2** Which factor most significantly influenced STMicroelectronics' decision to invest in the Catania facility?

**A** Government incentives.
**B** Proximity to customers.
**C** Control over complete manufacturing process.
**D** Lower operating costs in Sicily.

**Q 12.3** STMicroelectronics' partnership with Tesla primarily focused on?

**A** Joint research and development.
**B** Supply of SiC power devices for vehicle inverters.
**C** Battery technology development.
**D** Manufacturing facility sharing.

**Q 12.4** The transition from 150 mm to 200 mm wafer production primarily aims to?

**A** Increase manufacturing efficiency and reduce costs.
**B** Improve device performance.
**C** Meet Tesla's specific requirements.
**D** Compete with Asian manufacturers.

**Q 12.5** A SiC wafer manufacturing line produces 150 mm wafers with a yield rate of 85%. If the facility transitions to 200 mm wafers, calculate the required yield rate to maintain the same effective production area per month, considering these facts:

- Current monthly production (150 mm): 1000 wafers
- Area of 150 mm wafer: $\pi(75\,\text{mm})^2$
- Area of 200 mm wafer: $\pi(100\,\text{mm})^2$
- Current yield rate: 85%
- Target monthly production (200 mm): X wafers

**Formula:**

$$\text{Effective production area} = \text{Number of wafers} \times \text{Wafer area} \times \text{Yield rate}$$

**Calculate:**

**A** The effective production area with 150 mm wafers.
**B** The number of 200 mm wafers needed.
**C** The required yield rate for 200 mm wafers to maintain the same effective area.

## 12.4 Conclusion and Key Takeaways

The rapid evolution of SiC technology has positioned it as a transformative material in power electronics. SiC offers significant advantages over conventional silicon-based semiconductors. Its superior thermal conductivity, wide bandgap, and high breakdown voltage enable enhanced efficiency, reduced system size, and improved reliability across various high-power applications. These benefits make SiC particularly valuable in electric vehicles, renewable energy systems, industrial automation, and data centers, where energy efficiency and performance are paramount.

Market trends indicate strong and sustained growth for SiC technology, with a projected CAGR exceeding 30% from 2019 to 2024. This expansion is driven by increasing demand for high-performance power semiconductors, advancements in wafer production, device fabrication, and industry-wide efforts to standardize and optimize SiC-based solutions. As manufacturing costs continue to decline, broader adoption is expected across both existing and emerging applications.

Despite its advantages, challenges remain in material quality, device reliability, and scalability. Ongoing research and development efforts, along with collaborations between industry leaders and research institutions, are addressing these challenges by refining production techniques and developing more robust characterization methodologies. As a result, SiC technology continues to mature, and paving the way for its widespread integration into next-generation power electronics.

In summary, SiC is revolutionizing power electronics by delivering unparalleled performance in high-power and high-efficiency applications. With continuous advancements in technology, cost reductions, and expanding market adoption, SiC is well-positioned to play a critical role in shaping the future of electrification and energy management across multiple industries.

## 12.5 Discussion Points

The advancement and adoption of SiC technology present several key discussion points that influence its trajectory in the power electronics industry. These considerations encompass technological developments, economic feasibility, industry-wide challenges, and future market trends.

1. **Material and Manufacturing Challenges:** SiC has better properties, especially electrical and thermal; but the manufacturing process has to be more complex and expensive than silicon. What will further drive down the cost and scaling of optimal output? For example, through ongoing innovations in wafer manufacturing, defect reduction and device packaging?
2. **Performance and Reliability Considerations:** SiC devices are operated at higher voltages, temperatures and frequencies than their silicon counterparts. How do these factors impact the reliability of the device in the long run, and how must one test to guarantee the performance remains consistent through these extremes?
3. **Market Adoption and Industry Integration:** The adoption of SiC in electric vehicles, renewable energy, and industrial power systems is accelerating. What are some of the primary barriers that could hinder widespread implementation? Technological, or economic or regulatory? What role do government policies, subsidies, and sustainability goals play in promoting SiC-based technologies across various industries?
4. **Competitive Landscape and Future Innovations:** Key SiC players are ramping up R&D efforts to capture the fast growing SiC market. What competition will emerge between established semiconductor companies and emerging startups?

Discussion of these points in light of the other anticipated SiC market challenges and opportunities can further help industry leaders, engineers, researchers and policy makers to work together to ultimately get this transformative technology adopted on a large scale.

## 12.6 Limitations of the Case Study

1. **Scope of Market Analysis:** The study showed SiC's commercial usage in the most visible sectors such as electric vehicles, renewable energy, and industrial applications. This could also mean that the emerging sectors like aerospace and quantum computing might not have gone far enough yet since not much data is publicly available.

   Tech-specific market forecasts depend on existing trends and currently available reports. They may also vary based on any unexpected technological breakthroughs, policy changes, or events.
2. **Data Availability and Accuracy:** The study leverages industry data, academic publications, and corporate disclosure, which whilst credible are not necessarily providing real-time insights into the rapidly developing technologies.

   With the proprietary attributes of semiconductor manufacturing, certain technical and financial insights might be heavily redacted by key market players, which limiting the depth of the case study.
3. **Focus on SiC Compared to Alternative Technologies:** While SiC's advantages over traditional silicon-based semiconductors are well documented, this case study does not extensively compare SiC with other technologies that may offer competitive or complementary benefits. Future hybrid materials or entirely new semiconductor technologies may significantly change the player in the field, hence more research than this study could be needed.
4. **Economic and Policy Influences:** Government subsidies, trade policies and supply chain disruptions are important forces that impact various factors in the SiC market. The externalities however are highly volatile and hard to forecast with absolute certainty.

   Infrastructure, research funding and industrial penetration could vary between regions that affect the SiC deployment. A deeper examination might be needed to look at market penetration by geography at regional level.
5. **Long-Term Reliability and Performance Data:** While SiC is recognized for its superior thermal and electrical properties, long-term operational data for large-scale deployments is still emerging. This study primarily references existing research and commercial implementations, but ongoing real-world applications may yield additional insights in the future.

Within these limitations, the case study provides valuable insights into the current state and future potential of SiC technology. With continued research, industry collaboration and data collection, all of this together can contribute in improving market strategies and propelling more innovation in power electronics.

## References


1 Di Giovanni, F. "Silicon carbide: physics, manufacturing, and its role in large-scale vehicle electrification," *Chips*, 2(3), 209–222, 2023.

2 Boomer, K., Lauenstein, J.M. and Hammoud, A. "Body of knowledge for silicon carbide power electronics," (No. GRC-E-DAA-TN35760), 2016.

3 "STMicroelectronics annual report," 2023, https://investors.st.com/financials/annual-and-semi-annual-reports (accessed March 2, 2026).

4 Robertsson, S. and Karlsson, O. "Securing supply of silicon carbide semiconductors-a case study at a large automotive OEM," Master's Thesis, Chalmers University of Technology, 2022.

5 Pessoa, R.S. and Fraga, M.A. "The versatile horizon: SiC power semiconductors in electric vehicles, renewable energy, aeronautics, and space systems," *Journal of Aerospace Technology and Management*, 16, e3424, 2024.

6 Ruys, A.J. *Silicon Carbide Ceramics: Structure, Properties and Manufacturing*, Elsevier, 2023.

7 Choyke, W.J., Matsunami, H. and Pensl, G. *Silicon Carbide: Recent Major Advances*, Springer Science & Business Media, 2003.

8 Choyke, W.J. "Silicon carbide: presolar SiC stardust grains and the human history of SiC from 1824 to 1974," in *Handbook of Silicon Carbide Materials and Devices*, ed. Z.C. Feng, 3–28. Springer Science & Business Media, 2023.

9 Zekentes, K., Veliadis, V., Ryu, S.H., Vasilevskiy, K., Pavlidis, S., Salemi, A. and Zhang, Y. "SiC and GaN power devices," in *More-than-Moore Devices and Integration for Semiconductors*, ed. F. Iacopi and F. Balestra, 47–104. Springer International Publishing, 2023.

10 Bauer, D.J., Khazdozian, H., Mehta, J., Nguyen, R.T., Severson, M.H., Vaagensmith, B.C., Toba, L., Zhang, B., Hossain, T., Sibal, A.P. and Smith, B.J. "Critical materials strategy," (No. INL/RPT-23-72323-Rev. 001). Idaho National Laboratory (INL), 2023.

11 Lübben, M., Cüppers, F., Mohr, J., von Witzleben, M., Breuer, U., Waser, R., Neumann, C. and Valov, I. "Design of defect-chemical properties and device performance in memristive systems," *Science Advances*, 6(19), p.eaaz9079, 2020.

12 Karki, U., Gonzalez-Santini, N.S. and Peng, F.Z. " Effect of gate-oxide degradation on electrical parameters of silicon carbide MOSFETs," *IEEE Transactions on Electron Devices*, 67(6), 2544–2552, 2020.

13 Karelin, V.A., Strashko, A.N., Sazonov, A.V. and Dubrovin, A.V. "Obtaining the fine-grained silicon carbide, used in the synthesis of construction ceramics," *Resource-Efficient Technologies*, 2(2), 50–60, 2016.

14 Castellazzi, A., Fayyaz, A., Romano, G., Yang, L., Riccio, M. and Irace, A. "SiC power MOSFETs performance, robustness and technology maturity," *Microelectronics Reliability*, 58, 164–176, 2016.

**15** Wang, Y., Liu, Y., Chen, Z., Liu, Y., Guo, J., Zhang, W., Rao, P. and Li, G. "Recent progress in the pore size control of silicon carbide ceramic membranes," *Ceramics International*, 48(7), 8960–8971, 2022.

**16** Cheng, L., Palmour, J.W., Agarwal, A.K., Allen, S.T., Brunt, E.V., Wang, G.Y., Pala, V., Sung, W.J., Huang, A.Q., O'Loughlin, M.J. and Burk, A.A. "Strategic overview of high-voltage SiC power device development aiming at global energy savings," *Materials Science Forum*, 778, 1089–1095, 2014.

**17** Baliga, B.J., "Silicon carbide power devices: progress and future outlook," *IEEE Journal of Emerging and Selected Topics in Power Electronics*, 11(3), 2400–2411, 2023.

**18** Binner, J., Porter, M., Baker, B., Zou, J., Venkatachalam, V., Diaz, V.R., D'Angio, A., Ramanujam, P., Zhang, T. and Murthy, T.S.R.C. "Selection, processing, properties and applications of ultra-high temperature ceramic matrix composites, UHTCMCs–a," *International Materials Reviews* 65(7), 389–444, 2020.

**19** Rafin, S.S.H., Ahmed, R. and Mohammed, O.A. "Wide band gap semiconductor devices for power electronic converters," in *Fourth International Symposium on 3D Power Electronics Integration and Manufacturing (3D-PEIM)*, 1–8. IEEE, 2023.

**20** Sun, Q., Guo, H., Zheng, Z., Zhang, F., Hong, F., Zhang, Y., Wang, Y. and Jiang, S. "Influence of pinch effect on the lifetime of a 2MW silicon carbide photoconductive semiconductor switch," *IEEE Transactions on Electron Devices*, 71(3), 2018–2023, 2024.

**21** Pushpakaran, B.N., Subburaj, A.S., Bayne, S.B. and Mookken, J. " Impact of silicon carbide semiconductor technology in photovoltaic energy system," *Renewable and Sustainable Energy Reviews*, 55, 971–989, 2016.

**22** Yole Development, "Power SiC: materials, devices and applications," 2020, https://medias.yolegroup.com/uploads/2020/11/YDR20119_Power_SiC_Materials_Devices_Applications_2020_Flyer.pdf (accessed March 2, 2026).

**23** Mamun, A.A., McGarrity, M., Kim, J.H. and Zhao, F. "Silicon carbide-based DNA sensing technologies," *Micromachines*, 14(8), 1557, 2023.

**24** Kimoto, T. and Cooper, J.A. "Fundamentals of Silicon Carbide Technology: Growth, Characterization, Devices, and Applications," IEEE Press, 2014.

**25** Khan, M.Y.A. "Enhancing electric vehicle performance: a case study on advanced motor drive systems, integration, efficiency, and thermal management," *Control Systems and Optimization Letters*, 3(1), 20–27, 2025.

**26** Agarwala, A., Tahsin, T., Ali, M.F., Sarker, S.K., Abhi, S.H., Das, S.K., Das, P., Hasan, M.M., Tasneem, Z., Islam, M.M. and Islam, M.R. "Towards next generation power grid transformer for renewables: technology review," *Engineering Reports*, 6(4), e12848, 2024.

**27** Kainth, S., Sharma, P., Diwan, P.K. and Pandey, O.P. "Shaping the future: innovations in silicon wafer production and finishing," *Silicon*, 16(18), 6479–6497, 2024.

**28** Veliadis, V., Su, M., Di Giovanni, F., Veliadis, V., Su, M. and Di Giovanni, F. "Overview of silicon carbide," in *SiC Technology: Materials, Manufacturing, Devices and Design for Power Conversion*, ed. M. Di Paolo Emilio, 1–23. Springer Nature, 2024.

**29** Coriat, N. and LucchinI, N. "Integrating R&D with manufacturing to build world-wide competitiveness: SGS-Thomson integrated circuits," in *Europe's Next Step: Organisational Innovation, Competition and Employment*, ed. L. Andreasen, B. Coriat, F.D. Hertog, R. Kaplinsky, 151–166. Routledge, 1995.

**30** Emilio, M.D.P. " SiC market," in *SiC Technology: Materials, Manufacturing Devices and Design for Power Conversion*, ed. M. Di Paolo Emilio, 287–297. Springer, 2024.

**31** Borille, L. "Success and risk factors in strategic alliances: the case of alliances after the semiconductor shortage in the automotive industry," Master's Thesis, Università Ca' Foscari, 2023.

**32** STMicroelectronics. "STMicroelectronics to build integrated silicon carbide substrate manufacturing facility in Italy," 5 October 2022, www.newsroom.st.com/media-center/press-item.html/c3124.html (accessed March 2, 2026).

**33** Rafin, S.S.H., Ahmed, R., Haque, M.A., Hossain, M.K., Haque, M.A. and Mohammed, O.A. "Power electronics revolutionized: a comprehensive analysis of emerging wide and ultrawide bandgap devices," *Micromachines*, 14(11), 2045, 2023.

**34** Goue, O.Y. "Characterization of structural defects in wide band-gap compound materials for semiconductor and opto-electronic applications," Doctoral Dissertation, State University of New York at Stony Brook, 2016.

**35** STMicroelectronics. "STMicroelectronics to build the world's first fully integrated silicon carbide facility in Italy," 31 May 2024, www.newsroom.st.com/media-center/press-item.html/c3262.html (accessed March 2, 2026).

**36** Logožar, K. "Advantages and disadvantages of strategic alliances in international business," in *International Scientific Conference ITEMA Recent Advances in Information Technology, Tourism, Economics, Management and Agriculture*, 2022.

**37** Lelis, A.J., Habersat, D.B., Green, R., and Mooro, E. "Key reliability issues for SiC power MOSFETs. *ECS Transactions*, 58(4), 87, https://doi.org/10.1149/05804.0087ecst, 2013.

**38** Gajewski, D.A., Ganguly, S., Lichtenwalner, D.J., Van Brunt, E., Hull, B., Allen, S. and Palmour, J.W. "Reliability and standardization for SiC power devices," *Materials Science Forum*, 1092, 179–186, 2023.

**39** Transparency Market Research. "SiC power device market, market outlook 2031," 2022, https://www.transparencymarketresearch.com/silicon-carbide-power-devices-market.html (accessed March 2, 2026).

**40** Sky Questt. "Silicon carbide (SiC) power devices market size, share, and growth analysis," 2023, https://www.skyquestt.com/report/silicon-carbide-sic-power-devices-market (accessed March 2, 2026).

# 13

# Ericsson's 5G Infrastructure: Mobile to Connected Future

Technical Innovation, Strategic Positioning, and the Road to 6G

## 13.1 Synopsis

Picture this: internet connectivity is so fast and seamless that physical distance doesn't matter, and communication feels as natural as speaking to someone standing next to you. This ambitious vision was like a distant dream until telecommunications technology was introduced and, particularly 5G technology, made it very possible. This communication is not only between humans but also between things such as "Internet of Things." 5G is an advanced 5th generation communication technology that aims to eliminate network bottlenecks and enabling ultra-fast communication for billions of devices. At the forefront of this transformation is Ericsson, a global telecom leader in the deployment of 5G infrastructure to accelerate connectivity.

Driven by the increasing demand for faster and more reliable networks, from autonomous vehicles to smart cities, Ericsson has been pioneering innovations in network slicing, Massive MIMO solutions, and energy-efficient base stations. Scaling 5G, on the other hand, however, is a rather intricate engineering task, including infrastructure investment to spectrum allocation and cybersecurity. In this case study, we look at how Ericsson is tackling these challenges and leveraging technology as well as partnerships to guide the future of global communication networks.

Ericsson's advancements in 5G infrastructure are characterized by a multifaceted approach, encompassing technological innovation, strategic partnerships, and a deep understanding of the evolving telecommunications landscape. The company's focus extends beyond simply increasing data rates; it aims to create a versatile and adaptable network, capable of supporting a wide array of services, from enhanced mobile broadband to mission-critical applications [1].

As the world embraces 5G, research and development efforts are already underway for the next frontier: the sixth generation (6G) networks. Predicted to be widely deployed in the 2030s, 6G will take wireless communications to the limit with terabit-per-second speeds and sub-one-millisecond latency alongside artificial intelligence and quantum computing integration. Terahertz communication, AI-powered network automation and the use of advanced holographic displays are some of the main technologies expected to define 6G. Ericsson, as a pioneer in mobile technology, is leading the charge in next-generation connectivity that will power immersive extended reality (XR), seamless human–machine collaboration, and completely autonomous smart infrastructures. The past is the past as history demonstrates that each generation of mobile technology adds to its foundation and with Ericsson leading the advance, 6G will change the digital dimensions once more.

## 13.2 Background and Case Scenario

The evolution of mobile telecoms has been a story of major technical progress from simple analog voice services over to the multi-giga application digital networks that we use today. Ericsson has played a key part in this evolution, developing and deploying every generation of mobile technology.

Before the advent of 5G, mobile telecommunications underwent a remarkable evolution, progressing through successive generations from 1G to 5G, where each G stands for "generation" [2]. These successive generations of technology have today created a completely new world, fundamentally changing how we communicate and consume information. These critical phases illustrate the extraordinary evolution of technology and establish today's hyper-connected global landscape.

### 13.2.1 Historical Context of 1G–6G

The 1G network, which is said to be the birth of mobile communication, came out first in the late 1970s to early 1980s. Analog systems like Advanced Mobile Phone System (AMPS), and Nordic Mobile Telephone (NMT) were released to simply support the most elementary voice calls [3]. They had low capacity, poor voice quality, and almost no security features. Though constrained as they were, 1G networks were nothing short of a revolution from the previous systems of mobile radio telephone, at least in offering some first steps of mobile telephony.

The motivation behind developing 1G technology stemmed from the growing demand for mobile communication that was not confined to fixed locations. Prior to 1G, early mobile radio telephone services were primitively bulky, and lacked the capacity to support a broad user base [3]. The introduction of 1G used the analog frequency modulation (FM). It was rolled out to solve the problems through a more rational and scalable infrastructure for mobile communications and enabled people to communicate over the air while on the move [4].

The Advanced Mobile Phone System and Nordic Mobile Telephone 1G networks used the range of radio frequencies such as 450 MHz and 900 MHz bands [5]. These 450 MHz and 900 MHz frequency bands permitted the transmission of voice calls. Radio waves in these frequency bands were allocated by regulatory bodies for mobile communications. These bands were selected due to their favorable propagation characteristics, which enables coverage over a reasonable distance.

Additionally, 1G networks had a 30 kHz channel bandwidth [4]. The width or range of frequencies spectrum allocated for a single communication channel is known as channel bandwidth. Then more data or information could be transmitted in the wider channel bandwidth. The 1G network's channel bandwidth was as small as approximately 30 kHz and so it could not effectively carry high-speed data services [4].

The limitations of analog technology still shackled 1G networks. This resulted in poor voice quality, limited capacity due to the narrow bandwidth of FM signals, and lack of security features such as encryption [6]. The maximum data rate that could be achieved in 1G was less than 2.4 Kbps, which was excruciatingly slow by today's standards. To put this in perspective, a 2.4 Kbps connection probably would not even be able to load something as simple as a plain text-based webpage. The analog nature of 1G networks also made them susceptible to interference and signal degradation, especially in densely populated areas.

With all that said though, 1G networks were still a leap from the evolution of mobile communication that led to our digital era today. There were already millions of subscribers of 1G worldwide by the late 1980s and this testified to the demand for mobile communication being on the rise [7]. The transition from 1G to 2G networks in the 1990s marked a pivotal moment in the history of mobile telecommunications, introduced digital transmission, enhanced security, and support for mobile data services.

As illustrated in Figure 13.1, with the improvement of technology each succeeding generation expanded on what its predecessors had done, adding transformative features that fundamentally changed global communication. The 1990s saw the

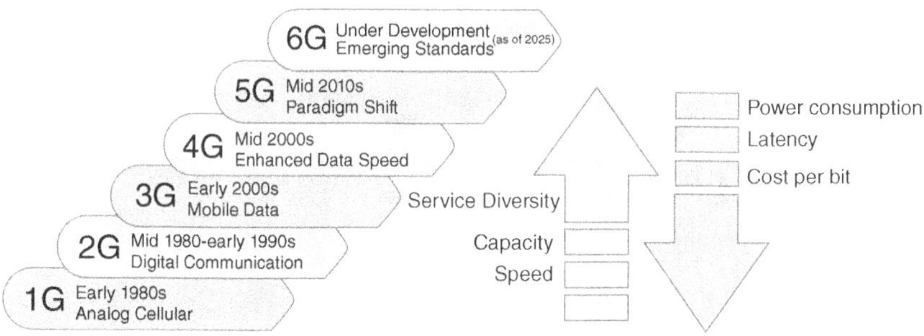


**Figure 13.1** Evolution from 1G to 5G and key factors that increased and decreased. *Source:* Author's own work.

emergence of second-generation (2G) networks. The shift from analog to digital transmission led to Short Message Service (SMS) and thus transformed mobile communication completely.

And this not only improved the efficiency of mobile networks but also ultimately led to global standardization. With the rapid increase in mobile adoption across geographies, interoperability and international roaming became essential for all mobile operators. To solve these problems, the telecommunication industry was looking for an answer that eventually gave rise to Global System for Mobile Communications (GSM). The launch of GSM was a turning point in mobile communications history, heralding an era of worldwide compatibility and enhanced functionality.

Ericsson helped to shape GSM, which was identified as the most globally prevalent mobile standard in the world. Over 80% of global mobile markets were using GSM by the early 2000s. The digital technology revolutionized communication engineering in terms of spectral efficiency, security, and call quality. The most important innovation was the introduction of Time Division Multiple Access (TDMA), breaking each of the frequency channels down into timeslots, so that a given frequency band can be shared by multiple users without making different bands specially for them. It enabled higher network capacity and reduced the operational costs for telecom operators.

GSM added encryption and authenticity mechanisms such as A5 ciphering algorithm. Subscriber Identity Modules (SIM) made security even better but also allowed easier user identification [8]. Another key feature was its standardized SMS, arguably at the cornerstone of communication, being used for person-to-person messaging as well as machine-to-machine messaging. See Figure 13.2 for GSM network infrastructure components. The infrastructure consists of three main

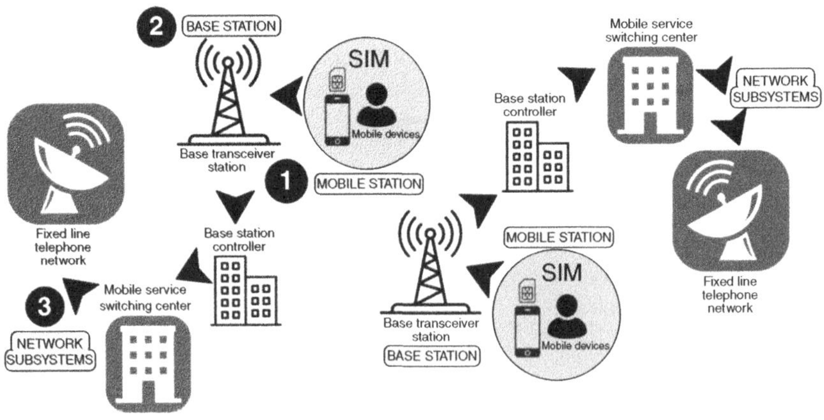


**Figure 13.2** GSM network infrastructure. *Source:* Author's own work.

subsystems: the Mobile Station (including user devices and SIM cards), Base Station Subsystem (covering base transceiver stations and controllers), and Network Subsystem (managing call-switching and routing). This approach of modular design facilitated easy communication by linking mobile devices via base stations to switching centers. These switching centers then routed calls to fixed telephone networks while providing both security and user mobility across different network locations.

2G evolution to digital networks paved the way for early mobile data services, including General Packet Radio Service (GPRS) and Enhanced Data rates for GSM Evolution (EDGE) that performed internet browsing, and multimedia messaging [9]. GSM networks by the end of the decade were capable of handling millions of SMS messages daily. Global mobile data traffic exceeded about 2 petabytes per month, kicking in the wireless evolution of 3G and more [10]. Over 250 billion SMS messages were sent annually worldwide by the year 2002, clearly indicating a rapid roll out for this new feature [11].

The early 2000s ushered in the era of third generation (3G) networks, which prioritized data connectivity alongside voice communication. The breakthrough came with the implementation of packet-switched data transfer. Networking became much more efficient and led way for faster, and smoother mobile internet as we know it today. Data rates were improved through technologies like Wideband Code Division Multiple Access (WCDMA) and High-Speed Packet Access (HSPA). It enabled users to enjoy reliable video calls, more refined multimedia messaging and enhanced web browsing in a real-time manner [12]. Not only this pushed mobile communication, but also laid the foundation of mobile internet. Figure 13.3 illustrates the technological evolution from 1G to 6G networks.

Mobile switching technology, especially subsequent generations of the AXE system by Ericsson, were critical for the optimization of call and data handling in 3G networks with continuous innovation. The AXE system with modular and scalable architecture increased the utilization of mobile exchanges. It helped in dynamically by allocating resources according to the demands of users and guaranteeing optimum performance even in the case of high load [13]. This system allowed telecom operators to send bursts of voice and data traffic, reliably and with low latency.

Moreover, Ericsson took part in the implementation of Universal Mobile Telecommunications System (UMTS). It was a key element of 3G and necessary to

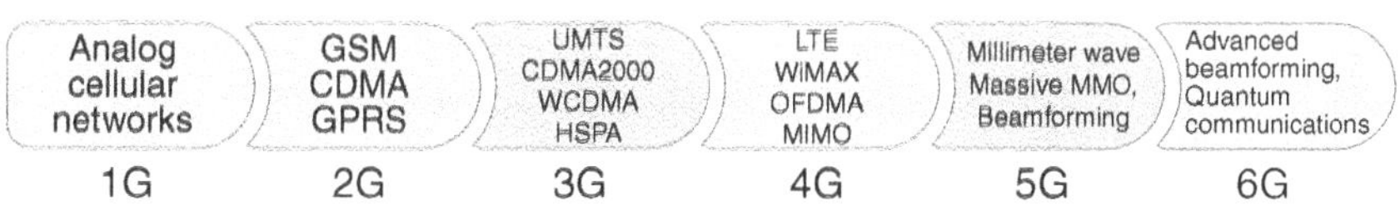


**Figure 13.3** Technological milestones in mobile networks from 1G to 6G.
*Source:* Author's own work.

increase network capacity, improve spectrum efficiency, and increase quality-of-service overall [14]. UMTS enhanced the radio access technology at a higher level, being Wideband Code Division Multiple Access supported with higher data rates and more concurrent links than its predecessors [15]. High-Speed Packet Access (HSPA), created faster data throughput by extending on UMTS with High-Speed Downlink Packet Access (HSDPA) and High-Speed Uplink Packet Access (HSUPA).

HSPA+ is an enhanced type of HSPA networking, designed to improve system capacity and data throughput for mobile broadband [16]. It utilized techniques like Multiple-Input Multiple-Output (MIMO) antennas, 64 Quadrature Amplitude Modulation (64-QAM) and more sophisticated error correction to increase data speeds and the spectrum efficiency. Thanks to those developments, HSPA+ could achieve peak data rates of up to 42 megabits per second (Mbps), that enabled continuous video streaming, fast web surfing and data-hungry applications [17]. Megabits per second (Mbps) is a unit of measurement that indicates how much digital data can be transmitted within one second. With the speed of 42 Mbps, users could download large files, use HD streaming and run all sorts of bandwidth-heavy applications without much interruption or buffers on mobile internet.

To put this perspective in real terms, downloading a high-resolution photo (around 5 MB) would take around 1 second. While a standard definition movie (around 700 mb) would download under two minutes if network conditions are ideal. Table 13.1 compares the download speed and estimated download time for 1G to 5G. It is clear to see that the advancement of mobile technologies led to significant reductions in download time, enhancing user experience and enabling more data intensive applications. These download times are just estimates to illustrate an example. Exact figures may vary based on network conditions, device capabilities, and other factors.

HSPA+, was much faster in contrast to the 2G networks previously, where a single image download can take minutes. This was a giant leap for mobile connectivity and much closer to setting internet users around the world one step closer to

**Table 13.1** Comparison of estimated download times for a 5 MB high-resolution photo across different mobile network generations.

| Network Generation | Typical Download Speed | Estimated Download Time |
|---|---|---|
| **1G** | ~2.4 Kbps | ~4 hours 45 minutes |
| **2G** | ~0.1 Mbps | ~6 minutes 40 seconds |
| **3G** | ~3.1 Mbps | ~13 seconds |
| **4G** | ~15 Mbps | ~2.7 seconds |
| **5G** | ~400 Mbps | ~0.1 seconds |

*Source:* Author's own work.

having a web that works in real time. More specifically, in 2010 the global monthly mobile data traffic was about 190 petabytes, a feature reflecting the adoption of 3G, and the need for ever more internet data via mobile connection [18].

3G brought mobile browsing, email access, and video streaming that redefined personal communication and businesses as well. Quickly, the 3G networks became commonplace. It helped shape the mobile economy making apps ecosystems and e-commerce or cloud computing into a reality. This shift not only improved consumer experience, but it also marked the beginning of the high-speed and low-latency connectivity requirement for the 4G and 5G era.

4G networks, introduced in the 2010s led to an important milestone in mobile technologies with Long-Term Evolution (LTE), as the dominant standard [3]. LTE provided up to 300 Mbps downlink peak speeds and 10 ms response time, which is critical for live applications such as for HD video streaming, online gaming, and cloud computing [19].

LTE, in comparison to 3G, had a spectral efficiency four times better by simply transmitting more data over the same bandwidth. Most of this innovation stemmed from technologies such as Orthogonal Frequency-Division Multiple Access (OFDMA), Multiple-Input Multiple-Output (MIMO) antennas and Evolved Packet Core (EPC) [20]. The explosion of mobile data traffic reached 7.2 exabytes/month by 2015, playing a key role in reshaping the economics and user experience of the digital economies. So, what exactly are these technologies?

OFDMA, a multi-user version of Orthogonal Frequency-Division Multiplexing (OFDM) was later deployed to increase spectral efficiency. Here, frequency bands were sliced up into subcarriers, that enabled multiple users to draw from the same data simultaneously [21]. The reduced interference and increased network capabilities made it a fundamental LTE component. Qualcomm and Ericsson were some of the major stakeholders in creating and fine tuning OFDMA for mobile networks.

MIMO improved data throughput and reliability using multiple antennas at both the transmitter and receiver ends. MIMO increased signal strength and reduced interferences by employing multiple data streams over different spatial paths [22]. Nokia, Huawei, and Cisco among others, led the charge of developing MIMO to be incorporated in LTE as part of wireless infrastructure enhancements.

EPC architecture replaced the circuit-switched core of legacy generations with a flat all-IP network that improved efficiencies for data movement and network operations [23]. It facilitated easy frictionless new technologies such as VoLTE (Voice over LTE) and Internet of Things (IoT). EPC was instrumental in advancing LTE network rollout, with major telecoms players such as Ericsson, Nokia, and Huawei.

The adoption of an all-IP (internet protocol) architecture in 4G networks represented a fundamental shift from traditional circuit-switched networks, which were originally designed for voice communication, to a fully packet-switched network. Particularly, datagram packet switching and hybrid approaches are optimized to achieve heavy data transfer. 4G, being an all-IP network in which voice and data had no segregation, was planned from the ground up. It enabled high-speed data transfer for all modes of communication like voice, video, as well messaging seamlessly.

Increased network resource utilization, better scalability, and reduced costs for mobile operators are some of the factors that this technology has supported. The all-IP architecture facilitated the addition of new technologies like VoLTE (better quality voice calls than legacy networks), IoT applications (that need low latency), and high-reliability connectivity. This important transition from 3G to 4G and the subsequent 5G evolution was made possible by the significant contributions of Ericsson, Nokia, and Huawei in rolling out all-IP networks [24].

The present decade is defined by the advent of fifth generation (5G) networks, which offer lightning-fast data speeds, ultra-low lag, and massive device connectivity. 5G is different from its predecessors in the sense that this generation is planned to serve diverse sets of use cases. It ranges from enhanced mobile broadband (eMBB) to a range of mission-critical applications for autonomous vehicles, smart cities, and industrial automation [25]. 5G can deliver peak data rates of up to 20 Gbps, 100 times faster than 4G, and decreases latency to 1 millisecond, nearly-instant [26]. Furthermore, 5G supports a massive number of device densities, even one million per square kilometer – an absolute necessity for the expansion of IoT and Industrial Internet of Things (IIoT) [27].

As a leader in 5G innovation, Ericsson introduced important 5G advancements including Massive MIMO, network slicing, and edge computing [28]. Massive MIMO extends beyond traditional MIMO, employing multiple antennas each at base station (64 to 256) [29]. This enables beamforming which directs the signals to individual users, rather than broadcast everywhere. It leads to spectral efficiency and reduced interference. Massive MIMO dynamically steers signal paths to increase network capacity and reliability for the busiest urban areas. Massive MIMO has been pioneered by operators such as Huawei, Nokia, and Ericsson, leading up to a five-times gain on spectral efficiency vs. LTE.

One of the breakthroughs in 5G is network slicing, which is operator-defined multi-tenancy on shared physical infrastructure [30]. Slices are optimized for a particular application and provide differentiated performance characteristics for different use cases. For example, communication between the autonomous vehicles use ultra-reliable low latency slices, and high-bandwidth slices are for augmented reality (AR) and virtual reality (VR) [31]. Flexibility allows telecom operators to manage their network resources with more agility and efficiency, thus balancing performance costs. According to GSMA, network slicing can reduce operational expenses by up to 30% while enhancing service quality [32]. Figure 13.4 illustrates 5G network slicing, which creates multiple virtual networks on a shared physical

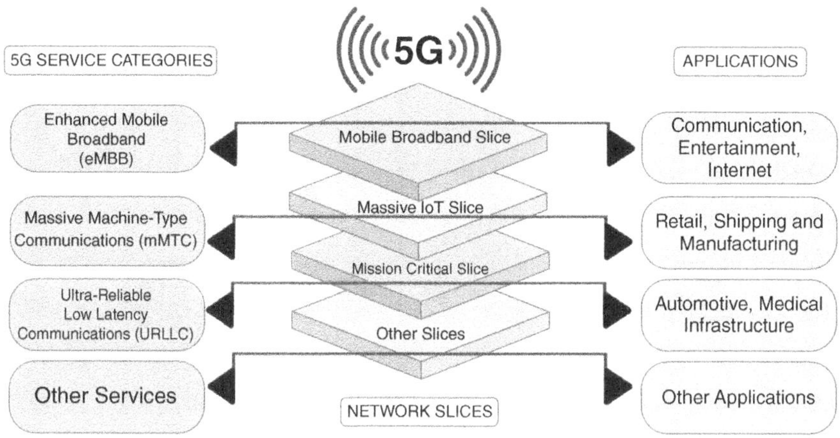


**Figure 13.4** The 5G network slicing and supported industry applications.
*Source:* Author's own work.

infrastructure, and each tailored to specific service requirements. The service categories shown on the left side aligns with the slices of the 5G network as seen in the center of the figure. On the right, various applications are linked to their corresponding network slices, demonstrating how 5G efficiently allocates resources to meet diverse industry needs.

Figure 13.5 displays the typical architecture of edge computing. The centralized cloud data centers are connected to the intermediate edge gateway servers, and many IoT devices to combine data processing and lower latency. The cloud data center is the node with vast storage and computation runs. It collects data on various gateway servers to further process and store over the long term. As data moves from IoT devices to the edge gateway servers, it effectively acts as the middleman between cloud and IoT devices. The servers provide local data processing, which in turn reduces the necessity to send large volumes of data over to the centralized cloud data center. Edge computing also cuts down on latency and running the data closer to the end user instead of bringing it all the way to distant cloud servers [33].

5G can be used in applications that need to respond in real time, such as industrial automation, remote health care, and immersive gaming by placing the computing resources closer to the edge of network, e.g., base stations or local data centers. Studies show that edge computing can reduce latency by up to 90%, ensuring seamless user experience [34]. The edge computing solutions to complement 5G are enabled in large part by players like Ericsson, Nokia, and Qualcomm. This architecture underscores the shift toward decentralized computing, where strategic placement of processing resources optimizes performance, scalability, and reliability across various sectors.

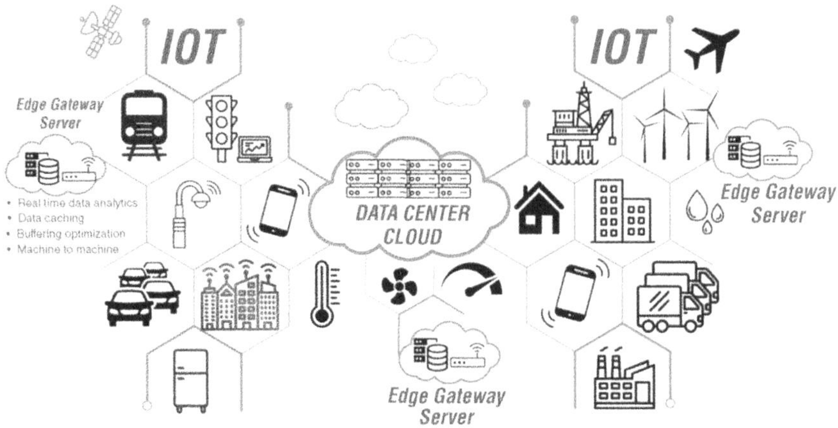


**Figure 13.5** Edge computing architecture. *Source:* Author's own work.

The innovations of 5G technologies help with better spectrum efficiency, customized network services and shortest data processing lag times for continuous connection in a digital world. The forecasted contribution of 5G is much bigger. Estimated to contribute USD$1.3 trillion to the global GDP by 2030, as more countries adopt it and turn it from ultra-convenience to future industry and a life changer [35]. Upgrades in 5G networks are broad changes to the wireless mobile network [36]. So, 5G new radio comes with a completely new air interface layout, innovative antenna techniques, and high-performance radio transmission providing significant improvements in data rates.

Throughout each generational shift, Ericsson's commitment to technological advancement and global connectivity has profoundly shaped the telecommunications landscape. The evolution from 1G to 5G demonstrates how mobile networks have progressed from simple voice calls to complex intelligent systems, capable of transforming industries and societies. As 5G continues to evolve, Ericsson remains a driving force behind the development of next-generation wireless networks, paving the way for future innovations in mobile technology. The COVID-19 pandemic underscored the importance of these technologies, accelerating the demand for remote management, intelligent services, and robust communication networks [37].

### 13.2.2 Ericsson's 5G Deployment Initiatives

Although the global market for 5G equipment does have a handful of competitive players like Nokia, Huawei, and Samsung, Ericsson's technological advantages are so unique that it cements the company as the market leader in both

Radio Access Network (RAN) and Core network technologies. These competitive strengths extend beyond standard 5G implementations, offering operators unique capabilities that differentiate Ericsson's solutions in the increasingly crowded telecommunications infrastructure market.

In the RAN space, Ericsson's leadership becomes apparent through its advanced Massive MIMO implementations and spectrum optimization technologies. The company's Massive MIMO solutions utilize sophisticated beamforming techniques that enable user-specific signal direction, significantly improving spectral efficiency and network capacity compared to traditional antenna systems [38]. This technology allows Ericsson's base stations to serve multiple users simultaneously without interference, creating narrow, focused beams that follow individual devices as they move through the network coverage area. The usage of dual-polarized beamforming also contributes to improved performance as it generates wider 'synchronization signal block' beams and it aims at enhancing the coverage patterns around specific cell shapes [39].

The solution is truly groundbreaking as Ericsson's Dynamic Spectrum Sharing (DSS) solves some of the most pressing issues in 5G rollout, spectrum scarcity and costs. Unlike other vendors that force operators to purchase additional spectrum bands for their 5G solutions, Ericsson's spectrum sharing technology makes 4G LTE and 5G new radio dynamically coexist in the same spectrum [40]. This feature enables telecom operators to roll out 5G immediately on their pre-existing spectrum holdings, which in turn slashes the capital expenditure associated with acquiring new spectra. The approach enables dynamic real-time bandwidth sharing between 4G and 5G networks. It uses access load prediction, combined with a specially designed evolution access control mechanism, balancing the load between different network types and guarantees consistent quality-of-service for both mobile generations [41].

In core network technologies, Ericsson maintains a market position with scale and reliability that few competitors can match. The company's 5G Core solutions already address more than 500 million subscribers in more than 130 live networks worldwide, showing unmatched deployment scale and operational experience [42]. This gives Ericsson unparalleled knowledge of network tuning, performance, and troubleshooting that can be obtained only from very large real-world implementations. The firm's cloud-native 5G Core design allows operators to make a seamless migration away from legacy systems, which is a necessity that operators have been moving toward without much disruption to their existing services.

The integration advantage that Ericsson offers through its comprehensive end-to-end portfolio further strengthens its competitive position. Whereas other vendors concentrate on elements of the network, Ericsson's RAN, Core and Transport solutions can be deployed together, integrated seamlessly, and avoiding the common interoperability issues encountered with multi-vendor solutions. This type of integration is especially useful in multi-vendor 5G scenarios, where performance bottlenecks can be introduced, or a significant amount of integration is needed to effectively interwork. The company operates a global services organization employing thousands of technical experts experienced in multi-vendor 5G deployments. These experts provide implementation and optimization services that transform the company's technology advantages into practical solutions. This approach enables successful network deployments across diverse geographic and regulatory environments.

Building on this competitive advantage, Ericsson's efforts on 5G demonstrate how the company converts its technical superiority into transformative real-world use cases. Just imagine downloading a high-definition full movie within seconds instead of waiting for several minutes. Or imagine conducting medical surgery remotely with zero delay. That's what 5G can bring, and Ericsson is set to spearhead that future. From being rather basic 1G voice calls to 2G text messaging, mobile communication turned into the digital infrastructure of hyper-connected highways. 5G is like upgrading the roads to highways and now another level up to hyper-connected superhighways, with an increasing number of smart devices and data-hungry applications which require faster networks with greater bandwidth. Ericsson's dominance in 5G is reinventing the whole industry and day-to-day life of normal people [43]. 5G is not just an upgrade from 4G; it is a fundamental paradigm shift in mobile communications.

To support the high-speed and high-capacity demands of 5G, Ericsson has leveraged Massive Multiple-Input Multiple-Output (Massive MIMO) technology [44]. Figure 13.6 compares 4G/LTE and 5G networks, highlighting the improvements of Massive MIMO and small cells that make up the 5G system. In 4G LTE, the conventional MIMO arrangements primarily work with small number of antennas in base station that led to limitations on the network capacity and data rate. Unlike traditional systems that only have a handful of antennas, Massive MIMO on the other hand, uses hundreds of antennas at a base station. This increases spectral efficiency and allows more users to connect at the same time without network congestion. It easily enables faster data speeds, and better coverage and reliability, especially in densely populated urban areas.

In addition to Massive MIMO, Ericsson has deployed small cells, which are compact, low-power base stations, designed to enhance 5G coverage in areas where macro towers may be insufficient [45]. Small cells have lower transmitter powers, usually in the 2–20 watts range and in particular bands (2.4 GHz or 5 GHz capable small cells for example) to deliver localized coverage and capacity densely [46]. The small cells act as stitches to combine seamless connectivity in hotspot zones like a stadium, an airport or city centers to allow users to have a continuous service and let applications like AR and VR work without interruptions. The strategic combination of Massive

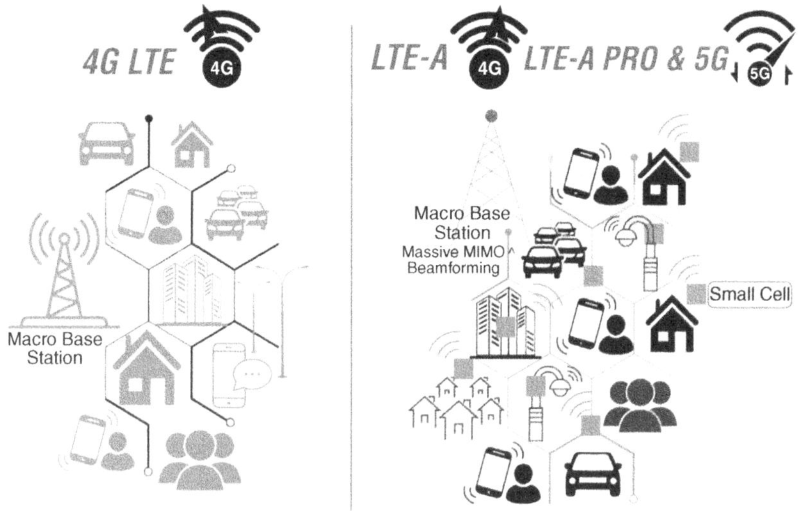


**Figure 13.6** 4G LTE vs. 5G network with Massive MIMO and small cells. *Source:* Author's own work.

MIMO and small cells in Ericsson's 5G networks represents a substantial technological advancement over 4G LTE infrastructure, directly implementing the competitive advantages that distinguish Ericsson's solutions in the global marketplace.

One of the key challenges of 5G deployment is increased energy consumption, given the growing number of connected devices and network nodes [47]. To combat this, Ericsson inserted AI and machine learning to facilitate network optimization in terms of power, performance, latency, and resource allocation in real time. AI-powered Ericsson solutions monitor real-time traffic patterns; they optimize power by serving the appropriate transmission powers to reduce energy consumption up to 30% during off-peak times while ensuring top-notch performance [48]. It improved sustainability by decreasing the operational costs for the telecom operators. For instance, Ericsson has shown 20%+ energy efficiency improvements for 5G networks vs. legacy 4G with its network energy management solutions. This new 5G New Radio (NR) solution by Ericsson can offer high-speed data transfer, low latency, and an enhanced network capacity [49].

In addition, Ericsson has added intelligent sleep modes for 5G base stations which take low power when load is low. This requires machine learning (or other) models across base stations to train on historical usage habits and predict future demands. These base stations perform efficiently and handle data traffic through the AI-based predictive analytics. Moreover, the next generation and optimized caching technology can further reduce the energy of sensors and their data transmission frequency.

5G's distinction is the potential to deliver ultra-low latency. It lowers the response times to merely 1 millisecond, which is nearly 50 times quicker than 4G networks [50]. The high-speed data link is required for use cases to the extent of autonomous vehicles, smart cities, industrial automation, and remote healthcare. For autonomous transportation, 5G makes real-time information exchange with red lights, and pedestrians, which significantly mitigates the chance of accidents with other road users [51]. In healthcare industry, particularly in emergency hospitals, low-latency feature of 5G enables surgeons to give real-time directions to remote robotic-assisted procedures [52].

To support the explosion of IoT devices and edge computing applications, Ericsson has developed cloud-native 5G core networks [53]. Unlike traditional network architectures, a cloud-native approach enables greater flexibility, scalability, and efficiency by running network functions as software-defined services [54]. This architecture facilitates network slicing, a technique that allows telecom providers to allocate customized network resources for different applications.

For example, a self-driving car fleet may require an ultra-reliable low-latency slice, while a smart factory may need a high-bandwidth slice for machine-to-machine communication [55]. By implementing cloud-native solutions, Ericsson ensures that 5G networks can efficiently handle diverse workloads, from consumer broadband services to mission-critical industrial applications.

Since the inception of 5G, Ericsson has been the technology leader, driven globally by the innovations in Massive MIMO, small cells, AI for energy-efficient solutions, ultra-low latency, and cloud-native 5G core networks. Partnering with leading telecom operators like Verizon, AT&T, Deutsche Telekom, and China Mobile, Ericsson is clearly planting trees for a hyper-connected future. With the widespread adoption of 5G driven solutions in future, the role of Ericsson will be crucial in enabling automation and beyond. The ultimate goal is to establish ubiquitous, high-speed connectivity as the standard across all sectors from healthcare to transportation and transforming what is currently exceptional into the standard infrastructure.

### 13.2.3 Challenges and Influencing Factors

5G networks' infrastructure development and deployment also gave Ericsson several challenges [56]. As networks become faster and more complex, engineers and telecom providers must overcome several obstacles that influence the speed, cost, and security of 5G adoption worldwide.

Obtaining regulatory approval in various parts of the world is one of Ericsson's biggest challenges in 5G deployment. Spectrum allocation, which dictates where in the spectrum the 5G waves can travel, is controlled by governments and regulatory bodies. Regulations, which often differ greatly between regions, may discourage 5G from global standards and mass expansion. Spectrum management in the US is under the Federal Communications Commission (FCC), and the Electronic Communications Committee (ECC) is followed by the European Union. In Asia, China, Japan, and South Korea have regulations governing 5G spectrum usage as well. In some regions, compliance is extremely high and that can result in infrastructure lagging and potentially hindering the launch of 5G services. Ericsson faces a formidable challenge in deploying a unified 5G network on a global scale, as it tries to navigate these nuanced and very often clashing regulatory environments.

Another challenge is spectrum allocation. In contrast to its predecessors, 5G encompasses low-band, mid-band and high-band as opposed to mere high frequency in mmWave (millimeter wave) [25]. This mmWave receives the fastest speeds and lowest latency but have very short range and do not penetrate buildings or walls well. It results in telecom operators needlessly strategizing frequency plans for coverage, peak-speed, and reliability. 5G with low-band frequencies, for example, offer excellent reach and penetration, but power is an issue compared to mid- and high-band spectrum. In turn, 5G technology needs to optimize these fragmented and largely overlapping spectra to allow tens of thousands of connected people in an urban environment to have access in rural places. If not, the logistical and financial factors of infrastructure development for 5G have to manage radio frequency crunch, inter-antenna sync, and more [57].

Figure 13.7 illustrates a typical 5G network architecture demonstrating how multiple frequency bands are used for different connectivity needs. Particularly, it delineates the three primary frequency bands utilized in 5G network. In advanced 5G technology, macro cells provide wide-area coverage via normal cellular towers. Small cells (including picocells and femtocells) support to increase network density in indoor areas. The combination of Massive MIMO (Many-input, Many-output) antennas are helping to improve spectral efficiency by enabling simultaneous data streams to multiple users. Moreover, the fiber-optic backhaul enables high-speed data transfer between network nodes and the use of edge computing. It decreases latency by processing data more on the end-user side. The 5G core network is built on cloud-native architecture as orchestrating network slicing. It provides dynamic bandwidth provision to multiple types of applications.

Infrastructure costs are a major hurdle as well. Traditional 4G networks relied primarily on macro towers and major base stations for coverage. In contrast, 5G technology incorporates extensive small-cell deployment alongside these larger installations. This hybrid approach ensures seamless, high-speed connectivity across all areas. However, the small cells bring a forest of base stations, fiber-optic backhaul, and edge computing farms that can add costs to implementation. The average cost range of a small-cell network depends on whether the installation is in rural or urban areas. In general, rural installations are typically more expensive due to resource limitations. According to AT&T, the cost of installing 5G network

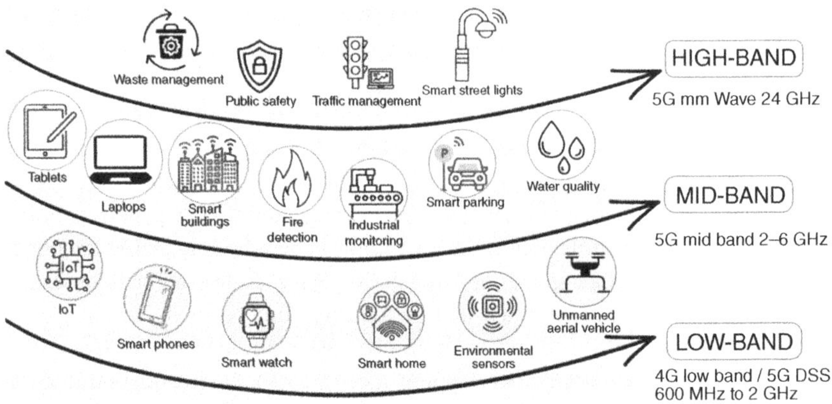


**Figure 13.7** 5G frequency bands and applications. *Source:* Author's own work.

infrastructure ranges from USD$ 6.8 million to USD$ 60 million [58]. Mapping small cells on streetlights, buildings, and other public assets is harder in these regions as it requires municipalities and private entities to get involved for permission and access.

Security is a growing concern too. 5G introduces new security risks including cybersecurity, data privacy, and the security of critical infrastructure. The more connected things are, the larger the cyber security surface area is, and different targets can be threatened. Otherwise, when you have network slicing in 5G, this can be an attack surface as there is no proper security. Network slices may implement varying security configurations and access control mechanisms. Access to user data like locations, browsing history or financial information over an unsecured network slice may be available for malicious actors to misuse. Safeguarding highly sensitive data with proper encryption (advanced encryption standard and Elliptic Curve Cryptography) is the needed to avoid prospective breaches that could cause severe service disruption. Beyond these regulatory and economic factors, Ericsson engineers face several technical challenges in optimizing 5G networks for performance, energy efficiency, and scalability.

The high-density small-cell deployment needed for 5G networks is a major concern. 5G does not need to have the same macro-cell towers that provide ultra-large area coverage for 4G but is about high-density base stations to support seamless coverage [59]. Smaller base stations need to be optimally located and wirelessly connected to the backhaul network. This has to be managed in a way that keeps the network always very stable without interference. Thus, it is important to minimize gaps in coverage, via designing the backhaul of intelligent site selection to algorithms for managing interference from neighboring small cells. It can also use radio resources efficiently by leveraging next-generation beamforming and dynamic spectrum sharing at the radio access network level. It provides a boost to the 5G small-cell network performance as well as energy efficiency.

Another challenge is the use of energy. Massive MIMO and beamforming systems in 5G equipment consume a lot of power, which may lead to higher operational expenditure (also higher environmental impact) [60]. Power-hungry hardware designs and low-traffic or idle-speed-reducing software algorithms are essential for energy-efficient hardware [61]. This can lower costs and increase the sustainability of 5G networks overall, and also through the implementation of cutting-edge power amplifiers, smart sleep modes, and renewable power-integration technologies.

There is also the interoperability of 5G and 4G networks. Most of 5G deployments are with non-standalone (NSA). In which, 5G services coexist in user equipment with an existing LTE infrastructure for the core network functions [62]. Engineers must deliver low latency, higher data rates, and seamless transition capabilities from 4G to 5G networks. This requires complex signaling protocols and powerful network management systems. Additionally, sophisticated interoperability between 4G and 5G layers is essential to maintain optimal end-user experience. It is particularly important for applications such as video streaming, online gaming, and mobile IoT services that require persistent connections.

Furthermore, low-latency and high-reliability are essential for mission-critical applications such as autonomous vehicles, industrial automation, and remote surgeries. Ultra Reliable Low Latency Communication (URLLC), needs completely new network architecture, edge computing and AI empowered traffic management for faster and immediate data mobility [63]. Achieving the goal of low latency communication, also with a high degree of reliability, requires advanced techniques like mobile edge computing (MEC), dynamic scheduling, and real-time network monitoring and optimization. To address these engineering challenges, Ericsson has explored several alternative solutions to improve the efficiency, flexibility, and reliability of 5G networks.

Open RAN (open-radio access network) being one of the approaches that allows for renting virtualized radio haul infrastructure from various vendors to create standardized interfaces between network functions [64]. It opens a disaggregated network architecture, where hardware and software are separated and lowers vendors lock-in with a broader ecosystem of equipment and software providers. With the O-RAN, telecom operators can choose network parts made by different vendors from open and interoperable specifications. It makes the 5G supply chain more flexible and competitive. This modular design deploys and operates at a lower cost for operators, as they are not locked to a specific vendor's proprietary solutions. Finally, the open interfaces and disaggregated components of O-RAN put 5G network infrastructure in practice in a more flexible and cost-effective manner. It means these O-RAN solutions are giving operators more choices regarding their suppliers in the entire telecom value chain [65].

Another promising solution is AI predictive maintenance, which leverages machine learning to evaluate network health in real time. This can help to detect possible failures before it happens and thus automatically prevents downtime as well optimizing network health. AI algorithms shift through real-time data, data history, traffic patterns, and hardware health to foresee when a component might fail [66]. Through proactive measures, operators can prevent beyond costly repairs from happening and maintain steady network performance.

Ericsson also developed hybrid cloud-native network management to scale better. The model unites cloud and on-premises services that allow telecom operators to dynamically assign resource types according to demand. Cloud-native 5G cores and edge computing lowers the latency and enabling real-time processing for apps.

By addressing these challenges through continuous innovation, Ericsson remains at the forefront of global 5G deployment, paving the way for next-generation wireless communication and future advancements in 6G technology. Currently, Ericsson is a leading provider of information and communication technology, with approximately 40% of the world's cell phone traffic passing through networks built by the company [67].

## 13.3 Think and Solve

The following questions summarize the core elements of the 5G Ericsson case study. Considering value proposition, challenges, and technical solutions pertaining to 5G network deployment. They are intended to assess understanding and familiarity of knowledge in mobile technologies, and the value in new technologies for network efficiency and tactical solutions.

**Q 13.1** What is one of the key advantages of 5G over 4G LTE?

**A** Lower frequency usage.
**B** Higher latency.
**C** Faster speeds and lower latency.
**D** Increased reliance on copper cables.

**Q 13.2** Which technology helps 5G networks optimize signal direction?

**A** Frequency modulation.
**B** Beamforming.
**C** Time-division multiplexing.
**D** Carrier aggregation.

**Q 13.3** Why does 5G require more small-cell deployments?

**A** 5G signals have a shorter range than previous generations.
**B** Governments require more towers for licensing.
**C** Small cells improve battery life on mobile devices.
**D** Small cells are cheaper than macro towers.

**Q 13.4** What role does AI play in Ericsson's 5G network optimization?

**A** Increasing power consumption.
**B** Reducing latency and managing network traffic.
**C** Encrypting data at the user level.
**D** Replacing physical infrastructure with cloud-only solutions.

**Q 13.5** Which of the following is NOT a challenge in 5G deployment?

**A** High energy consumption.
**B** Spectrum allocation.
**C** Decreasing data transfer speeds.
**D** Regulatory compliance.

**Q 13.6** Small-Cell Density Challenge

Imagine Ericsson is tasked with rolling out 5G in a densely populated urban area. Due to zoning restrictions, only 60% of the planned small-cell sites can be installed. What alternative methods could be used to maintain coverage and capacity? How can Massive MIMO and beamforming help compensate for fewer small cells?

**Q 13.7** Energy Efficiency Dilemma – A telecom operator finds that its 5G base stations consume 50% more power than expected. What energy-efficient solutions can be implemented to reduce power consumption while maintaining performance? How could AI-driven network optimization improve efficiency?

## 13.4 Conclusion and Key Takeaways

5G deployment represents a fundamental paradigm shift in telecommunications infrastructure. It delivers exceptional performance gains, ultra-low latency, and the capacity to connect billions of devices simultaneously. This technological advancement enables diverse automation applications across industrial, healthcare, and next-generation entertainment sectors. However, the road to a 5G everywhere is accompanied by significant challenges.

Challenges such as the need for dense small-cell infrastructure, spectrum allocation, and energy efficiency. There are also logistical hurdles such as regulatory approvals, and the cost of deployment that add even more challenges to the process of rolling it out. These issues ought to push everyone toward strategic planning and collaboration to unlock the full potential of 5G.

Real-world case studies show both the ways 5G can be deployed as well as what some challenges are encountered. Ericsson's implementation of 5G in industrial manufacturing demonstrated substantial operational savings and highlighted the tangible benefits of this technology. While experiences from different geographies show that infrastructure readiness, regulatory issues, and environmental concerns can drive higher adoption success rate.

In conclusion, although the road to comprehensive 5G adoption is complex, the benefits and economic growth is great. It is essential to work on the wide spectrum of problems together if we are to truly unleash the full capabilities of 5G networks.

## 13.5 Discussion Points

The deployment of 5G technology introduces a spectrum of opportunities and challenges that warrant comprehensive examination. Key discussion points include but are not limited to the following:

1. **Infrastructure Challenges and Solutions of Small-cell Deployment and Spectrum Allocation:** The dense small-cell networks and spectrum resource management to support 5G bands while accommodating high-frequency signals, presents challenges related to logistics and policy. Policymakers should weigh the interests of all stakeholders to optimize usage and minimize interference.
2. **Energy Efficiency Considerations:** 5G base stations may consume more power than anticipated. Implementing energy-efficient hardware and AI-driven network optimization can help manage and reduce energy consumption.
3. **Regulatory and Compliance Issues:** For standardization these are important. Adhering to international standards and local regulations is essential

for seamless 5G deployment. Collaboration among governments, industry bodies, and service providers is necessary to navigate these complexities.

4. **Security Concerns:** With increased data transmission and connected devices, robust security measures are must to protect against potential vulnerabilities and threats.

Addressing these discussion points through collaborative efforts among stakeholders can be pivotal in overcoming challenges and fully realizing the transformative potential of 5G technology.

## 13.6 Limitations of the Case Study

While the case study provides valuable insights into the deployment of 5G technology, several limitations must be acknowledged:

1. **Geographical Specificity:** The case study is illustrative of deployment patterns in urban regions. As a result, the findings might not be suitable for rural areas where infrastructural challenges and deployment-based assumptions might be different.
2. **Technological Evolution:** Given the rapid advancement of 5G technology, the case study may not encompass the latest developments or emerging solutions that have been introduced since its completion.
3. **Regulatory Variations:** The case study does not address regulatory challenges in detail. Regulatory environments can vary widely across different regions and countries, affecting the generalizability of the findings.
4. **Economic Considerations:** Economic factors, such as the cost of deployment and return on investment, are influenced by local market conditions. The case study does not discuss economic analyses in depth.

Recognizing these limitations is essential for contextualizing the case study's findings and for informing future research and deployment strategies in diverse settings.

## References


1 Lema, M.A., Laya, A., Mahmoodi, T., Cuevas, M., Sachs, J., Markendahl, J. and Dohler, M. "Business case and technology analysis for 5G low latency applications," *IEEE Access*, 5, 5917–5935, 2017.

2 Jiang, W. and Han, B. *Cellular Communication Networks and Standards: The Evolution from 1G to 6G*, Springer Nature Switzerland, 2024.

3 Jiang, W. and Han, B. "Evolution to first-generation (1G) mobile cellular communications," in *Cellular Communication Networks and Standards: The Evolution from 1G to 6G*, 7–21. Springer Nature, 2024.

4 Hodara, H. and Skaljo, E. "From 1G to 5G," *Fiber and Integrated Optics*, 40(2–3), 85–183, 2021.

5 Fadhil, H.M. and Dawood, Z.O. "Evolutionary perspective of mobile communication technologies," in *International Conference on Computer and Applications (ICCA)*, 80–84. IEEE, 2018.

6 Goyal, J., Singla, K., Akashdeep, and Singh, S. "A survey of wireless communication technologies from 1G to 5G," in *International Conference on Computer Networks and Inventive Communication Technologies*, 613–624. Springer International, 2019.

7 Telephone World. "Analog cellular (AMPS-1G)," 2025, www.telephoneworld.org/cellular-phone-history/analog-cellular-amps-1g (accessed March 2, 2026).

8 Kukliński, S., Czerwiński, M., Bieniasz, J., Kaminska, K. H., Paczesny, D. and Szczypiorski, K. "Evaluation of privacy and security of GSM using low-cost methods," in *World Conference on Computing and Communication Technologies (WCCCT)*, 95–105. IEEE, 2020.

9 Božanić, M. and Sinha, S. *Mobile Communication Networks: 5G and a Vision of 6G*, Springer, 2021.

10 Adler, R. and Charles, M. "Preparing for a 5G world," The Aspen Institute, Washington, DC, USA, 2016, https://www.aspeninstitute.org/publications/preparing-5g-world (accessed March 2, 2026).

11 Henson C., "SMS celebrates 20th anniversary," *The Wall Street Journal*, 3 December 2012, www.wsj.com/articles/BL-TEB-4980 (accessed March 2, 2026).

12 Holma, H. and Toskala, A. *WCDMA for UMTS: HSPA Evolution and Lte*, John Wiley & Sons, 2010.

13 O'Regan, G. "Overview of telecommunication," in *World of Computing: A Primer Companion for the Digital Age*, 227–237. Springer International Publishing, 2018.

14 Kumar, A. and Bajaj, R. "Internship in designing and optimisation Domain at Ericsson Global India Ltd.," Jaypee University of Information Technology, Waknaghat, 2020.

15 Sanchez, J. and Thioune, M. *UMTS*, John Wiley & Sons, 2007.

16 Tamgno, J.K., Imane, S.T.A., and Lishou, C. "Optimization of HPSA+ capacity of mobile networks," in *20th International Conference on Advanced Communication Technology (ICACT)*, 714–719. IEEE, 2018.

17 Holma, H., Toskala, A. and Tapia, P. *HSPA+ evolution to release 12: performance and optimization*, John Wiley & Sons, 2014.

18 Crotty B. "Cisco: Visual Networking Index: Global mobile data traffic forecast update, 2012–2017," Cisco White paper, 2013, https://s3.amazonaws.com/media.mediapost.com/uploads/CiscoForecast.pdf (accessed March 2, 2026).

**19** Farajidana, A., Chen, W., Damnjanovic, A., Yoo, T., Malladi, D. and Lott, C. "3GPP LTE downlink system performance," in *GLOBECOM 2009 IEEE Global Telecommunications Conference*, 1–6. *IEEE*, 2009.

**20** Rehman, M.U. and Safdar, G.A. *LTE communications and networks: femtocells and antenna design challenges*, Wiley, 2018.

**21** Katambire N.V. "Advancements in multiple input multiple output (MIMO) and orthogonal frequency division multiple access (OFDMA) as LTE-advanced enabler in 4G network," Master's Dissertation, University of Rwanda, 2016.

**22** Najam, A.I., Duroc, Y. and Tedjini, S. "Multiple-input multiple-output antennas for ultra wideband communications," in *Ultra Wideband: Current Status and Future Trends*, ed. M. A. A. Matin, 209–215. IntechOpen, 2012.

**23** Olsson, M., Mulligan, C., Rommer, S., Sultana, S. and Frid, L. *SAE and the Evolved Packet Core: Driving the mobile broadband revolution*, Academic Press, 2009.

**24** Glimstedt, H. "The dynamics of innovation in the wireless telecom industry during two eras of technological convergence, 1995–2015," ISI Growth, 2017, http://www.isigrowth.eu/2017/06/14/the-dynamics-of-innovation-in-the-wireless-telecom-industry-during-two-eras-of-technological-convergence-1995-2015 (accessed March 2, 2026).

**25** Vora, M. *5G-Advanced Technologies*, Educohack Press, 2025.

**26** Barb, G. and Otesteanu, M. "4G/5G: A comparative study and overview on what to expect from 5G," in *43rd International Conference on Telecommunications and Signal Processing (TSP)*, 37–40. IEEE, 2020.

**27** Gohar, A., and Nencioni, G. "The role of 5G technologies in a smart city: the case for intelligent transportation system," *Sustainability*, 13(9), 5188, 2021.

**28** Rafique, W., Barai, J. R., Fapojuwo, A.O. and Krishnamurthy, D. "A survey on beyond 5g network slicing for smart cities applications," *IEEE Communications Surveys & Tutorials*, 27(1), 595–628, 2024.

**29** Chataut, R. and Akl, R. "Massive MIMO systems for 5G and beyond networks—overview, recent trends, challenges, and future research direction," *Sensors*, 20(10), 2753, 2020.

**30** Zhang, S. "An overview of network slicing for 5G," *IEEE Wireless Communications*, 26(3), 111–117, 2019.

**31** Al Harthi, F.R.A., Touzene, A., Alzeidi, N. and Al Salti, F. " Implications of augmented reality applications for vehicle-to-everything (V2X)," in *Immersive Virtual and Augmented Reality in Healthcare*, 119–143. CRC Press, 2023.

**32** GSMA. *Network Slicing: Use Case Requirements*, version 2.0, January 2020, www.gsma.com/solutions-and-impact/technologies/networks/wp-content/uploads/2020/01/2.0_Network-Slicing-Use-Case-Requirements-1.pdf (accessed March 2, 2026).

**33** Nain, G., Pattanaik, K.K. and Sharma, G.K. "Towards edge computing in intelligent manufacturing: Past, present and future," *Journal of Manufacturing Systems*, 62, 588–611, 2022.

34 IndMALL. "How much faster is edge computing?" 2025, www.indmall.in/faq/how-much-faster-is-edge computing (accessed March 2, 2026).

35 PWC. "The global economic impact of 5G," PWC Global, technology industry publications, 25 January 2021, www.pwc.com/gx/en/industries/technology/publications/economic-impact-5g.html (accessed March 2, 2026).

36 Berzin, O., Ansay, R., Kempf, J., Sheikh, I. and Hendel, D. "The IoT exchange," arXiv preprint arXiv:2103.12131, 2021.

37 Sharifi, A., Khavarian-Garmsir, A.R. and Kummitha, R.K.R. "Contributions of smart city solutions and technologies to resilience against the COVID-19 pandemic: a literature review," *Sustainability*, 13(14), 8018, 2021.

38 Ericsson. "Massive MIMO explained: unlocking 5G efficiency," https://www.ericsson.com/en/ran/massive-mimo (accessed March 2, 2026).

39 Ericsson. "Broad beamforming technology in 5G massive MIMO," 2024, https://www.ericsson.com/en/reports-and-papers/ericsson-technology-review/articles/2024-issue-1 (accessed March 2, 2026).

40 Ericsson. "Reusing the same spectrum for 5G and 4G deployments," 2022, https://www.scribd.com/document/767691251/reusing-the-same-spectrum-for-5g-and-4g-deployments (accessed March 2, 2026).

41 Devopedia. "5G dynamic spectrum sharing," 2024, https://devopedia.org/5g-dynamic-spectrum-sharing (accessed March 2, 2026).

42 Ericsson. "5G core network solutions," 2025, https://www.ericsson.com/en/core-network/5g-core/solutions (accessed March 2, 2026).

43 Alkhalaileh, M., Calheiros, R.N., Nguyen, Q.V. and Javadi, B. "Data-intensive application scheduling on mobile edge cloud computing," *Journal of Network and Computer Applications*, 167, 102735, 2020.

44 Patzold, M. "It's time to go big with 5G [mobile radio]," *IEEE Vehicular Technology Magazine*, 13(4), 4–10, 2018.

45 Alimi, I. *5G Fixed Wireless Access: Revolutionizing Connectivity in the Digital Age*, Springer, 2025.

46 Muirhead, D., Imran, M.A. and Arshad, K. "A survey of the challenges, opportunities and use of multiple antennas in current and future 5G small cell base stations," *IEEE Access*, 4, 2952–2964, 2016.

47 Prasad, K.S.V., Hossain, E. and Bhargava, V.K. "Energy efficiency in massive MIMO-based 5G networks: opportunities and challenges," *IEEE Wireless Communications*, 24(3), 86–94, 2017.

48 Mobile Magazine. "How Ericsson's new 5G radios boost performance & efficiency," Mobile Magazine, 13 February 2025, www.mobile-magazine.com/sustainability/how-ericssons-new-5g-radios-boost-performance-efficiency (accessed March 2, 2026).

49 Dahlman, E., Parkvall, S. and Skold, J. *5G NR: The next generation wireless access technology*, Academic Press, 2020.

**50** Akbarzadeh, O. "Evaluating latency in a 5G infrastructure for ultralow latency applications," Doctoral Dissertation, Politecnico di Torino, 2022.

**51** Pawar, V., Zade, N., Vora, D., Khairnar, V., Oliveira, A., Kotecha, K. and Kulkarni, A. "Intelligent transportation system with 5G vehicle-to-everything (V2X): architectures, vehicular use cases, emergency vehicles, current challenges and future directions," *IEEE Access*, 12, 183937–183960, 2024.

**52** Qureshi, H.N., Manalastas, M., Ijaz, A., Imran, A., Liu, Y. and Al Kalaa, M.O. "Communication requirements in 5G-enabled healthcare applications: review and considerations," *Healthcare*, 10(2), 293, 2022.

**53** Bergström, H. and Fredriksson, O. "Detecting network partitioning in cloud native 5G mobile network applications," Master's Thesis, Chalmers, 2022.

**54** Huang, S.Y., Chen, C.Y., Chen, J.Y. and Chao, H.C. "A survey on resource management for cloud native mobile computing: opportunities and challenges," *Symmetry*, 15(2), 538, 2023.

**55** Biswas, A. and Wang, H.C. "Autonomous vehicles enabled by the integration of IoT, edge intelligence, 5G, and blockchain," *Sensors*, 23(4), 1963, 2023.

**56** Agiwal, M., Roy, A. and Saxena, N. "Next generation 5G wireless networks: a comprehensive survey," *IEEE communications Surveys & Tutorials*, 18(3), 1617–1655, 2016.

**57** Cengiz, K. and Aydemir, M. "Next-generation infrastructure and technology issues in 5G systems," *Journal of Communications Software and Systems*, 14(1), 33–39, 2018.

**58** Fortune Business Insights. " Small cell 5G network market size, share, trends & industry analysis, by operating environment (indoor and outdoor), by application (residential, commercial, and industrial), by small-cell type (femtocell, metrocell, picocell, and microcell)," 21 July 2025, www.fortunebusinessinsights.com/industry-reports/5g-small-cell-market-101600 (accessed March 2, 2026).

**59** Khan, S.A., Shayea, I., Ergen, M. and Mohamad, H. "Handover management over dual connectivity in 5G technology with future ultra-dense mobile heterogeneous networks: a review," *Engineering Science and Technology, An International Journal*, 35, 101172, 2022.

**60** Salh, A., Audah, L., Mohd Shah, N.S. and Anuar Hamzah, S. "Energy-efficient power allocation with hybrid beamforming for millimetre-wave 5G massive MIMO system," *Wireless Personal Communications*, 115, 43–59, 2020.

**61** Taleb, T., Benzaïd, C., Lopez, M.B., Mikhaylov, K., Tarkoma, S., Kostakos, P., Mahmood, N.H., Pirinen, P., Matinmikko-Blue, M., Latva-Aho, M. and Pouttu, A. "6g system architecture: a service of services vision," *ITU Journal on Future and Evolving Technologies*, 3(3), 710–743, 2022.

**62** Frank, H. "Private 5G networks: multi-connectivity driven aggregation for services, deployment and management," Doctoral Dissertation, University of Bristol, 2024.

**63** Khan, B.S., Jangsher, S., Ahmed, A. and Al-Dweik, A. "URLLC and eMBB in 5G industrial IoT: a survey," *IEEE Open Journal of the Communications Society*, 3, 1134–1163, 2022.

**64** Holma, H. and Toskala, A. *5G technology: 3GPP evolution to 5G-advanced*, John Wiley & Sons, 2024.

**65** Pongratz, S. "Open RAN market opportunity and risks," Dell'Oro Group report, 2021, https://www.delloro.com/wp-content/uploads/2021/11/DellOro-Group-Article-Open-RAN-Market-Opportunity-and-Risks.pdf (accessed March 2, 2026).

**66** Kroll, J.A. and Berzins, V. "Understanding, assessing, and mitigating safety risks in artificial intelligence systems," *Technical Report*, 2022, https://apps.dtic.mil/sti/html/trecms/AD1191920/ (accessed March 2, 2026).

**67** Telefonaktiebolaget LM Ericsson and Centre for Business History. "Ericsson history – how did we get here?" Ericsson, 2025, www.ericsson.com/en/about-us/history (accessed March 2, 2026).

14

# Huawei's Autonomous Networks: AI Innovation Amid Global Tensions

Self-Optimizing Systems, Predictive Analytics, and Geopolitical Consequences

## 14.1 Synopsis

Imagine managing a telecom network as orchestrating a massive symphony, where every instrument must play in perfect harmony, even as new instruments are added mid-performance. This mirrors the challenge that modern telecom operators face as they navigate an increasingly complex composition of proliferating networks, rising data demands, and seamless connectivity requirements. Traditional network management seems like a human conductor manually directing each orchestra section, requiring constant attention and precise timing. But, Huawei's AI-powered network optimization transforms this paradigm – evolving from a traditional conductor to an intelligent, self-conducting orchestra that adapts and harmonizes in real time, while ensuring perfect performance even as the musical complexity grows.

Global telecommunication leader Huawei has been a pioneer in the AI and network optimization integration. Using machine learning, big data analytics, and cloud, Huawei's system allows telecom operators to proactively anticipate network issues, optimize dynamic resource allocation and even have self-repairs in their network. This innovation not just improve network availability and performance but also cuts down operational costs and power consumption.

This case study illustrates the transformation of the telecom industry, some of the most important challenges such as network congestion, and latency network saturation that Huawei AI-driven solution is putting into place. It provides some insights into the technical backbone behind this system, its external manifestation and implications to bridge the gap on AI in telecommunications.

## 14.2 The Growing Complexity of Telecom Networks

Over the last few decades telecommunication networks have fundamentally changed. Whether it is the origins of analog based voice networking or into the growing need for ever "faster connectivity" in 5G and beyond, the increasing complexity of these networks is driven by several factors including the rise of mobile devices, the proliferation of Internet of Things (IoT) devices, and the increasing demand for bandwidth intensive applications such as video streaming and cloud computing [1].

### 14.2.1 Historical Context of Telecom Networks

Telecom networks in the 1980s and into the 1990s were relatively simple as they supported basic voice networks [2]. Yet the ubiquitous increase in the complexity of mobile devices; online use; and data hungry applications (e.g., video watching, cloud computing, IoT) have all multiplied this massively [3]. Gathering and processing such massive data from billions of devices, base stations and datacenters has been a massive challenge to manage complex telecom networks. Delivering continuous connectivity, efficient resource utilization, latency minimization, and quality of service under variable scenario demands more sophisticated, and smart ways of network management [4].

Thus, today's telecom networks are composed of complex web of connected devices, base stations, and data centers, each generating colossal amounts of data. Network management is always not just about interconnectivity. It is about efficiency, scalability, security, fault tolerance, being resourceful and reliable with lowest latency due to extreme amount of data transfer between interconnected nodes in the network. This complexity has pushed the traditional network management to its limit. Redesigning networks to match the "Quality of Service" demands from numerous applications is a key challenge, as mobile network traffic is foreseen to only increase [5].

Historically, network optimization was a full-fledge process of bricks and mortar, and rule-based ad hoc algorithms. Engineers looked at the performance metrics of a network, figured out where the bottlenecks are and adjust based on parameters hard coded into each system [6]. This traditional approach worked effectively in less complicated networks. However, network scale has expanded dramatically over time, and the speed of network growth has accelerated significantly. Heavy data traffic loads have also increased substantially. Consequently, the traditional approach is no longer sustainable for modern telecommunications infrastructure [7]. Human intervention is known to be slow, biased, and unable to scale efficiently to the dynamic needs of modern telecom networks.

This huge amount of data and increased complexity in today's telecom networks simply put manual optimization methods at their break point, driving the need for smarter automated solutions. Traditional approaches lacked the ability to efficiently handle massive data volumes. Particularly, modern telecom networks generate massive amounts of data every single second. These complex networks incorporate billions of end nodes and intermediate nodes worldwide. Such complex network topologies create several data transmission related challenges between source and destination [8]. This limitation provided the path for AI-based optimization methods. It makes use of machine learning and advanced analytics to automate the whole optimization algorithm and appropriate network conditions in real time [9].

Traditional network management techniques have difficulties managing networks data in real time based on the current industrial demands. These approaches rely heavily on reactive problem-solving, which involves addressing network issues only after they have interrupted service. This reactive model causes customer dissatisfaction and creates potential safety risks too. For instance, during the 2023 Optus outage in Australia, a routine software upgrade resulted in a 14-hour service disruption. Due to this, 2,000 individuals were prevented from making emergency calls [10]. The incident not only showed how reactive maintenance is hazardous but also a $12 million fine for non-compliance of the emergency call regulations came as the result of this.

This reliance on manual processes inhibits the ability to fix network issues faster and implement scalable solutions that reduce the time for a resolved outage, as well as higher quality service endurance. The complexity of contemporary telecom networks makes it necessary to have smarter and more automated solutions for network management and their self-optimization. AI-based strategies hold the potential of proactive detection of issues, and real-time self-optimization that can greatly improve network performance, reliability, and efficiency [11].

Scalability is another critical challenge. As networks expand, the volume of data required for analysis becomes overwhelming for manual processes. Also, network complexity increases significantly due to legacy equipment, and years of corporate mergers and acquisitions. This kind of infrastructure makes it extremely difficult for telecommunications service providers to maintain comprehensive security and to patch every access point on their networks. As a result, networks become vulnerable to cyberattacks from threat actors. For example, Volt Typhoon is a foreign state-sponsored group that has successfully infiltrated telecommunications infrastructure by exploiting unpatched devices and compromised routers [12].

Volt Typhoon, also known by aliases such as Vanguard Panda, Bronze Silhouette, Dev-0391, UNC3236, Voltzite, and Insidious Taurus, is a foreign-sponsored cyber group. This threat actor has been charged with targeting critical infrastructure sectors including communications (telecom), energy, and transportation systems. These threat actors typically exploit unpatched network devices and legacy systems to establish persistent access [13]. Their operational strategy focuses on identifying vulnerable entry points within network infrastructure and maintaining long-term, stealthy presence once initial access is gained.

Security authorities have raised alarms about the widespread and evolving nature of this kind of foreign-sponsored attack on telecommunications networks. The Cybersecurity and Infrastructure Security Agency (CISA), along with the National Security Agency (NSA) and the Federal Bureau of Investigation (FBI), assess that cyber actors are positioning themselves within IT networks. They could potentially launch disruptive or destructive cyberattacks against critical infrastructure during major crises or conflicts [13].

Government authorities have taken steps to disrupt and response to stealthy threats. For example, in January 2025, authorities announced countermeasures against foreign-sponsored telecommunications cyber intrusions, emphasizing the seriousness of these infiltrations and the need for robust cybersecurity measures [14]. This underscores the immediate need for telecom providers to strengthen their cybersecurity measures. Given vulnerabilities related to legacy equipment, proactive patch management and advanced threat detection systems are essential for protecting against any cyber risks like Volt Typhoon [15]. All of this emphasizes the immediate need for telecom providers to increase their own cybersecurity. In the face of vulnerabilities related to legacy equipment, the use of advanced threat detection systems is a key in protecting against cyber risks [16]. This proactive strength is necessary to protect the integrity and resilience of telecom networks to ensure uninterrupted services to citizens and to reduce the risk of potentially disastrous cyberattacks [17].

Rule-based systems have serious limitations in their adaptability as well. These systems usually have difficulty in adapting to more rapidly changing network conditions such as emergencies and sudden surges in traffic for example [18]. This is the problem that we are starting to see with the UK and Europe having a recent delay in deploying 5G. Dispersed telecommunications sectors and fierce competition have discouraged infrastructure investment, to deliver less than 50% of 5G coverage in Europe compared to 90% in the USA, and 95% in major Asian economies [19]. This shackles the capacity for more hyper-speed connections that are being demanded.

Automated optimization does not require much in terms of human resources and manual intervention. For instance, poor mobile connectivity in rural Australia has generated significant frustration among residents, highlighting the urgent need for substantial infrastructure investment to improve telecommunications services [20]. The federal government's Independent Telecommunications Review Committee has recommended modernizing the universal service obligation and utilizing satellites to enhance coverage, but these initiatives require considerable financial commitment.

Despite these attempts to address the issues, traditional network management approaches remain very challenging with high maintenance costs, slow response times, and lack of flexibility in feeding onto changing paradigms [21]. These constraints highlight the pressing need for more intelligent, automated network management that is instrumental in enabling the advent of artificial intelligence (AI) and machine learning (ML) solutions. Telecom operators use AI/ML to predict network disruptions and manage massive piles of data in real time with greater automation and proactive forecasts. It helps network operations to handle unplanned events and cut down operational costs. This change is critical to improve reliability, security, and customer experience in this modern telecommunications landscape.

This was the first paradigm shift of telecom networks with AI. AI gave the capability to consume large data in real time, learn patterns, predict possible issues, and optimize network performance independently. An AI enabled system could allocate resources, decrease energy usage, and even conduct self-correction operations without human intervention.

With AI-powered solutions being critical for telecom network management, the leadership role of major telecom infrastructure providers was instrumental to push innovation. Among them, Huawei has played a major part in the automation in telecommunication networks, partly with the rising complexity of artificial intelligence-driven network optimization [22]. Huawei was founded in Shenzhen, China in 1987 by Ren Zhengfei, a former People's Liberation Army telephony engineer with an initial focus on supplying China's countryside telecom equipment [23]. Over time, the company emerged as a leader in AI-driven network evolution. Huawei involves network infrastructure development, consumer electronics, and cloud computing services. In addition, their automated solutions focused on scale, speed, and security for modern telecom infrastructures.

Then by mid-2010s, Huawei started to boost its AI research and development, as it realized the transformative potential in telecom networks. Creating AI-based solutions puts that company in the very forefront role of network automation. However, Huawei's developments have been seen as a double-edged sword by some. Despite its global dominance in telecom infrastructure, the involvement of Huawei has been controversial as many western governments worry about security risks, and about its ties to the Chinese state. The duality of perception suggests wider geopolitical tensions surrounding Huawei's role in shaping the modern telecom industry.

The deployment of AI in telecom networks became essential due to the escalating complexity and data volumes generated by modern infrastructure [24]. Today's network traffic is expanding dramatically because of the quick development of smart devices like smartphones, smart homes, and smart automobiles [25]. Within a few decades, Huawei grew into one of the largest telecommunications companies in the world, with operations in over 170 countries and a workforce of more than 190,000 employees as of 2023 [26].

The company's rise was driven by its commitment to innovation and investment in research and development (R&D) [27]. Huawei allocates around 10–15% of its annual revenue to R&D, making it one of the top R&D spenders globally [28]. Its focus on cutting-edge technologies, including 5G, AI, and cloud computing, has positioned it as a key player in the global telecom market [29]. Despite facing geopolitical challenges, Huawei continues to lead in AI-driven telecom solutions, with a strong emphasis on creating intelligent, efficient, and sustainable networks.

### 14.2.2 Huawei's AI-Driven Telecom Network Optimization System

In 2018, Huawei introduced its "Autonomous Driving Network or ADN" concept. This system blended AI, big data and cloud computing to get networks that are self-optimizing, self-healing, and able to operate with a high degree of intelligence [30]. Huawei's AI solution has been widely used by many telecom operators globally, especially in Asian countries, Europe and Middle East, where they have demonstrated significant improvements in network efficiency and reliability.

Autonomous Driving Network (ADN) is a different piece of the puzzle in Huawei's Communications Network 2030, that is aimed at driving intelligent transformation of Communication Service Providers (CSPs) faster. ADN is an AI-driven automation layer that relieves self-healing and self-optimizing networks across multiple network technologies including 5G cellular networks, Wi-Fi networks, and other wireless infrastructure, delivering seamless connectivity and network operations.

The solution is based at core of single domain autonomy, cross-domain partnership, and value-driven network transformation. ADN gives broader business value, rapid service provisioning, proactive resolution enhancing customer experience, and predictive AI-driven maintenance leading to operation enhancement. It also improves allocation of resources and energy, decreasing network congestion and enhancing sustainability.

Huawei ADN Level 4 (L4) solutions are based on innovative technologies including telecom foundation models, converged awareness, and digital twin simulations from Huawei. It includes industry-specific AI copilots and scenario-optimized agents. This augments workforce capabilities, network performance optimization and enable digital transformation use cases for people in their industries. ADN delivers zero-wait, zero-touch, and zero-trouble experiences enabling highly autonomous networks that ensure sharp service quality and operational excellence.

The ADN Level 4 (L4) solution architecture enhances network intelligence by delivering business value and enabling new applications. Business Impact is spread over four main areas: (1) Monetization: Enables faster service provisioning and creates pathways to Network as a Service (NaaS), (2) Customer Experience: Delivers AI-powered enhancements to improve service quality while reducing complaint frequency, (3) Operations & Maintenance (O&M) Efficiency: Implements

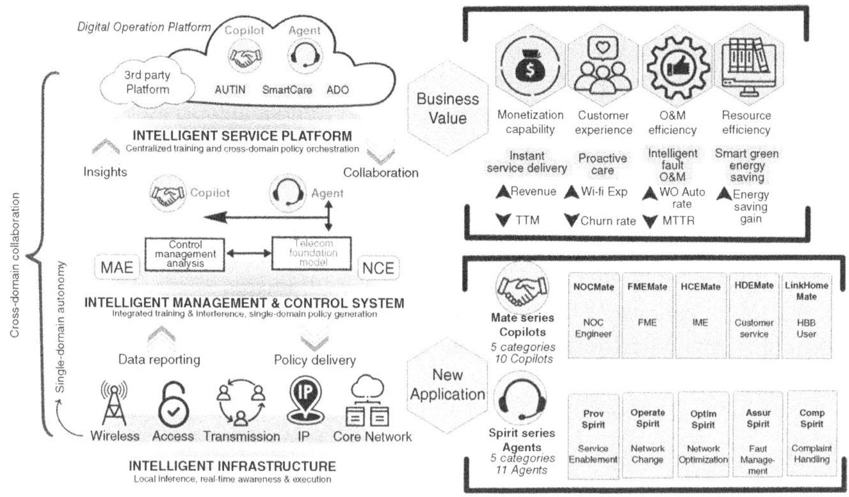


**Figure 14.1** ADN solution overview. *Note:* FME = Field maintenance engineer; IME = Installation and maintenance engineer. *Source:* Adapted from [31].

predictive maintenance and automated fault management to modernize operations, and (4) Resource efficiency: Optimizes network utilization through intelligent resource allocation and smart capacity management.

Figure 14.1 shows this in terms of new applications; ADN introduced the Mate series copilots and Spirit series agents to ensure various operational needs. The Mate series has five categories and ten copilots including NOCMate, FMEMate, HCEMate, HDEMate and LinkHome Mate. While the Spirit series offers five categories and eleven agents like ProvSpirit, OperateSpirit, OptimSpirit, AssurSpirit and CompSpirit, enabling scenario of specific automation and optimization across diverse network infrastructures including 5G cellular, Wi-Fi, and other wireless technologies. Together, these innovations move CSPs toward highly autonomous networks on a scale. It delivers a massive boost in efficiency, reliability and digital transformation.

Huawei's system distinguishes itself by processing real-time data at rates up to 1.2 terabits per second (Tbps), accommodating the demands of millions of connected devices and network nodes. This capability was demonstrated in 2019, a trial with BT Ireland. In which, Huawei's OptiXtreme series oDSP (optical Digital Signal Processor) chips and high-performance optical transmission modules enabled ultra-high-speed interconnection between two Dublin data centers [32]. The OptiXtreme series incorporates Channel-Matched Shaping (CMS) and AI neuron modules. It allows them to sense key optical channel parameters, compensate for link impairments, and optimize transmission spectrum efficiency. Additionally, these chips support single-wavelength transmission rates ranging from 200 to 600 Gbps, significantly enhancing the capacity and reliability of modern optical networks [33]. Huawei's Super-C line system further extends its capabilities, achieving a single-fiber capacity of 48 Tbps. It is six times the capacity of conventional 100G wavelength division multiplexing (WDM) systems [34]. These advancements position Huawei at the forefront of next-generation high-speed optical transmission solutions, ensuring efficient data processing and seamless connectivity in complex telecom environments.

The system can predict network issues with over 90% accuracy using AI and deep machine learning by dynamically allocating resources and automating decision levels [35]. In contrast to the passive reactive systems, Huawei solution helps telecom operators proactively identify and solve issues in a timely order to prevent user impacting events.

Its end-to-end automation is a key innovation in Huawei's approach [36]. While many competitors offer AI-enhanced tools for specific aspects of network management, Huawei's system spans the complete network lifecycle from planning and building through optimization and maintenance as whole. Federated learning is used in the field so that AI models can be trained over multiple data sources, and data privacy is a key here for telecommunication companies.

As mentioned earlier in this chapter, the major components in this advanced system are predictive analytics, self-healing networks, real-time optimization, and energy efficiency [37]. By using predictive analytics, machine learning can be used to predict traffic patterns and anomalies. It can impact the network performance with an accuracy over 85% for both [34]. Self-healing networks detect and resolve faults autonomously, reducing 90% of alarms [31]. This is often done quickly, minimizing service interruptions and 30% of the faults can be resolved before users file complaints. The key characteristics are dynamic optimization of system parameters like frequency allocation, power control and load balancing in real time with an efficiency gain of 20–40% under certain scenarios. Cloud-native architecture allows for easy scaling and integration with existing network architecture.

Figure 14.2 illustrates the transition from manual to automated network solution. This shows how advanced technologies are integrated to streamline the customer complaints process, and creating more efficient and effective workflows. Before automation was implemented, the manual method involves classification and basic analysis. After automation was implemented, the network system pre-process data to extract the keywords by utilizing Bidirectional Encoder Representations from Transformers (BERT) model into seven high-frequency scenarios for proper complaint classification. This shift directly increases the accuracy of the classification and automates the complaint process. This also reassigns analysis from the Core Network Department to the Monitoring department.

Moreover, Figure 14.2 provides a glimpse of the specific larger model signaling queuing. It leverages a three-layer signaling parsing system with natural language

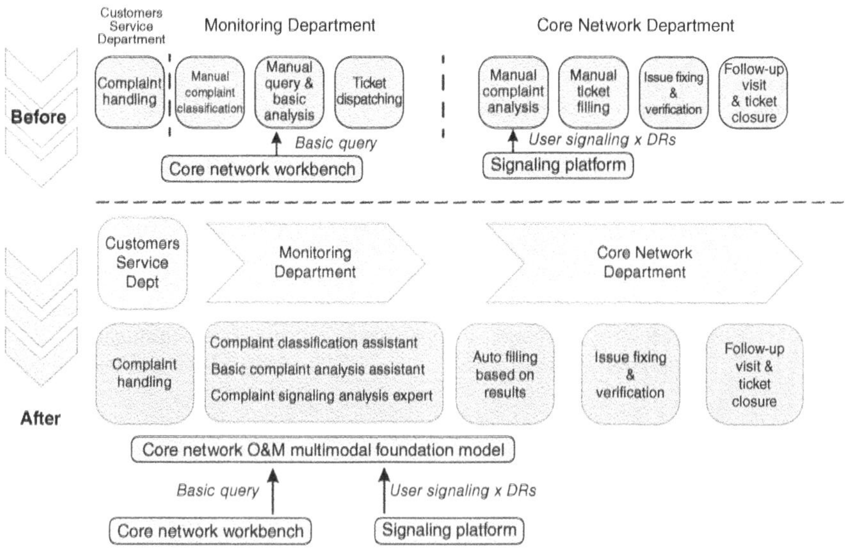


**Figure 14.2** Core network's transition from manual to automated customer complaint-handling. *Source:* Adapted from [31].

Q&As for faster and more precise analysis of signaling analysis. The model brings down single-phase analysis to five minutes instead of four hours. It enables comprehensive root cause identification and case recommendation like the one offered by experienced O&M experts. Implemented at China Mobile Zhejiang, this solution has integrated more than 30 experienced digital employees into the team. It reshaped the complaint-handling process and initiated a positive cycle of transformation in network operations and management.

This success story at China Mobile Zhejiang highlights the transformative impact of AI-driven signaling analysis on network operations. Not only does the network able to do more timely root cause analysis, but it is reducing the time required for analysis extensively and transforms existing complaint-handling workflows efficiently. Further evidence from other regions that this milestone has been achieved underscores the robustness and effectiveness of the system in different types of network deployments. Table 14.1 shows some of the key projects and their measurable outcomes, showcasing the widespread impact of Huawei's AI-optimized telecom solutions.

These deployments demonstrate the system flexing its muscles across different network setups and delivering tangible benefits. For instance, in China, Huawei's system helped China Mobile manage one of the world's largest telecom networks, significantly improving latency and user satisfaction. Etisalat in the UAE used the

**Table 14.1** Key projects with Huawei's AI-optimized telecom system.

| Telecom Operator | Region | Project Details | Results |
|---|---|---|---|
| China Mobile | China | Implemented across 31 provinces, managing over 10 million base stations and optimizing traffic for 900 million users | Reduced network latency by 30%; improved user experience ratings by 20% |
| Etisalat | United Arab Emirates | Focused on optimizing 5G networks in high-density urban areas | Achieved 35% improvement in network throughput; reduced energy consumption by 12% |
| Telefonica | Spain | Deployed across Spain's 5G network, with a focus on reliability during high-demand events like football matches and concerts | Improved network reliability by 25%; reduced operational costs by 15% |
| MTN Group | Africa | Optimized networks in 21 countries, particularly in rural and underserved regions | Reduced network downtime by 40%; improved data speeds by 15% in remote areas |

*Source:* [31, 38, 39]

system to improve 5G networks in urban areas, leading to throughput gains and energy efficiency improvements. Similarly, Telefonica in Spain utilized Huawei's AI-optimized solutions to maintain uninterrupted service during high-demand events, while MTN Group from Africa has focused on enhancing coverage and performance across rural areas throughout Africa.

Huawei's AI-driven telecom network optimization system is an important step for network operators in integrating technical innovation along with the practical applications [40]. The capacity of pre-empting and resolving issues, optimizing resources, and improving end-user experiences are the reasons this tool has become crucial for telecom operators. It navigates the challenges of modern networks.

This milestone becomes more important as telecom networks are growing complex and rich in data. Huawei also enhances network efficiency and has developed numerous special subsystems within the ADN (Autonomous Driving Network), such as specialized OptimSpirit, which is dedicated to the optimization of optical communication and transport networks. Based on AI-driven analytics and predictive automation, the subsystems deliver tailored solutions. It aimed at different network shortcomings and makes smooth experiences in different scenarios and networks.

As a subsystem within ADN of Huawei, OptimSpirit supports real-time wireless networks optimization as shown in Figure 14.3. By means of AI-analytics and adjusted resource management, it is at the heart of enabling high-speed, reliable and efficient optical transmissions. In contrast to conventional network management systems, it employs predictive maintenance of network and real-time fault detection to enhance network stability and spectral efficiency.

To illustrate one of the use cases, let us consider OptimSpirit implementation in the university campus. Figure 14.4 shows 24/7 assurance of campus network optimization agent. It is designed to help the campus network fault management

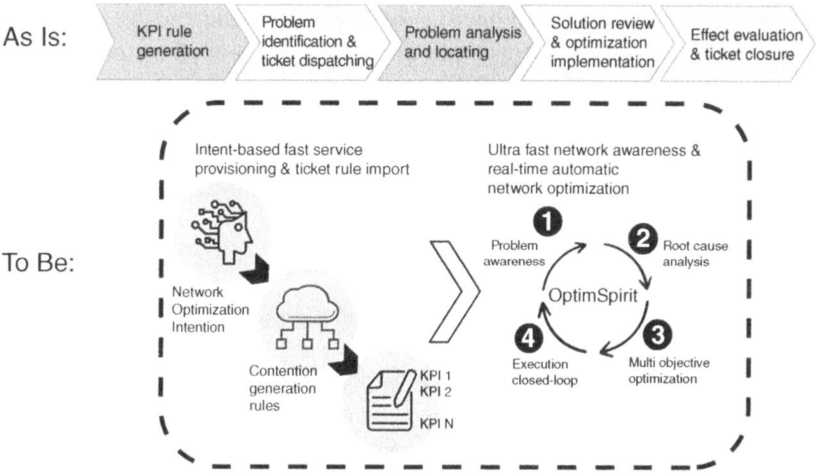


**Figure 14.3** Real-time wireless network optimization of OptimSpirit. *Source:* Adapted from [41].

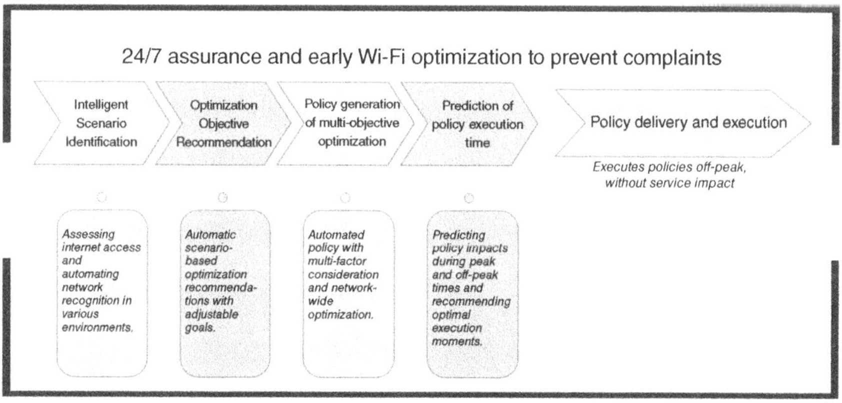


**Figure 14.4** 24/7 assurance campus network optimization agent. *Source:* Adapted from [41].

system and provides seamless Wi-Fi service during heavy data traffic. High-traffic Wi-Fi networks on campus are crucial for anyone doing remote education, watching videos, and using a lot of educational resources. Dormitory networks, on the other hand, suffer from interference from multiple neighboring buildings, unauthorized signal loiterers, and a large number of devices.

Huawei addressed this challenge with its campus network optimization agent based on telecom backboned model, and multi-objective Wi-Fi optimization algorithms. The agent automates network scenario identification using key performance indicators (KPI), recommends optimization objectives, generates multi-dimensional optimization policies, predict execution time, and promptly delivers real-time solutions.

The product proactively monitors network health in conjunction with fine-tuning connectivity before the user has any issues. A recent pilot study at a top Chinese university showed the efficiency of this technology as over 80% of non-fixable in terms of hardware issues were likely to occur before users felt any disruption. This resulted in a dramatic decrease in faulty tickets, while improving better network stability and user satisfaction.

### 14.2.3 Technologies, Challenges, and Key Enablers

Huawei Telecom Network Optimization System offers a next-generation solution powered with AI, Machine Learning, Big Data and Cloud. As Figure 14.5 presents, the core part of the system, such as ML, is formulated based on supervised learning, reinforcement learning and unsupervised algorithms for tuning network operations. With Supervised Learning, it makes predictive analysis. By analyzing the historical data, it gives the traffic pattern as well as identifying the potential network issues. It also helps in finding the hidden patterns from the big datasets.

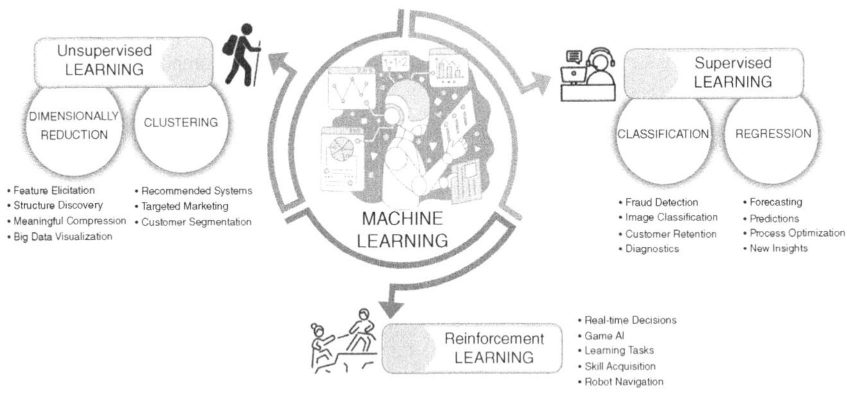


**Figure 14.5** The role of machine learning in autonomous telecom systems. *Source:* Adapted from [40, 41].

Unsupervised learning is generally employed, which is used for anomaly detection and orchestrates network behavior analysis. Reinforcement learning is for the system to fine tune its own network parameters and keep it updated based on refreshed feedback. Huawei systems integrate extra advanced algorithms and statistical models to improve decision-making and automating processes. It derives more value from large volumes of network data to realize a self-adapted intelligent management of telecom.

One of the innovative features of Huawei's system is its use of federated learning. This is an approach that trains AI model across the networks without the need to centralize the data [42]. Federated learning is very relevant in the telecom industry owing to the high importance of data privacy and security. This handles location-based information, communication patterns, and usage statistics entailed with user data carefully because they are sensitive information [43]. Federated learning allows users to train AI models at the devices and/or local servers where the data are, so that private data always stays where it belongs, at its source of origin. It is critically important to fulfill the strongest privacy and security law for data protection from General Data Protection Regulation (GDPR) [44]. This decentralized approach enhances privacy, lessens potential data breaches, and reduces costly data transmission back and forth from the centralized server data. In Huawei's framework, telecommunication operators combine their AI models together with federated learning, so that they can spawn smart AI algorithms without having users' data disclosed. Moreover, this approach boosts the scalability of the system to support higher levels of telecommunication network complexity in global telecoms with millions of connected devices.

However, AI-driven network management still faces data security risks, even with federated learning's privacy protection. While raw data remains localized, the shared model parameters and aggregated learning models can still reveal sensitive information about network infrastructure, traffic patterns, and operational characteristics. Additionally, the system remains vulnerable to model critical attacks where malicious nodes could corrupt the global learning model, and inference attacks that can extract sensitive information from shared model updates during the federated training process.

The system uses deep learning methods like convolutional neural networks (CNNs) and recurrent neural networks (RNNs) to process complex data streams from modern telecom networks. CNNs are perfect for spatial data, signal strength, interference patterns, and topology maps of the network [45]. By comparison, these networks can efficiently learn features from high-dimensional data, and work on problems such as areas of poor connectivity or optimization for base station placement. For instance, CNNs can take visualized network coverage and find out where signal degradations are happening then the operator can resolve in advance [46]. Also, CNNs perform extremely well in multi-dimensional data which makes the system capable of performing real time accuracy analysis of network conditions.

In the case of RNNs, they are designed to operate only on temporal data. Thus, it is best suited for forecasting fluctuations in traffic, latency problems, and anomaly of user behavior over time [47]. Therefore RNNs (in contrast to normal neural networks) that can have a form of memory to remember previous inputs, which makes it much more convenient to deal with sequential data. Huawei uses RNNs for predicting network congestion at peak hours, and when network devices might fail based on the historical data during major events for the network demand spikes from users [48]. To take as an example, RNNs can look at data from IoT devices connected to the network and forecast resource needs in certain places and at specific times. This dynamic resource allocation occurs automatically to prevent bottlenecks. The RNNs models have time-dependent relationships, so that the system has adaptive capability to respond to changes of network conditions and hence brings overall efficiency and reliability.

By leveraging CNNs, RNNs and federated learning, Huawei is running the system-focused network in true hybrid form. It helps in maintaining privacy-preserving data analysis, spatial optimization, and temporal forecasting. These technologies play a central role in the system to process data streams of large scale in real time. It also supports taking intelligent decisions to enhance network performance while at the same time being compliant with regulatory requirements.

Big data is central to Huawei's network optimization system. Some estimates put the amount of data generated by telecom networks as of today as massive as triple-digit gigabytes. For example, a single 5G cell could feed up to 10 GB a second (GSMA 2020). In the Huawei network, a data platform is built for real-time data processing and rapid analytics of massive amounts of data at scale to gain value from it. The system uses distributed computing frameworks, such as Apache Hadoop and Spark, to handle the sheer scale of data involved. It uses those metrics like re-queuing, signal strength, and user density to intelligently make appropriate decisions to optimize network performance.

Cloud computing is another important part of Huawei's network as well. The system is cloud-native, meaning it has been purpose-built on the architecture to enable easy scaling, and flexibility for integratoin with other telecom infrastructure. Cloud computing provides computational power for processing massive data sets and training complex neural networks in one place. It also allows the system to run on diverse network environments. The advantages of Huawei's cloud platform for edge computing enables data processing near the network edge and decreases response time. This is especially useful for applications such as autonomous vehicles and smart cities where milliseconds can make a significant difference.

Huawei's system, however technically advanced, encounters many difficulties and limitations. Telecom networks are big data handlers for user's private information and that is one of the most relevant issues for data privacy. Ensuring compliance with data protection regulations, such as the General Data Protection Regulation (GDPR) in Europe, requires robust encryption protocols and secure data storage practices. Huawei has therefore incorporated end-to-end encryption and anonymization solutions to mitigate this.

Another factor that is very significant is scalability. The system is designed for big networks, but networked devices are growing and the advent of 6G in the future will place additional demands on it. As the network becomes more complex, enabling real-time processing and decision-making over time needs constant improvements in AI algorithms as well as computational infrastructure. The system also has to be scalable and adaptable to local regulations that can impact its use in each jurisdiction.

The Huawei AI-driven telecom network optimization system is an example of an overarching solution to address the problems in network management that the modern telecom industry faces. By coupling up some intelligent AI algorithms with big data and cloud computing it can easily give impressive performance and flexibility. However, widespread adoption across the telecom industry still depends on further addressing key challenges such as data privacy, compliance, and scalability.

## 14.3 Think and Solve

**Q 14.1** Which of the following best describes federated learning in Huawei's telecom optimization system?

**A** Centralized training of AI models using all user data in one location.
**B** Training AI models across distributed datasets without sharing raw data.
**C** A method for compressing network data to reduce storage requirements.
**D** A technique for encrypting data during transmission between base stations.

**Q 14.2** What is the primary role of convolutional neural networks (CNNs) in Huawei's system?

**A** Analyzing spatial data, such as network topology and signal strength.
**B** Predicting traffic fluctuations and user behavior over time.
**C** Encrypting sensitive user data to comply with GDPR regulations.
**D** Managing real-time resource allocation across base stations.

**Q 14.3** How do recurrent neural networks (RNNs) enhance Huawei's telecom optimization system?

**A** By analyzing high-dimensional spatial data for network optimization.
**B** By training AI models across distributed datasets without centralization.
**C** By predicting temporal patterns, such as traffic spikes and latency issues.
**D** By reducing power consumption in 5G base stations.

**Q 14.4** What is a key benefit of federated learning in the telecom industry?

**A** It reduces the computational power required to train AI models.
**B** It ensures sensitive user data remains localized, enhancing privacy.
**C** It eliminates the need for AI in network optimization.
**D** It improves the physical infrastructure of telecom networks.

**Q 14.5** A telecom operator uses Huawei's AI-driven system to predict network traffic using RNNs. The system is tasked with forecasting the traffic load T($t$) (in Mbps) over time $t = 1$, modeled as: $T(t) = T0 + R \times t$

Calculate the traffic load after six hours.
Where:

- T0 = 500 Mbps (initial traffic load)
- R = 50 Mbps/hour (rate of traffic increase)
- $t$ is the time in hours.

## 14.4 Conclusion and Key Takeaways

Telecom network evolution has reached a critical turning point, driven by the increase of connected devices, growing data demands, and emerging technologies like 5G and beyond. Huawei's AI telecom network optimization system is a new paradigm in response to these evolving challenges. It combines next-generation research in federated learning, CNNs, RNNs, comparable data analytics, and cloud computing. Using these advancements, the system provides both real-time optimization of network performance as well as essential attributes like data privacy, scalability, and regulation compliance.

One of Huawei system's key contributions is its capability to manage tremendous amounts of data from modern telecom networks whilst at the same time executing federated learning to maintain user privacy. The decentralized architecture limits where the sensitive user data can be stored. It matches with legal regulation in tight data protection like GDPR. It also incorporates CNNs as well as RNNs for the system to be able to process spatial and temporal data streams. It results in fine predictions of network behavior, which might prevent future problems from being triggered. The advances collectively improve network resiliency, agility, and flexibility that are necessary in the enabling of new waves of applications like self-driving cars and/or smart cities, and industrial IoT (IIoT).

However, the execution of such a complex system has its own challenges. These include data privacy, compliance with different regulations, and scalability, as we move forward into a more automated future. Addressing these challenges require continuous innovation in AI algorithms and computational infrastructure. Also, there is a need for collaborative efforts from industry stakeholders and policymakers in tackling these challenges.

The key takeaway is AI's significant impact on telecom network optimization. The systems from Huawei prove that AI can take care of these complexities quite well. This solution can guarantee good connectivity and superior user experience. Big data and cloud computing integration shows that there is a need for further architecture improvements to address both scalability and flexibility to manage the increasing requirements of telecom networks. In the end, all such systems are dependent upon the fine balancing of techno-aided growth and ethical measures especially data privacy.

In conclusion, Huawei's AI-driven telecom network optimization system is the future of telecom infrastructure. By exploiting AI, big data, and cloud computing, the company positions itself as a pivotal enabler for next-generation networks. Telecoms is a rapidly evolving industry, and systems like this becoming increasingly important in creating a connected world that not only works well but is also reliable, safe, and sustainable.

## 14.5 Discussion Points

To expand on the subjects in this case study further, below are some discussion points to think about. These discussion points aim to foster a deeper understanding of the potential, challenges, and future directions of AI-driven telecom network optimization systems

1. How does AI improve the speed and reliability of telecom network optimization over conventional methods?
2. How effective is federated learning in ensuring data privacy while maintaining the accuracy of AI models?
3. How can AI-driven systems like Huawei's approach adapt to the growing complexity of telecom networks, especially with the rise of 5G and IoT?
4. How can telecom operators balance the need for innovation with compliance to data protection regulations such as GDPR? What ethical considerations should be prioritized when deploying AI-driven systems in telecom networks?
5. How can AI-driven systems contribute to energy-efficient telecom networks? What role does AI play in achieving sustainability goals in the telecom industry?

## 14.6 Limitations of the Case Study

Although the case study is about Huawei's AI telecom optimization system, it does not necessarily mean the solution works out-of-the-box for every telecom operator or network infrastructures. Unique configurations, challenges, and requirements of different networks may mean the approach and solutions of Huawei may not fit on different network environments. Below are some key limitations that are not fully discussed in this case study.

1. While the case study highlights state-of-the-art applications of deep AI models (CNNs, RNNs) etc. and federated leaning, it does not fully disclose the limitations of these models. AI models in general are extremely heavy computationally, requiring lots of training datasets and well-executed training, which could be challenging to achieve for smaller operators.
2. Huawei's system scalability to absorb the complexity of global telecom networks is not fully covered. Other networks with different geography, regulation, and technologies could lack the resources required to effectively adopt the system.
3. Although compliant with worldwide data protection regulations, the case study does not go into the intricacies of operating in disparate regional regulatory regimes. Different laws such as GDPR, CCPA and similar ambiguities could present substantial difficulties for actual implementation.
4. Although the system has well-described technological capabilities, the case study itself does not investigate the economic aspects of costs to deploy it, return on investment (ROI), or their viability for adoption by smaller telecom operators.
5. The Huawei system relies heavily on big data processing and cloud computing, which may not be available or affordable for smaller operators, due to non-existent digital infrastructure in some regions or high price of cloud service operations.
6. Though the system questions data privacy, it does not touch upon higher level ethical questions. For instance, algorithm bias or transparency of the telecom networks in case of AI decisions made by such networks.
7. This case study does not explore details of the integration of the Huawei system with telecom infrastructure from different vendors, in which, interoperability issues may arise, especially under multi-vendor environments.

By acknowledging certain limitations, this case study stands as a neutral view of Huawei's AI telecom optimization system allowing for both its possibilities and gaps in engineering to be seen.

## References


1 El Khatib, M.M., Al-Nakeeb, A. and Ahmed, G. "Integration of cloud computing with artificial intelligence and its impact on telecom sector—a case study," *IBusiness*, 11(01), 1–10, 2019.

2 Huurdeman, A.A. *The Worldwide History of Telecommunications*, John Wiley & Sons, 2003.

3 Wang, Z., Wei, G., Zhan, Y. and Sun, Y. "Big data in telecommunication operators: data, platform and practices," *Journal of Communications and Information Networks*, 2(3), 78–91, 2017.

4 Santos, J., Wauters, T., Volckaert, B. and De Turck, F. "Low-latency service delivery in a continuum of virtual resources: state-of-the-art and research directions," *IEEE Communications Surveys & Tutorials*, 23(4), 2557–2589, 2021.

5 Alfoudi, A.S.D., Newaz, S.S., Otebolaku, A., Lee, G.M. and Pereira, R. "An efficient resource management mechanism for network slicing in a LTE network," *IEEE Access*, 7, 89441–8945, 2019.

6 Bakhshi, T. "State of the art and recent research advances in software defined networking," *Wireless Communications and Mobile Computing*, 2017(1), 7191647, 2017.

7 Srinidhi, N.N., Kumar, S.D. and Venugopal, K.R. "Network optimizations in the internet of things: a review," *Engineering Science and Technology, an International Journal*, 22(1), 1–21, 2019.

8 Verma, S., Kawamoto, Y., Fadlullah, Z.M., Nishiyama, H. and Kato, N. "A survey on network methodologies for real-time analytics of massive IoT data and open research issues," *IEEE Communications Surveys & Tutorials*, 19(3), 1457–1477, 2017.

9 Umoga, U.J., Sodiya, E.O., Ugwuanyi, E.D., Jacks, B.S., Lottu, O.A., Daraojimba, O.D. and Obaigbena, A. "Exploring the potential of AI-driven optimization in enhancing network performance and efficiency," *Magna Scientia Advanced Research and Review*, 10(1), 368–378, 2024.

10 Taylor, J. "Optus fined $12m after thousands could not call triple zero during 2023 outage," *The Guardian*, 7 November 2024, www.theguardian.com/business/2024/nov/08/optus-000-triple-zero-fine-2023-outage-acma (accessed March 3, 2026).

11 Min, S. and Kim, B. "AI technology adoption in corporate IT network operations based on the TOE model," *Digital*, 4(4), 947–970, 2024.

12 Sabin, S. "The long, complex road to resuring U.S. telcos," *Axios*, 6 December 2024, www.axios.com/2024/12/06/telecom-cybersecurity-china-hack-us (accessed March 3, 2026).

13 Hung, H.T. "Multilateral cooperation in building critical infrastructure security and resilience: case of American deterrence of Chinese cyberthreats," *Journal of Cyber Policy*, 9(3), 1–26, 2025.

**14** Ogungbemi, O.S., Ezeugwa, F.A., Olaniyi, O.O., Akinola, O.I. and Oladoyinbo, O.B. "Overcoming remote workforce cyber threats: a comprehensive ransomware and bot net defense strategy utilizing VPN networks," *SSRN*, 4911878, 2024.

**15** Thomas, R. and Jain, E. "China state-linked group accused of hacking SingTel, Bloomberg News reports," *Reuters*, 5 November 2024, https://www.reuters.com/technology/cybersecurity/china-state-linked-group-accused-hacking-singtel-bloomberg-news-reports-2024-11-05 (accessed March 3, 2026).

**16** Bendiek, A., Bund, J. and Kerttunen, M. "The attribution dividend: Protecting critical infrastructure from cyber attacks," (no. 46/2024). SWP Comment, 2024.

**17** Vaidya, T. "2001–2013: Survey and analysis of major cyberattacks," arXiv preprint, arXiv:1507.06673, 2015.

**18** Costa, D.G., Peixoto, J.P.J., Jesus, T.C., Portugal, P., Vasques, F., Rangel, E. and Peixoto, M. "A survey of emergencies management systems in smart cities," *IEEE Access*, 10, 61843–61872, 2022.

**19** Powell, E. "Why are the UK and Europe falling behind on 5G?" *The Times*, 7 March 2025, www.thetimes.com/business-money/companies/article/uk-europe-5g-vodafone-three-telecoms-mobile-mn0z9j9ds (accessed March 3, 2026).

**20** BAI Communications. "Connectivity in regional Australia: challenges and solutions," 22 April 2021, www.baicommunications.com/blog/connectivity-in-regional-australia-challenges-and-solutions (accessed March 3, 2026).

**21** Modarressi, A.R. and Mohan, S. "Control and management in next-generation networks: challenges and opportunities," *IEEE Communications Magazine*, 38(10), 94–102, 2000.

**22** Ouyang, Y., Wang, L., Yang, A., Shah, M., Belanger, D., Gao, T., Wei, L. and Zhang, Y. "The next decade of telecommunications artificial intelligence," arXiv preprint, arXiv:2101.09163, 2021.

**23** Tao, L. "Huawei's successful transformations and revolutions," in *Huawei Wisdom: Develop Strategy to Execution at Critical Moments*, 1–32. Springer Nature, 2022.

**24** Samek, W., Stanczak, S. and Wiegand, T. "The convergence of machine learning and communications," arXiv preprint arXiv:1708.08299, 2017.

**25** Bhatti, U.A., Tang, H., Wu, G., Marjan, S. and Hussain, A. "Deep learning with graph convolutional networks: an overview and latest applications in computational intelligence," *International Journal of Intelligent Systems*, 2023(1), 8342104, 2023.

**26** Wu, J., "The rise of China: a key player in the supply chain," in *Global Trends in Manufacturing Supply Chains*, 29–61. Springer Nature, 2025.

**27** Lin, X., Liu, B., Han, J. and Chen, X. "Industrial upgrading based on global innovation chains: a case study of Huawei Technologies Co., Ltd. Shenzhen," *International Journal of Innovation Studies*, 2(3), 81–90, 2018.

**28** Li, L., Guo, B., Murmann, J.P. and Wu, D. "Huawei's R&D management transformation," in *The Management Transformation of Huawei: From Humble Beginnings to Global Leadership*, 292–346. Cambridge University Press, 2020.

**29** CSET Policy Brief. "The Huawei Moment," 2021, https://cset.georgetown.edu/publication/the-huawei-moment (accessed March 3, 2026).

**30** Dang, W., Huang, R., Yu, Y. and Zhang, Y. *Autonomous Driving Network: Network Architecture in the Era of Autonomy*, CRC Press, 2024.

**31** Huawei. "Autonomous driving network (ADN)," White Paper, 2024, www.file.huawei.com/-/media/corp2020/pdf/giv/striding-towards-the-intelligent-world/2024/intelligent_world_adn_2024_en.pdf?la=en (accessed March 3, 2026).

**32** "Huawei and BT Ireland jointly complete industry's first 1.2Tb/s transmission real-time trial in a live network," *TelecomTV*, 5 December 2019, www.telecomtv.com/content/optical/huawei-and-bt-ireland-jointly-complete-industrys-first-1-2tb-s-transmission-real-time-trial-in-a-live-network-37161 (accessed March 3, 2026).

**33** Huawei Tech. "Pushing bandwidth limits and moving towards optical networking 2.0," 2024, www.huawei.com/en/huaweitech/publication/87/pushing-bandwidth-limits-and-moving-towards-optical-networking (accessed March 3, 2026).

**34** Huawei Community. "News alert: Huawei completed industry's first 1.2Tb/s real-time data transmission," *Huawei Consumer Community*, 2024, https://consumer.huawei.com/en/community/details/News-Alert-Huawei-completed-Industry-s-First-1-2Tb-s-Real-Time-Data-Transmission/topicId_44277 (accessed March 3, 2026).

**35** Guo, Y., Wang, H., Zhang, H., Liu, T., Li, L., Liu, L., Chen, M., Chen, Y. and Lip, G.Y. "Photoplethysmography-based machine learning approaches for atrial fibrillation prediction: a report from the huawei heart study," *JACC: Asia*, 1(3), 399–408, 2021.

**36** Wang, C., Chen, M., Wang, Q., Fang, Y. and Qiu, L. "New product development paradigm from the perspective of consumer innovation: A case study of Huawei's integrated product development," *Journal of Innovation & Knowledge*, 9(2), 100482, 2024.

**37** Fang, H., Yu, P., Tan, C., Zhang, J., Lin, D., Zhang, L., Zhang, Y., Li, W. and Meng, L. "Self-healing in knowledge-driven autonomous networks: context, challenges, and future directions," *IEEE Network*, 38(6), 425–432, 2024.

**38** Allen T, "Enabling the Telco to Techco Transformation for Success in the Intelligent Era", *Huawei Technologies Co. Ltd*, https://www.huawei.com/en/huaweitech/cases/ai-for-techco (accessed March 20, 2026).

**39** MTN Group. "Annual Report: AI-Driven Network Expansion," 2023, https://www.mtn.com/wp-content/uploads/2024/04/MTN-Group-FY-23-Integrated-Annual-Report-1.pdf (accessed March 23, 2026).

**40** Alabi, M. "The impact of artificial intelligence on network optimization in telecommunications," *ResearchGate*, 1, 384664972, 2023.

**41** Huawei. "Striding Towards the Intelligent World 2024: Autonomous Driving Network (ADN)," 2024, https://www-file.huawei.com/-/media/corp2020/pdf/giv/striding-towards-the-intelligent-world/2024/intelligent_world_adn_2024_en.pdf?la=en (accessed March 3, 2026).

**42** Zhang, C., Liu, X., Zheng, X., Li, R. and Liu, H. "Fenghuolun: a federated learning based edge computing platform for cyber-physical systems," in *IEEE International Conference on Pervasive Computing and Communications Workshops*, 1–4. IEEE, 2020.

**43** Rahman, A., Hasan, K., Kundu, D., Islam, M.J., Debnath, T., Band, S.S. and Kumar, N. "On the ICN-IoT with federated learning integration of communication: concepts, security-privacy issues, applications, and future perspectives," *Future Generation Computer Systems*, 138, 61–88, 2023.

**44** Albshaier, L., Almarri, S. and Albuali, A. "Federated learning for cloud and edge security: a systematic review of challenges and AI opportunities," *Electronics*, 14(5), 1019, 2025.

**45** Khan, A., Sohail, A., Zahoora, U. and Qureshi, A.S. "A survey of the recent architectures of deep convolutional neural networks," *Artificial Intelligence Review*, 53, 5455–5516, 2020.

**46** Li, Z., Liu, F., Yang, W., Peng, S. and Zhou, J. "A survey of convolutional neural networks: analysis, applications, and prospects," *IEEE Transactions on Neural Networks and Learning Systems*, 33(12), 6999–7019, 2021.

**47** Gültekin, Y.B. "Traffic volume prediction using LSTM improved with rules obtained from anomalies," Master's Thesis, Middle East Technical University, Turkey, 2021.

**48** Mehri, H. "Traffic prediction in 5G networks using machine learning," Doctoral Dissertation, Boise State University, 2024.

# 15

# SpaceX Starlink: Revolutionizing Satellite Internet Through LEO Innovation

From 600 ms to 20 ms Latency – Bridging the Digital Divide While Crowding the Sky

## 15.1 Synopsis

What if the internet, a tool many of us take for granted, could reach every corner of the Earth, from bustling cities to the most remote villages? Starlink is SpaceX's attempt to solve this global connectivity challenge through the most ambitious satellite internet network ever imagined. Starlink is not just the epic story of technological progress, but a revolutionary attempt aimed at eliminating the digital boundaries. Satellite internet has, historically speaking, faced great obstacles to getting off the ground. Predictable geostationary satellites placed 36,000 kilometers in the sky act as echo chambers for communication, much like shouting across a valley and having to wait for the echo to return. This limitation has, for years, made satellite internet unviable to on-demand applications like video calls, online gaming, or autonomous systems.

SpaceX is the company that is sending thousands of Low Earth Orbit (LEO) satellites into space to give the whole world high-speed and low-latency internet. Starlink is at the heart of SpaceX's new conception of what satellites should be, and how low it can fly to provide much faster signal transfer time as well as giving superior end-user experience. This bold undertaking has tested more than just what engineering capability can achieve. It has sparked worldwide debate over the future of connectivity, space sustainability, and the ethics of mega-constellations.

The analysis of Starlink highlights how to reimagine old problems with new solutions, backed by innovation and perseverance to achieve success. This is a revolution of perspective that the use of satellite communications can chang humanity, economies, and entire countries. This case study discusses Starlink's design challenges and solutions as well as valuable takeaways for all readers.

## 15.2 From Outer Space to Every Place: The Development of Satellite Internet

The vision to bring the internet to the most distant corners of our planet, whether deep in valleys or even out into the middle of the ocean, that has been at the core of the development of satellite internet. Governments and corporations began using satellites for communication during the space race era with Sputnik 1 in 1957 and Telstar 1 in 1962 [1]. Early systems relied on geostationary satellites (GEO) which provided some kind of coverage but with high latency. Commercial satellite internet entered service in the 1990s, but didn't keep up with demand. In the late 2010s, things began to change with LEO satellites such as Starlink. It offers much faster, and low-latency internet to the world [2]. Satellite internet, which used to serve as a backup, is now the only lifeline for connectivity to the global web. It promises to bridge the digital divide and make the internet accessible to a broader audience.

### 15.2.1 Challenges of Traditional Geostationary Satellites

Geostationary satellites are the most common form of satellite communication utilized for decades, usually positioned 35,786 km above the Earth's equator [3]. This provides substantial geographic coverage and reliable broadcasting, but the high altitude comes with its own technical problems. The biggest of them is high latency [4]. Signals must travel a total round-trip distance exceeding 72,000 kilometers (or 44,738 miles). This results in a minimum one-way delay of about 250 milliseconds and a round-trip latency of 500 to 600 milliseconds. It is caused by the finite speed of electromagnetic waves which is approximately 299,792 kilometers per second in a vacuum [5]. This kind of latency interrupts real-time applications like online gaming, video conferencing, and financial trading [6].

GEO satellites face significant bottlenecks, particularly in bandwidth capacity. These latency and bandwidth constraints become increasingly important as demand for the high-speed internet continues to surge. Especially, it is challenging in urban areas, where the same bandwidth must be shared among numerous users [7]. The challenges are compounded by several factors such as limited adaptability to atmospheric, space condition effects, inefficient spectrum utilization due to their fixed orbital positions, substantial deployment, and maintenance costs, encompassing satellite construction, launch operations, and ground infrastructure. All these acts to limit the scalability and efficiency of GEO-based networks. These problems created a demand for novel approaches and methods to solve latency and bandwidth optimization to maximize the performance of satellite communication.

For example, the latency of GEO satellites (500 to 600 ms) is much higher than that of terrestrial fiber-optic networks which is around 50 milliseconds [8].

Fast response time is particularly important in real-time applications. If any delay experienced by the user for more than 100 milliseconds, it may cause perceivable lag, and a weaker user experience related to reaction times. Similarly, the high latencies of video conferencing do affect the synchronization of conversations as they introduce noticeable audio and video delays [9]. These limitations demonstrate the pressing necessity of alternatives to fulfill modern demands within communication systems.

Low latency connections are a must for financial transactions, especially for stock trading and high-frequency trading (HFT), as even milliseconds of difference can lead to substantial financial loss [10]. Furthermore, telemedicine and remote drone control are cited as other applications that GEO satellites cannot serve in an efficient way [11].

Another challenge of GEO satellites is their lack of coverage on high-latitude areas. Because these satellites are in equatorial orbit, their signal is geared toward mid-latitude areas [12]. This means their coverage at high latitudes, especially the polar areas, is quite poor. Above 70 degrees of latitude, signal strength decreases since the satellite is not raised sufficiently. It results in less elevation angle for the satellite and increasing atmospheric attenuation while reducing link reliability [13].

This restriction specifically hits hard at research stations in the extreme Arctic and Antarctic who need to have a dependable connection for scientific data transfer. This issue also affects communities in north Alaska, northern Canada, Greenland, and some parts of Scandinavia and Russian. In these geographical regions, the terrestrial broadband coverage is sparse or non-existent. The maritime and aviation sectors operating in polar routes also face challenges in maintaining seamless global connectivity for navigation, safety, and operational efficiency [14].

Moreover, the other constraint on GEO satellites arises from bandwidth restrictions. As shown in Table 15.1, the C-band (3.7–3.98 GHz), Ku-band (12–18 GHz) and Ka-band (26.5–40 GHz) are the operational bands for most GEO satellites. Other bandwidths may have greater susceptibility to atmospheric interference. This can result in temporary service degradation or complete signal loss [15]. In addition, space weather events like solar flares and geomagnetic storms can also potentially disrupt satellite signals, degrading performance and reliability [16]. There is increased complexity in establishing stable satellite communications due to these environmental variations [17]. Nevertheless, even with the advent of high-throughput satellites (HTS), the total available bandwidth is still restricted by the fixed spectrum allocation and frequency reuse strategies [18]. This effectively results in less data throughput and slower internet speeds, which further degrades the quality of service (QoS) during peak times. More users translates to a higher contention ratio, which directly correlates with poor performance. Moreover, the cost per megabit is still significantly high on satellite internet as spectrum efficiency needs to be balanced with affordable services.

**Table 15.1** Operational bands and atmospheric interference for GEO satellites.

| Frequency Band | Range (GHz) | Bandwidth | Susceptibility to Interference | Use Case |
|---|---|---|---|---|
| **C-Band** | 3.7–3.98 GHz | Narrow | Low (less affected by rain fade) | Broadcasting, basic internet |
| **Ku-Band** | 12–18 GHz | Moderate | Moderate (affected by rain fade) | Internet, broadcasting, VSAT systems |
| **Ka-Band** | 26.5–40 GHz | Wide | High (severe atmospheric attenuation) | High-speed internet, HTS systems |

*Source:* [19–21].

Deployment and maintenance also pose significant economic challenges. The launch of a single GEO satellite can cost up to USD$ 30,000 per kilogram or roughly around US$100–300 million, including satellite construction, launch vehicle expenses, insurance and an estimated operating cost for a 15 year lifespan [22]. GEO satellites are semi-permanent and limited to around a 15–20 years of operational life unlike low Earth orbit (LEO) constellations that can be quickly replaced and upgraded. When they reach the end of their lifespan, replacement means launching new satellites, which has a lengthy lead time and greater costs. This is why it is expensive to sustain GEO satellite networks, constraining their ability to scale rapidly in the face of growing global demand for broadband connectivity.

As low-latency, high-speed global connectivity becomes more vital, more alternative satellite solutions are being developed to overcome the very intrinsic weaknesses of conventional geostationary satellites. The most exciting innovation within satellite communications today are new LEO systems, such as SpaceX's Starlink and OneWeb. Such LEO systems could disrupt how we consume and interact with the use cases around GEO satellite-based services. LEO is boasting significant improvements in latency and overall performance compared to the traditional GEOs.

### 15.2.2 Development of MEO and LEO Satellites

Unlike GEO satellites that sit above a location in the sky at roughly 36,000 kilometers, LEO satellites are perched in a much lower orbit 500–1200 km [23]. Medium Earth Orbit (MEO) satellites, on the other hand, operate at altitudes between 2,000 km and 35,786 km, striking a balance between the advantages of both GEO and LEO. The satellites are much closer to Earth so latency is cut to around 20–40 milliseconds [24]. This ultra-low-latency of LEO satellites can be used only if and where applications need low-latency responsiveness like online gaming, video conferencing, communications to the autonomous spacecraft, and remote medical procedure [25].

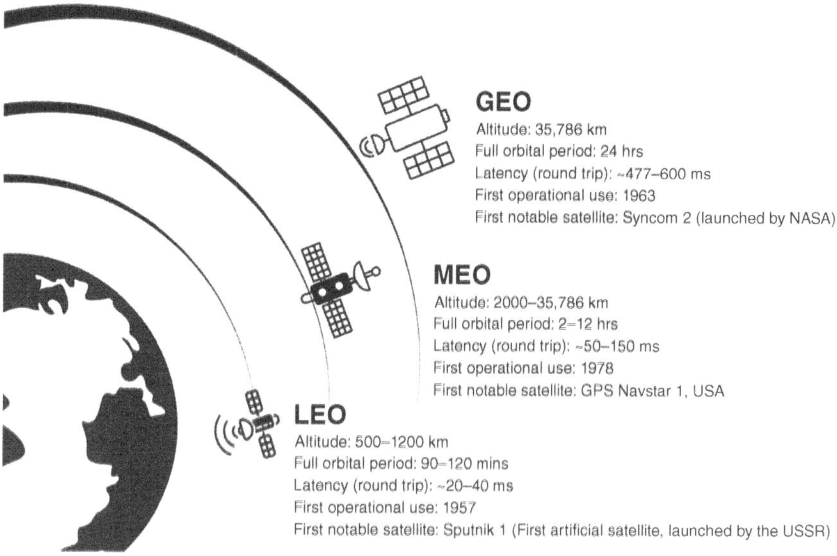


**Figure 15.1** Comparison of GEO, MEO, and LEO satellite characteristics, including altitude, orbital period, latency, and historical milestones. *Source:* Author's own work.

In contrast, GEO satellites, with their 500–600 milliseconds latency, cannot adequately support these time-sensitive applications [26].

When considering global coverage, LEO satellites provide better service in high-latitude and polar areas while GEO satellites face significant geographical limitations [27]. The three orbital classes have evolved since the 1950s, with each serving distinct purposes (Figure 15.1). MEO satellites, generally used in navigation systems (GPS, Galileo and GLONASS) and have moderate latency of 50 to 150 milliseconds. They are used for applications that need good coverage but at much lower latency than GEO, without the huge infrastructure costs of LEO constellations.

SpaceX Star Link is a game changer that provides the internet to research stations, like those in Antarctica, that are in hard to reach places [28]. For instance, Starlink started delivering high-bandwidth internet to McMurdo Station, a US facility in Antarctica in early 2023, so that scientists and staff can work with a stable connection for research and communication [29]. Previous connectivity in the region was slow and spotty, preventing scientific collaboration as well as operational efficiency. Starlink has radically increased data transfer speeds and is now allowing for real-time communication with global research teams. It greatly improved productivity in one of the most isolated regions in the world [30].

Table 15.2 gives a technical and economic comparison of LEO vs. MEO vs. GEO satellites. It emphasizes the factors like orbit height, latency, coverage area, costs, and use cases. While GEO is for wide coverage with a lifespan in decades, it has high latency and high deployment costs. On the other hand, while MEO satellites

**Table 15.2** Comparison of GEO, MEO, and LEO satellites.

| | LEO (Low Earth Orbit) | MEO (Medium Earth Orbit) | GEO (Geostationary Earth Orbit) |
|---|---|---|---|
| **Orbit Altitude** | 500–1,200 km | 2,000–35,786 km | ~35,786 km |
| **Orbital Period** | ~90–120 minutes | 2–12 hours | 24 hours |
| **Coverage Area** | Small; requires many satellites for global coverage | Larger than LEO; fewer satellites needed | Very large; one satellite covers ~1/3 of Earth |
| **Latency** | ~20–40 ms | ~50–150 ms | ~600 ms |
| **Data Throughput** | High due to proximity to Earth, but requires many satellites for consistent coverage | Moderate, used in navigation and some broadband applications | Lower compared to LEO/MEO, but stable for broadcasting |
| **Launch and Deployment Costs** | Lower cost per satellite, but requires more launches | Moderate cost, fewer satellites than LEO but more than GEO | High cost per satellite and launch |
| **Maintenance and Replacement** | Frequent replacements (~5–7 years lifespan) | Longer lifespan than LEO (~10–15 years) | Long lifespan (~15+ years), but expensive to replace |
| **Signal Strength and Power** | Strong, but requires frequent handoffs between satellites | Moderate, more stable than LEO | Weak, requires high-powered ground antennas |
| **Applications** | Earth observation, communications, internet (e.g., Starlink) | Navigation (GPS, Galileo), communications (O3b) | Broadcasting, weather monitoring, fixed satellite services |
| **Deployment Timeframe** | Rapid deployment: constellations can be expanded quickly | Moderate: typically used for navigation and broadband system | Slow: requires precise positioning and high investment |
| **Economic Considerations** | Lower initial investment but higher operational costs due to frequent replacements | Balanced cost vs. performance; suitable for navigation and specialized communication | High upfront investment, but long-term cost efficiency for broadcasting and fixed services |

*Source:* Author's own work based on data from [2, 31].

used in global positioning systems (GPS) and broadband applications, LEO ensures low-latency connectivity and faster turnaround but requires large constellations and frequent replenishment. This comparison intends to give an overview of their trade-offs for various applications.

LEO satellite constellations are on the rise, with private companies such as SpaceX, OneWeb, and Amazon's Project Kuiper are driving most of the movement. As of 2023, SpaceX is one of the main developers in this field with more than 4,000 Starlink satellites placed in orbit. The intended number may soon increase to 12,000 [32]. Starlink is keen to supply high-speed internet in every corner of the world, especially where there is no or very limited terrestrial infrastructure so tackling the digital divide which remains in some parts of world [33]. With download speeds like 50–200 Mbps, Starlink has positioned itself as a real broadband substitute for locations that lack or have spotty terrestrial infrastructure [34].

Meanwhile, OneWeb has focused on enterprise and government use cases with endeavors to close connectivity voids across aviation, maritime, and defense among others [35]. The company is partnering with different countries and entities to meet these targets. OneWeb, for example, has participated with the UK in the drive for enhanced rural broadband, especially in areas of Scotland, where it was not possible via traditional means [36]. This partnership has brought fast internet to the people and the commercial sector in these areas. It contributes to improving better economic status and quality of living in those regions.

Along with this, OneWeb planned to launch its satellites in alliance with Indian Space Research Organization (ISRO). And as of early 2023, the company had launched 618 satellites which are expected to complete their first-generation constellation [37]. This partnership has fueled OneWeb's mission and underscored an increasing international collaboration needed in satellite business. Table 15.3 highlights the key players and nations investing in different satellite orbits.

Leo constellations are a revolutionary change in satellite communication tech. Their capability of delivering low-latency, high-speed internet from a global footprint has paved new ways for applications to be feasibly used. LEO networks have enabled additional connectivity in developing countries to meet their remotest education and

**Table 15.3** Major companies and countries' usage of GEO, MEO, and LEO satellites.

| Orbit Type | Companies | Countries |
|---|---|---|
| **LEO** | SpaceX (Starlink), OneWeb, Amazon (Project Kuiper), Telesat, China SatNet | USA, UK, Canada, China, Russia |
| **MEO** | SES (O3b), Boeing, Lockheed Martin, China Satcom | USA, Luxembourg, China, Russia |
| **GEO** | Intelsat, SES, Eutelsat, Hughes, Viasat, China Satcom, Inmarsat | USA, France, UK, China, India, Japan |

*Source:* Author's own work.

telemedicine initiatives. This has allowed some regions that were never served before to have access to key services. Besides, LEO satellites are also being used for live data capture and analytics with some industries like agriculture, and environmental monitoring for faster decision-making with operational efficiency [38].

Although LEO satellite constellations like Starlink represent a major technological breakthrough, they bring significant security and reliability challenges [39]. While having their advantages, LEO systems have their fair share of downsides as well. Thousands of satellites being deployed could introduce concern on orbital congestion and collisions. The proliferation of objects in low Earth orbit has raised concerns about space debris, and the potential for collisions, which could lead to Kessler Syndrome, rendering certain orbits unusable [40]. For the successful and sustainable future of space, regulations and international cooperation are important. Furthermore, the high costs associated with launching and maintaining vast networks and advanced ground infrastructure create significant barriers to smaller players in the market.

To sum up, LEO satellites quite literally transformed the landscape in satellite technology. They opened new global connectivity and real-time doors for GEO satellite constraints in latency and coverage. SpaceX and OneWeb are proving the idea is sound with LEO constellations, while changing internet access and eliminating the digital divide. However, the long-term viability relies on further addressing technological, regulatory, and financial hurdles to make this technology globally accessible.

### 15.2.3 SpaceX and Reusable Rocket Technology

SpaceX, established by Elon Musk in 2002, started to dramatically lower the costs of space access via reuse of rocket technology. In 2008, the company hit a big milestone with Falcon 1 as the first privately developed liquid-fueled rocket to ever make it into orbit. That success opened the backbone behind all of SpaceX's satellite deployment missions [41].

Prior to the founding of SpaceX, many of the aerospace companies and government entities were undertaking satellite launches, space exploration and even using reusable rocket concepts. But none of them have set goals close to being a commercial entity focused on cost savings. Since the 1960s, NASA had been the primary player in space flight, managing the Space Shuttle from 1981–2011 [42]. While the Space Shuttle was partially reusable, refurbishment costs kept it from being economical [43]. Boeing and Lockheed Martin were major NASA contractors involved with the Saturn V rocket, the Space Shuttle, and the International Space Station (ISS) [44]. The Atlas and Titan rockets used for commercial satellites were developed by Lockheed Martin [45].

Arianespace, founded in 1980, led the market for commercial satellite launch services until SpaceX came. Similarly, Orbital Sciences Corporation, later acquired by Northrop Grumman, built small to larger launchers, including the Pegasus and Taurus rockets [46]. While the Russians used their well-established Soyuz rockets to launch satellites as well as astronauts into space, China CNSA launched its very first astronaut in 2003, marking the beginning of an expanding space program [47].

All of these were done with non-fully reusable rockets; however, no entity had perfected a design that would allow for effective reusability and with an aggressive cost reduction on launches. Through reusable rocket technology, SpaceX created a paradigm shift in how to get into space at much lower cost for orbital launches [48]. SpaceX wanted to fill the gap by building Falcon 9, the first true orbital class rocket to achieve vertical landing and reusability in 2015 [49]. This advancement drastically lowered the cost of launching satellites and opened up commercial spaceflight.

Figure 15.2 shows the complete journey of a Falcon 9 rocket by showcasing its unique reusable launch profile that sets it apart from the traditional expendable rockets. At the beginning of the liftoff, the Falcon 9 takes off from the launch site using its mighty Merlin engines. This is immediately followed by its main engine cutoff (MECO) and stage separation where the first stage (booster) detaches and drops off from the second stage. The second stage then ignites to insert the

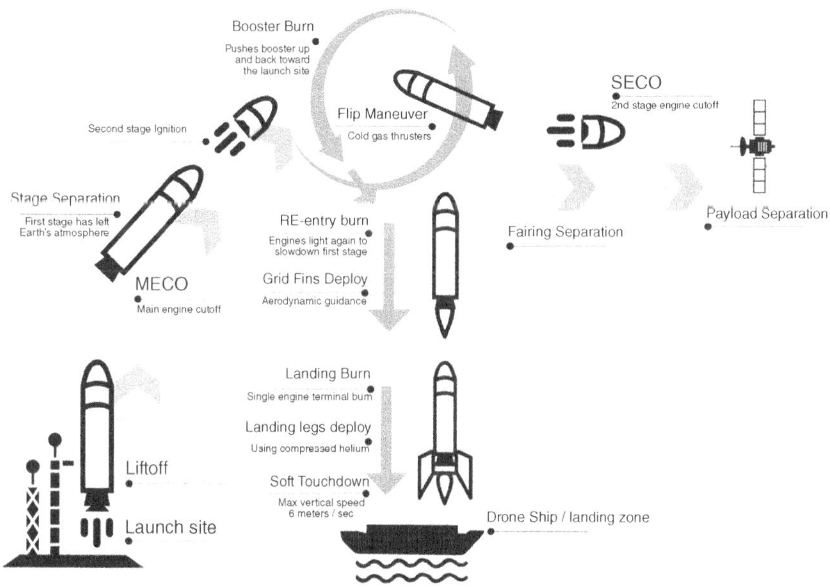


**Figure 15.2** Key phases in the launch and landing of SpaceX's Falcon 9 rocket. *Source:* Adapted from [49].

payload into orbit, while the payload fairing separates, exposing the satellite for orbital release. Second engine cutoff (SECO) signifies the end of the delivery of the payload into its assigned orbit, where it is then successfully deployed.

Simultaneously, the first-stage booster executes a groundbreaking sequence of maneuvers, so that its recovery and reusability become reality. Once separated, the booster performs a flip maneuver to align for Earth return and then executes a boost-back burn to guide itself to the landing pad or drone ship. During re-entry, the booster ignites its engines for an entry burn to slow descent and reduce aerodynamic heating and thermal stress, while grid fins deploy for precise flight control. Finally, the booster uses its engines to brake while its legs unfold like in airplanes before landing. It enables a soft touch down on land, standing up on the landing ship or landing platform as shown in Figure 15.2.

The difference between SpaceX's reuse method and traditional ways is that in the first stage on normal launches, everything else is either thrown in the ocean, or burned up on its way out, lasting only one mission. This is a wasteful and expensive route, having to build all new rockets for every mission.

The reusability of Falcon 9 means this changes the typical paradigm, to catch and refurbish the first stage booster. Using this innovation, SpaceX was able to reduce the costs to millions of dollars per launch. For instance, the price to manufacture a Falcon 9 booster is somewhere between USD$ 30–40 million but refurbishing and reusing it costs only a fraction of that price. They have also brought down the cost for encapsulating payload fairings, to about 6 million per pair. Not only does this reduce the costs but also speeds up the launch schedule for SpaceX, allowing it to become the leading commercial launch provider. This makes space more accessible for various applications, including satellite deployment, scientific research, and future crewed missions.

In 2015, SpaceX revealed its plan for Starlink, to deliver global satellite internet via a network of orbital constellations. The project started officially with the launch of two prototype satellites named Tintin A and Tintin B in 2018 [50]. The first operational batch of Starlink satellites were launched by SpaceX in May 2019, after testing was successful. Through 2025, early in the first half of the year, SpaceX had deployed more than 4,000 LEO satellites as the largest constellation in history [51].

With the scope of SpaceX operations continuing to grow, the company focuses on refining its satellite architecture and launch techniques. The next evolution of Starlink development is about pin-sharp configuration. The design and deployment of such innovative solutions can provide worldwide coverage in terms of good latency of broadband speeds and guarantee network competitiveness against legacy communication infrastructures [52].

### 15.2.4 Engineering Principles Applied

The satellite constellation design of Starlink is an engineering marvel, attempting to defeat the challenges of global internet accessibility. Starlink provides high-speed LEO satellite-based internet for every corner of the world, especially for rural and hard to reach places. The system is built on several key innovations, including dynamic routing through inter-satellite laser links, phased-array antennas for rapid signal acquisition and beam steering, and the integration of ground infrastructure to ensure seamless global connectivity [53].

Starlink satellites are in a different orbit (540–570km) as opposed to traditional geostationary satellites that circulate above Earth at altitude about 36,000km [54]. The nearness of the Earth significantly minimizes the time for signals to traverse from satellite to the ground. This makes delivery of latencies about 20 milliseconds, comparable to a fiber-optic network. Latency is very important for use cases like online gaming, video conferencing, and telemedicine, where any small delay can have severe effect on performance. SpaceX has now launched more than 4,000 operational satellites to date but intends to add a constellation of up to 42,000 more. AS of 2025, they plan to improve their coverage and further increase capacity [51].

Probably one of the most groundbreaking ideas about the design of Starlink is its inter-satellite links (ISL) that allow satellites to communicate with one another. These links provide data to go along with the fewest number of hops end-to-end without ever actually using a ground station. For instance: a signal across the ocean can use ISL to be transmitted from one continent to another and not use terrestrial infrastructure. Further, this capability minimizes latency and maximizes network effectiveness and resiliency. ISLs make sure there is continuous data flow by creating a mesh network and handling signals effectively even if there is congestion or satellite failure. This mesh topology helps to re-route signals without dynamic configuration [55]. Figure 15.3 illustrates this network architecture of Starlink.

Phase array antenna integration is involved in both the satellites and User Terminals (UTs). It is another important technological advancement of Starlink. It ensures reliable, and high-bandwidth communication. The antennas use several small elements to electronically steer the beam of signal rather than mechanically moving. Technology enables satellites to dynamically track and stay in contact with multiple ground stations and user devices simultaneously. On the user side, phased-array antennas enable seamless handoffs between satellites as they move across the sky, ensuring uninterrupted connectivity. This rapid signal acquisition and beam steering capability are key for maintaining consistent internet speeds, low latency, even in challenging environments [58].

Starlink uses a strategically placed ground infrastructure to supplement the satellite network, with the high-performance gateway stations on fiber as a

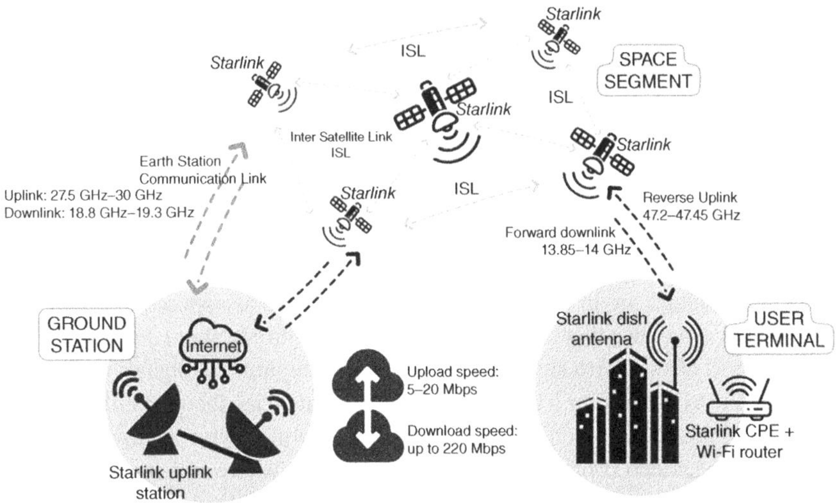


**Figure 15.3** Starlink system network architecture (internet). *Source:* Author's own work with data from [56, 57].

ground network [59]. The ground stations are the lifeline relay points for the terrestrial internet backbone and the space constellation. They provide optimal data transfer speeds and redundancy. Building and incorporating advanced ground-based infrastructure helps the network growth to be reliable. It also enables the ability to support a larger number of users at high speed.

By using these new technologies together, Starlink is redefining satellite internet and providing an alternative to mainstream broadband in areas where infrastructures are lacking. But, as remarkable as those innovations are at improving connectivity, they also unleash a cloud of new engineering problems associated with keeping latency low for global distributed network. The core issue lies in the constraints of ground-to-space communication, as traditional satellite networks have long struggled with high latency due to the vast distance signals must traverse. Solving this problem necessitates a satellite network that is founded on a radically different design approach.

While Starlink's method represents a breakthrough, alternative solutions have also been considered to combat latency and improve global internet access. Hybrid systems combining geostationary satellites with LEO constellations are one way to go at solving latency, and another involves operating a single type of orbit [60]. This hybrid model leverages orbits effectively, in terms of the trade-off between cost and performance. Geostationary satellites provide long-term, stable service over broad territory while LEO satellites enable the low latency with fast response times, suitable for applications requiring almost real-time response [61].

Another method to improve efficiency focuses on optimizing data transmission through advanced compression techniques. Compression algorithms can boost transmission speed by reducing the amount of data sent over network [62]. This has already become the norm for video streaming services, as they deliver high-definition content over limited bandwidths. If satellite communications followed the same techniques, bandwidth could be less of an issue and can help to improve network performance, especially where spectrum is scarce.

Figure 15.4 shows the flow of data compression in satellite communications which benefits improving network efficiency by reducing the data volume being transmitted. Therefore, this increases bandwidth utilization while also decreasing latency. Lossless compression algorithms (e.g., Huffman coding and Lempel–Ziv–Welch (LZW)) offer data integrity with compression efficiencies of around 50%. Loss compression approaches such as discrete cosine transform (DCT) or wavelet-based approaches offer much higher rates of up to ~90% for multimedia applications, particularly using lossy compression techniques [63]. By lowering bandwidth demand, compression reduces congestion in LEO satellite constellations. Starlink uses advanced compression algorithms, so that the data can be transmitted more efficiently on the available spectrum and resulting in throughput improvement, and service quality. This method is remarkably useful at combating the sort of challenges posed by bandwidth limits and high latency problems that are inherent in satellite-based internet networks.

Orbit design and arrangement of the satellites are also playing a vital role in latency reduction and coverage optimization [67]. The constellation is organized in several orbital planes into which Starlink disperses its satellites and makes sure that they are not densely congested on each orbit. Polar orbits, likewise, allow satellites to provide global coverage and this is extremely beneficial for remote areas. Careful orbital design improves the network and protects satellites from collision risks due to an increasing number of satellites being launched into orbit [68].

Despite its technological advancements, Starlink faces several operational challenges that prevent scalability and sustainability. Satellites are expensive to manufacture at scale and require a large initial investment in automation and supply

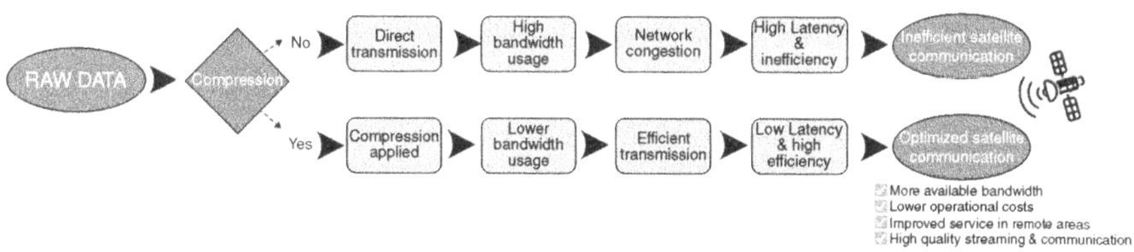


**Figure 15.4** Data compression impact on network efficiency in satellite communication. *Source:* Author's own work with data from [64–66].

chains to support mass production [69]. This was resolved by building satellites in-house at SpaceX to eventually scale down production costs [70]. The company has further improved the cost reduction through its reusable concept, for example, the Falcon 9 rocket, specifically the reusable booster, which makes launches more cost effective [71].

Also equally urgent is the problem of space debris. The growing crowd of satellites in space is a concern over likely congestion and collision, as SpaceX plan to deploy up to 42,000 satellites [56, 72]. Measures taken by SpaceX include automated collision avoidance systems and a managed deorbit process for when a satellite is no longer operational. However, ensuring sustainability and resilience of space operations remains a growing challenge. It requires industry-wide collaboration and regulatory oversight.

Overcoming these obstacles and building upon their efforts, Starlink aims to finally bridge the worldwide digital divide and connect everyone on Earth with the internet. Beyond improving global connection, the advent of low-latency satellite networks is planned to be the building block for future improvements on space-based connectivity. Nevertheless, while there have been some advancements in this space, high latency is still one of the biggest challenges to traditional satellite communications. Understanding the core issue of latency and the complexities of achieving global coverage without sacrificing performance is essential to developing an effective solution.

## 15.3 Think and Solve

**Q 15.1** What is the primary benefit of Low Earth Orbit (LEO) satellites over Geostationary Earth Orbit (GEO) satellites?

**A** Higher data throughput.
**B** Better signal strength.
**C** Lower latency.
**D** Longer operational life.

**Q 15.2** How does SpaceX's Starlink constellation aim to address the digital divide?

**A** By providing high-speed internet in remote areas.
**B** By reducing the cost of satellite launches.
**C** By enhancing the resolution of satellite imaging.
**D** By improving GPS accuracy.

**Q 15.3** What distinguishes Medium Earth Orbit (MEO) satellites from LEO and GEO satellites?

**A** They offer higher signal strength.
**B** They have a longer orbital period.
**C** They provide moderate coverage.
**D** They are cheaper to maintain.

**Q 15.4** Which application benefits most directly from the low-latency connections provided by LEO satellites?

**A** Online gaming.
**B** Weather monitoring.
**C** Fixed satellite services.
**D** TV broadcasting.

**Q 15.5** Calculate the round-trip latency for a connection using Starlink's LEO satellites.

**Given:**
Satellite Altitude: 550 km
Speed of Light: 299,792 km/s

**Use the formula:**
Latency = Round trip distance/ Speed of Light
Round trip distance = 2 × Satellite Altitude
Also, think about solving the same question, if you consider Earth's radius.

## 15.4 Conclusion and Key Takeaways

Starlink is the most transformative initiative in modern satellite communications technology today. By deploying a large constellation of LEO satellites, SpaceX intends to deliver high-speed, low-latency global internet that reaches the underserved regions. Think about these key takeaways:

1. **Technological Innovation:** For applications requiring minimum latency in communication like online gaming, remote controlled systems, and video conferencing, Starlink uses LEO constellation networks.
2. **Global Connectivity:** The project tackles the problem of providing ubiquitous and economically sustainable internet globally to increase education, healthcare possibilities and economic prospects worldwide.

3. **Overcoming Traditional Barriers:** LEO takes far less time to transmit data and has more frequency agility suited to different environmental conditions than geostationary satellites which have high latency and limited bandwidth.
4. **Economic and Environmental Considerations:** Although LEO constellations offer a cost and scalable solution, they also raise concerns about space junk, orbital waste, and the need for sustainable operations to prove the long-term viability of LEO.
5. **Future Implications:** The Starlink success may well open the floodgates to more commercial space activities, potentially leading to international partnerships and paving the way for new space sustainability regulations.

Starlink is a crucial chapter in satellite communications evolution, showing new and better possibilities to solve global connectivity problems. As this initiative continues to evolve, it will further reshape internet accessibility worldwide and transform how people connect with technology and information worldwide today.

## 15.5 Discussion Points

Although SpaceX and its Starlink constellations are major milestones in space and communication technology, there are still challenges that need to be further discussed. Technical, economic, environmental and regulatory factors have to be unpacked to realize the full implications for the future of Starlink. These discussion points can be used for future research and group discussions.

1. **Technological Advancements:** In what ways do LEO constellations offer advantages over traditional geostationary satellites? How does Starlink's uses advanced propulsion systems to reduce operational costs?
2. **Challenges:** What are the technical hurdles in deploying a massive LEO satellite constellation? How does Starlink plan to manage the costs associated with launching and maintaining a large number of satellites?
3. **Economic Implications:** How might Starlink's market entry affect traditional telecommunications tech providers? What potential revenue streams could arise from Starlink's services?
4. **Environmental Considerations:** What is the potential impact of increased space debris from Starlink's satellites? How effective are the measures proposed to mitigate space pollution?
5. **Regulatory and Policy Issues:** What are the key legal and regulatory challenges facing Starlink's global expansion? How can international collaboration be fostered to ensure the responsible use of space resources?

## 15.6 Limitations of the Case Study

Understanding the limitations of any case study is essential to grasp the extent of its findings and the areas that may require further exploration. The following outlines the key constraints that shape the analysis of the Starlink project:

- **Scope and Data Availability:** The case study primarily focuses on the technical and some economic aspects of Starlink and not being able to discuss in detail on user adoption rates, real-world service quality, and specific regional impacts.
- **Technological Constraints:** The case study might not have considered unanticipated technological hurdles such as long-term satellite reliability or system integration complexities.
- **Economic Factors:** This analysis may be unable to account for evolving global economic backdrop, and competitive market landscape or the long-term cost-competitiveness of a large LEO satellite constellation.
- **Environmental Impact:** While environmental impact assessments and regulations primarily focus on space debris and orbital sustainability, they often lack detailed environmental obligations and certain specific sustainable measures.
- **Regulatory and Policy Challenges:** What the case study could not address is about international space law and the legal implications of potential cyberattacks on government satellites or global cooperation in orbital resources.

## References

1 Slotten, H. R. *Beyond Sputnik and the Space Race: The Origins of Global Satellite Communications*, JHU Press, 2022.

2 García Bescós, A. "Study of the benefits and applications of VLEO (Very Low Earth Orbit) for communications and definition of space new business models: Starlink case," Bachelor's Thesis, Universitat Politècnica de Catalunya, 2021.

3 Caviglione, L., Celandroni, N., Collina, M., Cruickshank, H., Fairhurst, G., Ferro, E., Gotta, A., Luglio, M., Roseti, C., Abdel Salam, A. and Secchi, R. "A deep analysis on future web technologies and protocols over broadband GEO satellite networks," *International Journal of Satellite Communications and Networking*, 33(5), 451–472, 2015.

4 Beyaz, M. "Satellite communications with 5G, B5G, and 6G: challenges and prospects," *International Journal of Communications, Network and System Sciences*, 17(3), 31–49, 2024.

**5** Pavur, J., Moser, D., Lenders, V. and Martinovic, I. "Secrets in the sky: on privacy and infrastructure security in DVB-S satellite broadband," in *Proceedings of the 12th Conference on Security and Privacy in Wireless and Mobile Networks*, 277–284, 2019.

**6** Penttinen, J.T. (ed.) "Satellite systems: communications," in *The Telecommunications Handbook: Engineering Guidelines for Fixed, Mobile and Satellite Systems*, 555–602. John Wiley & Sons, 2015.

**7** Kim, B., Lee, G.M., Choi, G. and Roh, B.H. "Effective bandwidth based capacity request for real-time voice in GEO satellite system," in *MILCOM 2016-2016 IEEE Military Communications Conference*, 379–384. IEEE, 2016.

**8** Varrall, G. *5G and Satellite RF and Optical Integration*, Artech House, 2022.

**9** Meulenberg, P. "An investigation into the effectiveness of implementing video conferencing over IP," Doctoral Dissertation, Swinburne, 2005.

**10** Nocera, A. *High Frequency Trading and Financial Stability*, Universita Ca Foscari Venezia, 2020.

**11** Garg, P.K., "Potential of geospatial data in healthcare for society 5.0," *Geospatial Data Science in Healthcare for Society 5.0*, ed. P.K. Garg, N.K. Tripathi, M. Kappas, L. Gaur, 1–28. Springer, 2022.

**12** Su, Y., Liu, Y., Zhou, Y., Yuan, J., Cao, H. and Shi, J. "Broadband LEO satellite communications: architectures and key technologies," *IEEE Wireless Communications*, 26(2), 55–61, 2019.

**13** Pratt, T. and Allnutt, J.E. *Satellite Communications*, John Wiley & Sons, 2019.

**14** Hatchwell, J.A. "Concepts for data collection in the Arctic," 1972, https://apps.dtic.mil/sti/tr/pdf/ADA085970.pdf (accessed March 4, 2026).

**15** International Telecommunications Union. *Handbook on Satellite Communications*, John Wiley & Sons, 2002.

**16** Xue, D., Wu, L., Xu, T., Wu, C.L., Wang, Z. and He, Z. "Space weather effects on transportation systems: a review of current understanding and future outlook," *Space Weather*, 22(12), e2024SW004055, 2024.

**17** Kaushal, H. and Kaddoum, G. "Optical communication in space: Challenges and mitigation techniques," *IEEE Communications Surveys & Tutorials*, 19(1), 57–96, 2016.

**18** Yahia, O.B., Garroussi, Z., Bélanger, O., Sansò, B., Frigon, J.F., Martel, S., Lesage-Landry, A. and Kurt, G.K. "Evolution of high throughput satellite systems: vision, requirements, and key technologies," arXiv preprint arXiv:2310.04389, 2023.

**19** Panagopoulos, A.D., Arapoglou, P.D.M. and Cottis, P.G. "Satellite communications at Ku, Ka, and V bands: propagation impairments and mitigation techniques," *IEEE Communications Surveys & Tutorials*, 6(3), 2–14, 2004.

**20** NASA. "Chapter 6: electromagnetics,", https://science.nasa.gov/learn/basics-of-space-flight/chapter6-1/ (accessed March 15, 2025).

21 SES White Paper. "Satellite solutions in C-Band," www.ses.com/sites/default/files/2022-10/SES-Satellite-Solutions-in-C-band.pdf. (accessed March 4, 2026).

22 Withington, T. "Space on budget," Armada International, March 2020, www.armadainternational.com/2020/03/space-on-budget. (accessed March 15, 2025).

23 Kurunjimalar, R. and Palanivel, T.G. "Mobile satellite communication with cross layer architecture," *Indian Journal of Science and Technology*, 9, S1, 2016.

24 Vasisht, D., Shenoy, J. and Chandra, R. "L2D2: low latency distributed downlink for LEO satellites," in *Proceedings of the 2021 ACM SIGCOMM 2021 Conference*, 151–164, 2021.

25 Rinaldi, F., Suraci, C., Pizzi, S., Molinaro, A. and Araniti, G. "Reliable eHealth services delivery over 3D LEO-based non-terrestrial networks," in *ICC 2023-IEEE International Conference on Communications*, 4076–4081. IEEE, 2023.

26 Ravishankar, C., Gopal, R., BenAmmar, N., Zakaria, G. and Huang, X. "Next-generation global satellite system with mega-constellations," *International Journal of Satellite Communications and Networking*, 39(1), 6–28, 2021.

27 Al-Hraishawi, H., Chougrani, H., Kisseleff, S., Lagunas, E. and Chatzinotas, S. "A survey on nongeostationary satellite systems: the communication perspective," *IEEE Communications Surveys & Tutorials*, 25(1), 101–132, 2022.

28 Di Pippo, S. *Space Economy: The New Frontier for Development*, EGEA spa, 2023.

29 Wall, M., "SpaceX's Starlink internet service is now available in Antarctica," Space X, September 2022, www.space.com/spacex-starlink-internet-service-antarctica (accessed March 4, 2025).

30 Phillips, P. *The Application of Satellite-based Internet of Things for New Mobility*, SAE International, 2024.

31 Oluropo, O., Ukommi, U. and Udoh, R. "Comparative coverage and horizon plane analysis for LEO, MEO, GEO and HEO satellites," *Journal of Multidisciplinary Engineering Science and Research (JMESR)*, 1(2), 54–56, 2022.

32 Baccelli, F., Candel, S., Perrin, G. and Puget, J.L. "Large satellite constellations: challenges and impact," Doctoral Dissertation, Académie des sciences, 2024.

33 De Zoysa, S. and Siriwardana, D. "Final frontier of strategic information warfare and cybersecurity challenges to satellites and space infrastructure."

34 Michel, F., Trevisan, M., Giordano, D. and Bonaventure, O., "A first look at starlink performance," in *Proceedings of the 22nd ACM Internet Measurement Conference*, 130–136, 2022.

35 Penent, G. "European Space Programs and the Digital Challenge, ed. L. Nardon," Institut Francais des Relations Internationales (IFRI), 2017.

36 Ofcom. "Connected nations UK report," 2024, www.ofcom.org.uk/siteassets/resources/documents/research-and-data/multi-sector/infrastructure-research/connected-nations-2024/connected-nations-uk-report-2024.pdf?v (accessed March 4, 2026).

**37** Kyi, R. and Laishram, N. "Concerns, implications and possible barriers to licensing: a legal and operational analysis on setting up Starlink in India," *Jus Corpus Law Journal*, 3, 575, 2023.
**38** Malisuwan, S. "Challenges, opportunities, and future research in the integration of 5G/6G networks, LEO satellites, and IoT for environmental protection and sustainable development," *ECSTAR. Journal of Environmental Engineering and Science B*, 12(2023), 146–159, 2023.
**39** Yue, P., An, J., Zhang, J., Pan, G., Wang, S., Xiao, P. and Hanzo, L. "On the security of LEO satellite communication systems: vulnerabilities, countermeasures, and future trends," Authorea Preprints, 2022.
**40** Juarez, N.R. "Strategies to rapidly decommission information technology satellites to prevent low Earth orbit space debris," Doctoral Dissertation, Walden University, 2024.
**41** Seedhouse, E. *SpaceX: Starship to Mars–The First 20 Years*, Springer Nature, 2022.
**42** Neal, V. *Spaceflight in the Shuttle Era and Beyond: Redefining Humanity's Purpose in Space*, Yale University Press, 2017.
**43** Baiocco, P. "Overview of reusable space systems with a look to technology aspects," *Acta Astronautica*, 189, 10–25, 2021.
**44** Carou, D., "The aerospace sector," in *Aerospace and Digitalization: A Transformation Through Key Industry 4.0 Technologies*, 9–16, Springer, 2021.
**45** Dougherty, N., "Simulated altitude liquid rocket testing at the USAF AEDC: Atlas/Mercury, Titan/Gemini, Saturn/Apollo, Space Shuttle, ISS," in *46th AIAA/ASME/SAE/ASEE Joint Propulsion Conference & Exhibit*, 6982, 2010.
**46** Karasopoulos, H.A. "Supply, demand, and entrepreneurial ventures in the space launch industry," Doctoral Dissertation, Massachusetts Institute of Technology, 2001.
**47** Paladini, S. *The New Frontiers of Space*, Springer International Publishing, 2019.
**48** Peeters, W. "The paradigm shift of NewSpace: new business models and growth of the space economy," *New Space*, 12(3), 202–213, 2024.
**49** Bergan, B. *Space Race 2.0: SpaceX, Blue Origin, Virgin Galactic, NASA, and the Privatization of the Final Frontier*. Motorbooks International, 2022.
**50** Valmunen, K. "A comparative study of the maritime satellite internet services available to the Finnish shipowner," Bachelor's Thesis, Satakunta University of Applied Sciences, 2024.
**51** Caraveo, P. and Iacomino, C. "The third phase of the evolution of the space economy (2000–today) and its foreseeable developments," in *Europe in the Global Space Economy*, 19–41. Springer International Publishing, 2023.
**52** Vunnam R.V. "Evolution of space exploration: a statistical analysis of space launches including challenges and future outlook," preprint, 2024.
**53** Höyhtyä, M. *Satellite Communications and Networks*, Springer, 2024.

**54** Wei, S. "A new location method based on internet satellite beacon signal," *In 7th International Conference on Information Communication and Signal Processing (ICICSP)* (pp.8–12). IEEE, 2024.

**55** Wang, G., Yang, F., Song, J. and Han, Z. "Free space optical communication for inter-satellite link: Architecture, potentials and trends," *IEEE Communications Magazine*, 62(3), 110–116, 2024.

**56** Techplayon. "Space X – Starlink system architecture for internet," March 2025, www.techplayon.com/starlink-system-architecture (accessed March 4, 2026).

**57** Badikov, G.A. and Zaytsev, A.V. "Economic model of costs required to create and operate the Starlink fast internet system," in *AIP Conference Proceedings*, vol. 2549, no. 1. AIP Publishing, 2023.

**58** Dicandia, F.A., Fonseca, N.J., Bacco, M., Mugnaini, S. and Genovesi, S. "Space-air-ground integrated 6G wireless communication networks: a review of antenna technologies and application scenarios," *Sensors*, 22(9), 3136, 2022.

**59** Osoro, O.B. and Oughton, E.J. "A techno-economic framework for satellite networks applied to low Earth orbit constellations: assessing Starlink, OneWeb and Kuiper," *IEEE Access*, 9, 141611–141625, 2021.

**60** Ma, F., Zhang, X., Li, X., Cheng, J., Guo, F., Hu, J. and Pan, L. "Hybrid constellation design using a genetic algorithm for a LEO-based navigation augmentation system," *GPS Solutions*, 24, 1–14, 2020.

**61** Cui, G., Duan, P., Xu, L. and Wang, W. "Latency optimization for hybrid GEO–LEO satellite-assisted IoT networks," *IEEE Internet of Things Journal*, 10(7), 6286–6297, 2022.

**62** Azar, J., Makhoul, A., Barhamgi, M. and Couturier, R. "An energy efficient IoT data compression approach for edge machine learning," *Future Generation Computer Systems*, 96, 168–175, 2019.

**63** Kadhim, D.J., Mosleh, M.F. and Abed, F.A. "Exploring text data compression: a comparative study of adaptive Huffman and LZW approaches," in *BIO Web of Conferences*, vol. 97, 00035. EDP Sciences, 2024.

**64** Kondrateva, O. "Efficient communication for data-intensive applications in small satellite networks," Doctoral Dissertation, Mathematisch-Naturwissenschaftliche Fakultät, 2025.

**65** Adanvo, V.F., Mafra, S., Montejo-Sánchez, S., Tondo, F.A. and Souza, R.D. "Improving efficiency and reliability in information offloading for LEO satellite networks by inter-satellite communication techniques," *IEEE Transactions on Aerospace and Electronic Systems*, 61(2), 1336–1352, 2024.

**66** Esho, A.O.O., Iluyomade, T.D., Olatunde, T.M. and Igbinenikaro, O.P. "A comprehensive review of energy-efficient design in satellite communication systems," *International Journal of Engineering Research Updates*, 6(2), 13–25, 2024.

**67** De Micheli, C., "Optimizing satellite constellation orbits for low latency internet connections," Doctoral Dissertation, Politecnico di Torino, 2021.

**68** Yue, P., An, J., Zhang, J., Ye, J., Pan, G., Wang, S., Xiao, P. and Hanzo, L. "Low Earth orbit satellite security and reliability: issues, solutions, and the road ahead," *IEEE Communications Surveys & Tutorials*, 25(3), 1604–1652, 2023.

**69** Parishanmahjuri, H. "Development of the Global Value Chains in the Space Industry," Master's Thesis, Università Ca' Foscari, 2022.

**70** Seedhouse, E., *SpaceX: Making Commercial Spaceflight a Reality*, Springer Science & Business Media, 2013.

**71** Wall, M. "SpaceX launches first-ever used rocket in milestone flight," *Space.com*, March 2017, www.space.com/36375-spacex-reused-falcon-9-rocket-cost-savings.html (accessed March 4, 2026).

**72** Hiriart, T., Castet, J.F., Lafleur, J.M. and Saleh, J.H. "Comparative reliability of GEO, LEO, and MEO satellites," in *Proceedings of the International Astronautical Congress*, IAC-09 D (Vol. 1), 2009.

# Appendix

## Answers to Multiple Choice Questions

### Chapter 1

**Q 1.2** What was the primary engineering challenge in the BorWin2 project?

**B** Difficulty in integrating HVDC with the onshore AC grid.

**Q 1.3** Which of the following engineering principles was critical in designing the HVDC transmission for BorWin2?

**D** Both B and C.

**Q 1.4** What alternative solution could be considered if HVDC transmission were not used for BorWin2?

**C** Implementing an HVAC (High Voltage Alternating Current) transmission system.

### Chapter 2

**Q 2.1** Which of the following represents the primary engineering challenge in the GreenHyScale project?

**B** Stabilizing electrolyzer operation with intermittent wind power.

**Q 2.2** In the context of system integration, what is the main concern when scaling up to 100 MW?

**C** Power quality and grid stability.

**Q 2.3** When considering the hydrogen production system's response to wind power fluctuations, which parameter is most critical?

**B** Response time.

## Chapter 3

**Q 3.1** What is the primary challenge in integrating offshore wind energy with hydrogen production?

**B** Variability and intermittency of wind power.

**Q 3.2** What is the most used process for green hydrogen production in Ørsted's projects?

**C** Electrolysis

**Q 3.3** Why is hydrogen storage critical for Ørsted's hydrogen projects?

**A** Hydrogen must be stored to balance production fluctuations.

**Q 3.4** Which aspect of GFDK's design most directly influenced SeaH2Land's development?

**B** Multi-stakeholder governance model.

## Chapter 4

**Q 4.1** What was the primary advantage of CdTe thin-film solar cells compared to crystalline silicon solar cells?

**B** Lower production cost and better performance in high temperatures

**Q 4.2** What was one of the biggest challenges in scaling CdTe solar panel production?

**C** Limited availability of tellurium and supply chain dependency

**Q 4.3** How did First Solar address the efficiency gap between CdTe and silicon-based solar panels?

**A** By improving deposition techniques and reducing material defects

**Q 4.4** Why was cadmium telluride considered an environmentally sustainable alternative to traditional silicon solar panels?

**B** CdTe panels had a lower carbon footprint and required less energy to manufacture

## Chapter 5

**Q 5.1** Which of the following factors primarily influences the aerodynamic efficiency of a wind turbine blade?

**C** Blade length and shape optimization.

**Q 5.2** What is a major benefit of AI-driven predictive maintenance in wind farms?

**B** Reduced downtime and maintenance costs.

**Q 5.3** How does grid integration impact the efficiency of wind energy systems?

**A** It helps to balance supply and demand, preventing power fluctuations.

**Q 5.4** Which cybersecurity risk is most relevant to connected wind farms?

**B** Hacking of control systems leading to turbine malfunctions.

## Chapter 6

No multiple choice questions.

## Chapter 7

**Q 7.1** What was one of the major reasons for the increased need for grid-scale energy storage between 2010 and 2017?

**B** Increased integration of intermittent renewable energy sources like wind and solar.

**Q 7.2** What was a significant technical advantage of the Tesla Powerpack system used in the Hornsdale Power Reserve?

**C** Nickel-Manganese-Cobalt (NMC) chemistry for long cycle life and energy density.

**Q 7.3** What economic benefit did the Hornsdale Power Reserve provide within its first year of operation?

**A** Reduced South Australia's FCAS costs by 90%.

**Q 7.4** Which emerging technology could address some of the technical challenges faced by lithium-ion batteries in grid-scale applications?

**B** Solid-state batteries for improved energy density and safety.

## Chapter 8

**Q 8.2** As transistors shrink to smaller nodes such as 3nm, what is a key challenge engineers face?

**C** Managing short-channel effects and leakage current becomes increasingly difficult.

**Q 8.3** Why are FinFET transistors preferred over traditional MOSFETs for power-efficient semiconductor design?

**B** They provide better electrostatic control, reducing leakage current and power loss.

## Chapter 9

**Q 9.1** Power Efficiency vs. Connectivity: You are designing a battery-operated IoT device that must last at least five years while maintaining reliable connectivity. What is the best strategy to maximize battery life without sacrificing performance?

**C** Implement dynamic sleep modes, adjust signal strength based on network conditions, and optimize transmission intervals.

**Q 9.2** Transistor Geometry and Leakage: A conventional MOSFET-based IoT device design has a leakage current of 100 nA per transistor. If a FinFET-based alternative reduces leakage by 90%, what is the new leakage current per transistor?

**C** 10 nA

**Q 9.3** IoT devices, such as smart meters and health monitors, require long battery life while maintaining reliable connectivity. What is one of the biggest challenges in designing these devices?

**B** Reducing power consumption while ensuring effective wireless communication.

**Q 9.4** Which of the following best describes the role of dynamic sleep modes in an IoT device?

**B** They reduce power consumption by putting the device into a low-power state when it is not actively transmitting or receiving data.

## Chapter 10

**Q 10.1** What is the primary reason for semiconductor manufacturers adopted 3D stacking technology?

**B** To improve transistor density and performance while overcoming planar scaling limits.

**Q 10.2** Which of the following is a key advantage of Intel's Foveros technology?

**A** It allows heterogeneous integration of different process nodes.

**Q 10.3** One major challenge of 3D stacking technology is:

**C** Difficulty in managing heat dissipation within stacked layers.

**Q 10.4** Why is 3D stacking considered a paradigm shift in semiconductor engineering?

**B** It eliminates the need for Moore's Law by keeping chips two-dimensional.

## Chapter 11

**Q 11.1** What is a key differentiator in NVIDIA's approach to GPU technology compared to AMD?

**B** Focus on proprietary technologies like CUDA and ray tracing.

**Q 11.2** Which market segment has traditionally been dominated by NVIDIA?

**B** Professional visualization and discrete gaming GPUs.

**Q 11.3** Why is TSMC crucial to the GPU industry?

**A** It manufactures GPUs for multiple companies, including NVIDIA and AMD.

## Chapter 12

**Q 12.1** What was STMicroelectronics' primary strategic advantage in SiC production compared to its competitors?

**B** Vertically integrated supply chain.

**Q 12.2** Which factor most significantly influenced STMicroelectronics' decision to invest in the Catania facility?

**C** Control over complete manufacturing process.

**Q 12.3** STMicroelectronics' partnership with Tesla primarily focused on?

**B** Supply of SiC power devices for vehicle inverters.

**Q 12.4** The transition from 150mm to 200mm wafer production primarily aims to?

**A** Increase manufacturing efficiency and reduce costs.

## Chapter 13

**Q 13.1** What is one of the key advantages of 5G over 4G LTE?

**C** Faster speeds and lower latency.

**Q 13.2** Which technology helps 5G networks optimize signal direction?

**B** Beamforming.

**Q 13.3** Why does 5G require more small-cell deployments?

**A** 5G signals have a shorter range than previous generations.

**Q 13.4** What role does AI play in Ericsson's 5G network optimization?

**B** Reducing latency and managing network traffic.

**Q 13.5** Which of the following is NOT a challenge in 5G deployment?

**C** Decreasing data transfer speeds.

## Chapter 14

**Q 14.1** Which of the following best describes federated learning in Huawei's telecom optimization system?

**B** Training AI models across distributed datasets without sharing raw data.

**Q 14.2** What is the primary role of convolutional neural networks (CNNs) in Huawei's system?

**A** Analyzing spatial data, such as network topology and signal strength.

**Q 14.3** How do recurrent neural networks (RNNs) enhance Huawei's telecom optimization system?

**C** By predicting temporal patterns, such as traffic spikes and latency issues.

**Q 14.4** What is the key benefit of federated learning in the telecom industry?

**B** It ensures sensitive user data remains localized, enhancing privacy.

## Chapter 15

**Q 15.1** What is the primary benefit of Low Earth Orbit (LEO) satellites over Geostationary Earth Orbit (GEO) satellites?

**C** Lower latency.

**Q 15.2** How does SpaceX's Starlink constellation aim to address the digital divide?

**A** By providing high-speed internet in remote areas.

**Q 15.3** What distinguishes Medium Earth Orbit (MEO) satellites from LEO and GEO satellites?

**C** They provide moderate coverage.

**Q 15.4** Which application benefits most directly from the low-latency connections provided by LEO satellites?

**A** Online gaming.

# Index

Green Fuels for Denmark (GFDK) 60

Printed and bound by CPI Group (UK) Ltd, Croydon, CR0 4YY
17/09/2026
14964234-0005